Connectionist Modeling and Brain Function: The Developing Interface

Neural Network Modeling and Connectionism

Jeffrey L. Elman, Editor

Connectionist Modeling and Brain Function: The Developing Interface
Stephen José Hanson and Carl R. Olson, editors

Neural Network Design and the Complexity of Learning
J. Stephen Judd

Connectionist Modeling and Brain Function: The Developing Interface

edited by

Stephen José Hanson and Carl R. Olson

A Bradford Book

The MIT Press
Cambridge, Massachusetts
London, England

Printed and bound in the United States of America

Library of Congress Cataloging-in-Publication Data

Connectionist modeling and brain function.

(Neural network modeling and connectionism)
"A Bradford book."
Inclues bibliographical references.
1. Brain--Mathematical models--Congresses.
2. Connectionism--Congresses. 3. Cognitive science--Congresses. 4. Neural circuitry--Congresses. 5. Neural computers--Congresses. I. Hanson, Stephen José. II. Olson, Carl R.
QP376.C645 1990 89-63775
ISBN 0-262-08193-8 CIP

Contents

Part III: The Somatosensory System

Part IV: The Visual System

Series Foreword

The goal of this series, Neural Network Modeling and Connectionism, is to identify and bring to the public the best work in the exciting field of neural network and connectionist modeling. The series includes monographs based on dissertations, extended reports of work by leaders in the field, edited volumes and collections on topics of special interest, major reference works, and undergraduate and graduate-level texts. The field is highly interdisciplinary, and works published in the series will touch on a wide variety of topics ranging from low-level vision to the philosophical foundations of theories of representation.

Preface

To explore promising trends in the zone of interface between neuroscience and connectionist modeling, we recently organized at Princeton a two-day symposium involving presentations by pairs of neuroscientists and modelers working on identical or related problems. The talks covered a broad range with respect both to subject matter and approach. Collectively, the presentations provided a solid background on the major research areas where connectionist-inspired biology-based modeling is currently taking place. In the course of the presentations, questions concerning the theoretical advantages and biological plausibility of particular learning rules and network architectures were frequently raised and considered.

The present book, which arose from the symposium, should be regarded as a tutorial on the current state of the field. It will be of interest to a broad audience consisting of brain scientists, cognitive psychologists and those in related disciplines. It provides descriptions of specific experimental approaches and also confronts general issues related to learning, associative memory and sensorimotor development. It explores the computational nature of models in terms of learning rules and architectures and also considers issues of biological plausibility. The individual chapters, while retaining a level of detail and sophistication suitable to an audience of specialists, also illuminate the broad outlines of the field for a general audience.

Acknowledgements

This book owes its existence to a number of institutions and people aside from the contributors themselves. It would not have been possible without the symposium that brought most of the authors together initially and the symposium would not have been possible without financial support from the Langfeld Fund of the Princeton Department of Psychology, from the Cognitive Science Laboratory, and from the Human Information Processing Group. We would like to thank George A. Miller, in particular, for his support, guidance and participation in the symposium. We also are grateful to a number of people who worked behind the scenes to make things run smoothly and helped at critical moments: Shari Landes, Catherine Hanson, Christopher Hanson (who generously provided his colored chalk) and Carol Colby. Finally, most of the printing and layout of the book was done at at the Princeton Cognitive Science Laboratory, Bellcore, and Siemens Research Center. We would also like to thank Harry and Betty Stanton, Joanne Poole and Helen Osborne of MIT Press/Bradford for guidance and advice as the book developed.

List of Contributors

Terry Allard
Speech Motor Control Laboratories
Waisman Center
University of Wisconsin
Madison, WI 53705

José Ambros-Ingerson,
Center for the Neurobiology of Learning and Memory
University of California
Irvine, CA 92717

Phil Antón
Center for the Neurobiology of Learning and Memory
University of California
Irvine, CA 92717

Patricia S. Churchland
Department of Philosophy
University of California at San Diego
La Jolla, CA 92093

Leif H. Finkel
Department of Bioengineering
and David Mahoney Instintue of Neurological Sciences
University of Pennsylvania
Philadelphia, PA 19104

Richard Granger
Center for the Neurobiology of Learning and Memory
University of California
Irvine, CA 92717

Stephen Grossberg
Center for Adaptive Systems
Boston University
111 Cummington Street
Boston, MA 02215

Stephen José Hanson
Mathematics, Communications
and Computer Science Research
Bell Communications Research
Morristown, NJ 07960

Cogntive Science Laboratory
221 Nassau Street
Princeton University
Princeton, NJ 08542

Christof Koch
Computational and Neural
Systems Program
California Insititute of Technology
Pasadena, CA 91125

Ralph Linsker
IBM Research Division
Thomas J. Watson Research Center
Yorktown Heights, NY 10598

Gary Lynch
Center for the Neurobiology of Learning and Memory
University of California
Irvine, CA 92717

Kenneth D. Miller
Department of Physiology
University of California of San Francisco
San Francisco, CA 94143

Carl R. Olson
Department of Psychology
Princeton University
Princeton, NJ 08542

Christie L. Sahley
Department of Biological Sciences
Purdue University
West Lafayette, IN 47907

Terrence J. Sejnowski
The Salk Institute
10100 North Torrey Pines Road
La Jolla, CA 92037

Michael P. Stryker
Department of Physiology
University of California of San Francisco
San Francisco, CA 94143

Gerald Tesauro
IBM Research Division
Thomas J. Watson Research Center
Yorktown Heights, NY 10598

Dejan Todorović
Univerzitet u Beogradu
Laboratorija za Eksperimentalnu Psihologiju
Cika Ljubina 18-20
11000 Beograd
Yugoslavia

Connectionist Modeling and Brain Function: The Developing Interface

CHAPTER 1

Introduction: Connectionism and Neuroscience

Stephen José Hanson and Carl R. Olson

Introduction

These are revolutionary times for the brain and cognitive sciences. At the heart of the revolution is connectionism. Neural networks - systems consisting of simple neuronlike computing elements - have become a subject of intense interest to scientists spanning a broad range of disciplines including mathematics, physics, biology and psychology.

The wide appeal of neural networks is the result of several factors. In the first place, they are powerful. They are capable in principle of approximating any real valued function mapping and have been shown in practice to solve complex engineering problems in areas including speech recognition, visual pattern analysis and robotic control. It is likely in the future as further learning rules and architectures are investigated that powerful applications to classification, temporal processing and problem solving will be discovered. In the second place, connectionist networks solve problems in a simple and elegant way closely analogous to processes thought to occur in the brain. The behavior of a connectionist system emerges from structural changes driven by local learning rules. The arbitrary distinction between program and computer that is intrinsic to Artificial Intelligence and has restricted its biological relevance is absent in connectionism. In a fundamental sense, the connections of a network are both hardware and software.

This suggests another reason for interest in neural networks. By studying them we may be able to unravel basic features of brain function that have remained intractable to other approaches. The human brain is a uniquely complex organ composed of a hundred billion neurons connected by ten trillion synapses. It performs a vast array of ill-defined and overlapping functions by means of neural machinery that has evolved opportunistically over hundreds of millions of years. It would be unnatural, in light of its slow and no doubt tortuous evolution, for the brain to work in the way in which a machine designed by an engineer works. Most people who have thought for any length of time about the brain share with Lashley the feeling that to a considerable degree it must work by mass action, exhibiting what at one time was called equipotentiality and is now called distributed representation. By studying simulated neural systems based on known physiological and anatomical properties of the brain it should be possible to characterize the process by which masses of neurons interacting according to local rules undergo self-organization into efficiently functioning distributed ensembles.

The focus of this volume is on modeling studies of brain-inspired neural networks and on closely related studies of wet neuroscience. One of the themes is that there is a close and steadily increasing interaction between modeling experiments and studies of the real brain. But it will also become apparent that there is tension between studies of artificial and biological systems. The source of the tension is the gross disparity by just about any measure one can think of between the simplicity of the modeled systems and the complexity of the actual brain. The disparity is unavoidable because, even in the case of comparatively well understood brain systems, we are very far from having a thorough understanding of fundamental organizational details. This observation brings us to the issue of biological plausibility. Under what conditions can a model that does not capture all the details of a brain system be considered nevertheless to be plausible? Our general answer to this question is that even models with no relation to the brain beyond the fact that they are built from neuronlike elements may provide insight into the brain by showing that a particular function is in principle computable by neurons. The mere constraint that operations be carried out by synaptically interconnected neurons is incomparably stronger than any constraint accepted by other approaches, notably in artificial intelligence. Furthermore, in the interests of proportion, one must note that there is no convincing instance in mammalian studies of emergent network properties that depend critically on the precise parameters obtained from measurements of the brain.

In bringing together the chapters that make up this volume, we have deliberately created a couterpoise between biological and modeling studies. In some cases, pairs of authors were asked to write related chapters treating the same subject from two sides. In the remaining cases, brain studies and modeling are dealt with jointly by authors with dual expertise. Considered as a whole, the chapters demonstrate clearly the power of neural-net studies in explaining brain processes. Incidentally they demonstrate the remarkable diversity that exists within the field. It may help to provide orientation if we note in closing two of the key dimensions along which the studies described here differ.

One dimension is the level of the phenomenon that the model is expected to reproduce. This may range from being strictly neural to being strictly cognitive. The neural end of the continuum is occupied by studies of self-organizing sensory maps in the somatosensory system (Allard; Finkel) and in the visual system (Miller and Stryker; Linsker). The touchstone for success in these studies is essentially that the properties of units in the artificial system should match the properties of neurons in the brain. The cognitive end of the continuum is occupied by studies of spatial analysis of the body (Olson and Hanson) and of the brightness-from-luminance problem (Grossberg and Todorović). The criterion for success in these studies is that the artificial system should master an input-output function of which we as humans are capable. Occupying an intermediate point along the continuum are studies of learning (Sahley; Tesauro; Granger et al.) in which the distinction between properties of neurons and the emergent properties of behavior is not clearcut.

A second distinction concerns the mechanism of neural plasticity and the role which it plays. In the bulk of the studies described here the underlying process is competitive learning based on Hebb synapses and its role is essentially to let neurons sort out their functions by local interactions. This is a conservative approach because the Hebb rule is very simple and because learning is permitted to occur with no direction from the outside except in the form of sensory input. However there is a large class of learning phenomena that rely on feedback from the world. These phenomena depend on comparatively more complex patterns of wiring and may necessitate the use of learning rules (notably the delta rule) that possess greater power than the Hebb rule. Processes of this sort are dealt with in a few of the studies included here (Sahley; Tesauro;

Olson and Hanson).

Of necessity, the chapters composing this book represent a selected subset of approaches used in computational neuroscience. The broad outlines of the field from which they have been drawn are presented in the next chapter.

CHAPTER 2

Computational Neuroscience

T. J. Sejnowski, C. Koch and P. S. Churchland

Introduction

The ultimate aim of computational neuroscience is to explain how electrical and chemical signals are used in the brain to represent and process information. This goal is not new, but much has changed in the last decade: more is known about the brain because of advances in neuroscience, much more computing power is available for performing realistic simulations of neural systems, and new insights are available from simplifying models of large networks of neurons. Brain models are being used to connect the microscopic level accessible by molecular and cellular techniques with the systems level accessible by studying behavior.

Understanding the brain is a challenge that is attracting a growing number of scientists from many disciplines. Although there has been an explosion of discoveries over the last several decades concerning the structure of the brain at the cellular and molecular levels, we do not yet understand how the nervous system enables us to see and hear, to learn skills and remember events, to plan actions, and to make choices. Simple reflex systems have served as useful preparations for studying the generation and modification of behavior at the cellular level[1] In mammals, however, the relationship between

---

1. See, for example, work on stretch reflexes of the spinal cord (Houk and Rymer, 1981), the jamming-avoidance response in the electric fish (Heiligenberg, 1987), and the gill-withdrawal reflex of the sea slug Aplysia (Kandel et al., 1987).

perception and the activity of single neurons is more difficult to study, since the sensory capacities assessed with psychophysical techniques are the result of activity in many neurons from many parts of the brain. In humans, the higher-brain functions such as reasoning and language are even further removed from the properties of single neurons. Moreover, even relatively simple behaviors, such as stereotyped eye movements, involve complex interactions among large numbers of neurons distributed in many different brain areas (Lisberger et al., 1987; Lee et al., 1988, Goldman-Rakic, 1988).

Part of the reason why generating explanations for higher functions is so difficult is that nervous systems have many levels of organization between the molecular and systems levels, each with its own important functions. Properties not found in components of a lower level can emerge from the organization and interaction of these components at a higher level. For example, rhythmic pattern generation in some neural circuits is a property of the circuit, not of isolated pacemaker neurons (Selverston, 1983; Getting, 1983). The sources of such network properties are not accessible using methods suited to investigating single neurons. New experimental techniques and theoretical models are needed to understand the properties of these distributed systems (Churchland and Sejnowski, 1989a).

Neurons are organized in local circuits, columns, laminae, and topographic maps for purposes that are poorly understood (Knudsen et al., 1987; Barlow, 1986). Higher brain functions such as perception and attention may depend on coherent functional units distributed through several different maps and nuclei that would not be obvious from observations of anatomical boundaries (Mountcastle, 1979; Goldman-Rakic, 1988). Techniques for studying these intermediate levels of organization are currently being developed, including methods for simultaneously recording from multiple single units, optical recording of columnar organization in the cortex with voltage- and ion-sensitive dyes, and large-scale measurements of brain structure and activity with PET, MEG, 2-DG and MRI.[2] Though valuable, each of these new techniques has severe limitations on its

2. Multielectrode recording: Reitboeck, 1983; Llinas, 1985; Voltage- and ion-dependent dyes: Grinvald, 1985; Tsien and Poenie 1986; Positron Emission Tomography (PET): Raichle, 1986; Magneto-encephalography (MEG): Williamson et al., 1983; Magnetic Resonance Imaging (MRI): Bachus et al.,

spatial or temporal resolution (Sejnowski and Churchland, 1989). Assuming that there are emergent properties of networks, it is difficult to imagine a major advance in our understanding of brain function without a concomitant development in direct, efficient techniques for probing the mechanisms of distributed processing. Statistical methods must also be developed for analyzing and interpreting the mountain of information that will be collected (Gerstein and Aertsen, 1985, Abeles, 1982).

Modeling promises to be an important adjunct to these experimental techniques and is essential in addressing the conceptual issues that arise when studying information processing in the brain (Arbib, 1989; Koch and Segev, 1989; Schwartz, 1989; Cowan and Sharp, 1988; Reichardt and Poggio, 1981). The advantages of brain models are varied as follows: 1) a model can make more accessible to scrutiny the consequences of a complex, nonlinear brain system with many interacting components. 2) new phenomena may be discovered by comparing the predictions of a simulation to experimental results and new experiments can be designed based on these predictions; and 3) experiments that are difficult or even impossible to perform in living tissue, such as the selective lesion of particular channels, synapses, neurons, or pathways, can be simulated using a model.

What kind of a computer is the brain? Mechanical and causal explanations of chemical and electrical signals in the brain are different from computational explanations[3] The chief difference is that a computational explanation refers to the information content of the physical signals and how they are used to accomplish a task. This difference is easiest to see in simpler physical systems that compute. For example, a mechanical explanation for the operation of a slide rule includes the observations that certain marks are lined up, the slider is moved, and a result is read off. A computational explanation states that a slide rule computes products because the marks on the

---

1987; 2-deoxyglucose (2-DG): Sokoloff, 1984.

3. Marr and Poggio (1976) and Marr (1982) have emphasized the importance of this level of analysis.

sliding wood rules correspond to logarithms, and adding two logarithms is equivalent to multiplying the corresponding pair of numbers. Thus, the physical system carries out a computation by virtue of its relationship to a more abstract algorithm (Churchland et al., 1989). One of the major research objectives of computational neuroscience is to discover the algorithms used in the brain.

Unlike a digital computer, which is general purpose and can be programmed to run any algorithm, the brain appears to be a collection of special purpose systems that are very efficient at performing their tasks but are limited in their flexibility. The architecture of an efficient system like a slide rule, or a brain system, constrains the algorithm it implements in a fashion that does not occur in a digital computer (Mead, 1989). The clues from structure are particularly valuable because the nervous system is a product of evolution, not engineering design. The computational solutions evolved by nature may be unlike those that humans would invent at least for the reason that evolutionary changes are always made within the context of a design and architecture that already are in place (Jacob, 1982).

Classes of Brain Models

Realistic Brain Models One modeling strategy consists in using a large scale simulation that tries to incorporate as much cellular detail as is available. For brevity, we call these realistic brain models. Although this approach to simulation can be very useful, the realism of the model is both a weakness and a strength. As the model is made increasingly realistic by adding more variables and more parameters, the danger is that the simulation ends up as poorly understood as the nervous system itself. Equally worrisome, since we do not yet know all the cellular details, there may be important features that are being inadvertently left out, thus invalidating the results. Finally, realistic simulations are highly computation intensive. Present constraints limit simulations to tiny nervous systems or small components of more complex systems. Only recently has sufficient computer power been available to go beyond the simplest models.

An example of a realistic model at the level of a single neuron is the Hodgkin-Huxley model (1952) of the action potential in the squid giant axon. The action potential is a stereotyped, transient, electrical event that propagates along an axon and is used for communicating information over long distances. The action potential is a result of the voltage- and time-dependent properties of several types of membrane channels. The dynamics of the membrane channels were modeled by a set of coupled, nonlinear differential equations that were solved numerically. The velocity of the action potential predicted by Hodgkin and Huxley agreed to within 10% of the measured value. Two important lessons can be drawn from this example. First, the model was the culmination of a large number of experiments. In general, realistic models with a large number of parameters require a correspondingly large number of measurements to fit the parameters. Second, the voltage-dependent membrane channels postulated to account for the data were verified only much later with the introduction of single-channel recording techniques (Hodgkin, 1976). We should expect to make hypotheses that go beyond the immediate data.

An example of a realistic model at the network level is the Hartline-Ratliff model of the *Limulus* lateral eye (Ratliff, 1962). The photoreceptors in the retina have lateral inhibitory synaptic interactions with neighboring photoreceptors, and as a consequence the contrast in the firing rates at light intensity borders is enhanced. Because the interactions are approximately linear, a model of the network can be mathematically analyzed in great detail. Once all of the parameters are determined from experimental measurements, the model can accurately predict the firing rates of all the fibers in the optic nerve stimulated by an arbitrary spatio-temporal pattern of light falling on the retina (Brodie et al., 1978). Models of networks that are nonlinear and have feedback connections are much more difficult to analyze; thus only a qualitative mathematical analysis is possible and simulations of their behavior are essential. It is worth emphasizing once again the necessity of collecting a nearly complete set of experimental measurements before a realistic model can be attempted.

Simplifying Brain Models Because even the most successful realistic brain models may fail to reveal the function of the tissue, computational neuroscience needs to develop simplifying models that capture important principles. Textbook examples in physics that admit exact solutions are typically unrealistic, but they are valuable because they illustrate physical principles. Minimal models that reproduce the essential properties of physical systems, such as phase transitions, are even more valuable. The study of simplifying models of the brain can provide a conceptual framework for isolating the basic computational problems and understanding the computational constraints that govern the design of the nervous system.

The class of models currently being investigated under the general headings of connectionist models, parallel distributed processing models, and "neural networks" is of this second type, which we shall hereafter refer to as simplifying brain models. These models abstract from the complexity of individual neurons and the patterns of connectivity in exchange for analytic tractability (Hopfield and Tank, 1986). These models are being been investigated as prototypes of new computer architectures (Feldman and Ballard, 1982; Kohonen, 1988) or as models for psychological phenomena (Hinton and Anderson, 1981; Rumelhart and McClelland, 1986; Grossberg, 1988). Nonetheless, many of the results are applicable to the brain.

One of the best studied architectures is the class of layered feedforward networks. In this architecture, information is coded as a pattern of activity in an input layer of model neurons and is transformed by successive layers receiving converging synaptic inputs from preceding layers. These three findings are of significance for brain models: 1) Even systems with only a few intermediate layers have enormous power in representing complex nonlinear functions (Lippman, 1987; Lapedes and Farber, 1988), 2) The performance of a network in specific problem domains (such as visual and speech processing) depends critically on how the incoming information is represented by the neurons (such as the type of preprocessing) and the symmetries in the pattern of connections; and 3) For difficult problems, the processing units in the middle or "hidden" layers generally encode many different combinations of input variables using a semi-distributed type of representation (Churchland and Sejnowski, 1989b; Sejnowski and Rosenberg, 1987). By combining the power of these models with further constraints from neurophysiology and neuroanatomy, it may be possible to interpret some of the properties that have been observed from single-unit recordings, as we illustrate in the next section (Zipser and Andersen, 1988; Hurlbert and Poggio, 1988; Lehky and Sejnowski, 1988, Anastasio and Robinson, 1989).

These simplifying brain models also make an important bridge to computer science and other disciplines that study information processing. Issues such as convergence of the network to a stable solution, the amount of time needed for the network to achieve a solution, and the capacity of networks to store information are being investigated in simplifying models in ways that are not at present feasible with realistic models (Mitchison and Durbin, 1988). The scaling of these properties with the size of the network is crucially important for practical applications and the plausability of the model as a brain model (Abu-Mostafa, 1986; Tesauro and Janssens, 1988). Many of the current models do not scale well without additional constraints on the architecture, such as restricting the connectivity to local neighborhoods.

Technology for Brain Modeling Computational brain models are almost always simulated on digital computers (Koch and Segev, 1989). Computers are getting faster, but they must perform the many parallel operations in the brain one at a time and are many orders of magnitude too slow. Parallel computers with thousands of processors are being developed, but are still inadequate (Sejnowski, 1987). A new approach towards simulating biological circuitry is being pioneered by Mead (1989), who is constructing hardware devices whose components directly mimic the circuits in the brain. The severe physical restrictions imposed on the density of wires and the cost of communications in electronic circuits are similar to the constraints that are imposed on biological circuits. Fast hardware can deliver the computing power necessary to evaluate the performance of models in real time. This approach, which he terms "synthetic neurobiology" (Mead, 1989), allows for the rapid determination of the strengths and limitations of a theory.

Mead uses analog subthreshold cMOS VLSI technology (Very Large Scale Integrated circuit). Several chips that implement simplifying models of visual information processing have already been produced that are highly efficient. A "retina" chip computes the spatial and temporal derivative of arbitrary images projected onto a hexagonal array of 48 by 48 phototransistors – which are approximately logarithmic over five orders of magnitude of light amplitude – coupled via a horizontal resistive grid and injecting current into model "amacrine" cells that compute a temporal derivative (Sivilotti, Mahowald, and Mead, 1987). Similar circuits can be designed for computing optical flow in real time (Tanner and Mead, 1986; Hutchinson et al. 1988).

These VLSI chips and new techniques in optical information

processing may lead to a new computing technology, sometimes called artificial neural systems, or neurocomputing (Abu-Mostafa and Psaltis, 1987; Shriver, 1988). This technology for performing massively-parallel computation could have a major influence on the next generation of research in computational neuroscience. For example, an analog VLSI model of a neuron that included conductance mechanisms, synaptic apparatus, and dendritic geometry could be produced in great quantities. These chips could be used as coprocessors in a conventional digital computer to greatly increase the speed of realistic simulations. If this technology is developed now, it should be possible to simulate our visual system in real time by the twenty-first century (Sejnowski, 1987).

Specific Examples of Brain Models

In the following, we will discuss several different models, demonstrating the great variety of different levels of structure, analysis and measurement existing in contemporary computational neuroscience. It is clearly impossible to discuss within the confines of this article even a small fraction of the models in the literature that address some problem in neurobiology, so we will limit ourselves to a few examples drawn from the invertebrate and vertebrate vision literature. This choice reflects the idiosyncrasies and research interests of the authors and in no way implies that other areas within neurobiology have not developed equally relevant models[4].

The modeling of learning and memory is an important area not covered here. Evidence for neural plasticity is accumulating at the cellular and molecular levels in a variety of systems (Kandel, et al., 1987; Alkon, 1987; Thompson, 1986; Brown et al., 1988, Merzenich

4. Models in motor control and vision: Crick, 1984; Hildreth and Hollerbach, 1987; Ullman, 1986; Models of olfaction: Freeman, 1975; Lynch, 1988; Wilson and Bower, 1989; Models of sensory-motor integration: Robinson; 1981; Arbib, 1989; Pellionisz and Llinas, 1985; Grossberg and Kuperstein, 1986; Krauzlis and Lisberger, 1989.

and Kaas, 1982). The conditions for plasticity can be incorporated into large-scale network models whose properties can be explored in simulations of realistic models (Traub, et al, 1987; Wilson and Bower, 1989, Lynch, 1988; Finkel and Edelman, 1987) and analytically for simplifying models (Hopfield, 1982; Kohonen, 1984; Palm, 1982; Amari, 1983). A separate review would be required to summarize all the interesting models of learning and memory in brain circuits that are now being explored (Sejnowski and Tesauro, 1988; McNaughton and Morris, 1987; Donnegan, et al., 1988; Brown et al., 1989).

Detecting and computing motion Visual motion is a fundamental source of information about the world (Hildreth and Koch, 1987; Nakayama, 1985) and motion detection is the first stage of motion processing. The best known motion detection scheme was first proposed 30 years ago. Behavioral data, gained on the basis of open- and closed-loop experiments performed on beetles and flies (Hassenstein and Reichardt, 1956; Reichardt, 1957), indicated that a sequence of two light stimuli impinging on adjacent photoreceptors is the elementary event evoking an optomotor response. The relation between stimulus input to these two photoreceptors and the strength of the optomotor output follows the rule of algebraic sign multiplication (see also Varju and Reichardt, 1967)[5]

The correlation model of motion detection follows from these observations (Hassenstein and Reichardt, 1956). As illustrated in figure 1a, the output of one photoreceptor is high-pass filtered before being multiplied by a low-pass filtered signal from a neighboring receptor. The product is then integrated in time, which is equivalent to the auto-correlation of the visual input (Reichardt, 1961). Since the low-pass filter can be thought of as a delay, this direction-selective subunit (figure 1a) will respond to motion in one direction but will yield little or no response in the opposite direction. This theoretical model has a number of nontrivial properties, such as phase invariance and dependence on contrast frequency (Reichardt, 1961), that have

5. A fly eye contains about 24,000 visual receptors, (Franceschini, 1985) and four layers of neurons between the photoreceptors at the input level and motoneurons at the output level (Hausen and Hengstenberg, 1987).

Figure 1: Movement detection and figure-ground segregation. (a) The correlation model (Hassenstein and Reichardt, 1956; Reichardt, 1957) In its essential form, the two outputs from two neighboring photoreceptors are multiplied after one signal is delayed with respect to the second. If the product is averaged in time and subtracted from the time- averaged product from the other branch, the overall operation is equivalent to cross-correlating the intensity on the retina. The model is symmetrical from an anatomical point of view, but is functionally antisymmetrical. The correlation model was proposed on the basis of behavioral experiments carried out in insects (Reichardt, 1969; Varju and Reichardt, 1967). (b) The scheme proposed by Barlow and Levick (1965) to account for direction-selective ganglion cells in the rabbit retina. The output of one channel is gated in the null direction by a signal from a neighboring channel. In the preferred direction, the delayed signal from the neighboring receptor arrives too late to veto the excitatory signal. (c) A detailed cellular implementation of Barlow and Levick's scheme of direction selectivity (Koch et al., 1982), based on the proposal (Torre and Poggio, 1978) that the nonlinear interaction between excitatory and inhibitory synaptic induced conductance changes of the silent or shunting type can approximate a multiplication. Detailed computer simulations of the cable properties of dendrites show that shunting inhibition can effectively veto excitation as long as the inhibitory synapses are close to the excitatory synapses or between excitation and the cell body (Koch et al., 1986). (d) The neural circuitry underlying pattern discrimination in the visual system of the house fly *Musca Domestica*, as proposed by Reichardt and his group (Reichardt, and Poggio, 1979; Reichardt, Poggio and Hausen, 1983; Egelhaaf, 1985; Reichardt and Guo, 1986). In each eye, two overlapping sets of retinotopic arrays of elementary movement detectors of the correlation type (Hassenstein and Reichardt, 1956) respond selectively to either front-to-back (+) or back- to-front (-) motion. All movement detectors feed into two pool neurons (S_R and S_L). One of these cells (*pr*) is excited by motion in the front-to-back direction, while the second is excited by the opposite motion. The pool cells are also coupled with their homologues in the contralateral hemisphere. The output of both pool neurons shunt the output of the movement detectors (the inhibitory synaptic reversal potential is close to the resting potential of the cell), before they either excite or inhibit the two output cells (X_R and S_L). The final motor output is controlled by the X cells. The key elements of this model (X and S cells) have been identified with neurons in the third optic ganglion of the fly (Reichardt, Poggio and Hausen, 1983; Egelhaaf, 1985). Filled synapses are excitatory, open synapses are inhibitory and dotted synapses are shunting. Synapses represented by combined circle and triangle are subject to shunting. Courtesy of W. Reichardt, (Figure shown in adjoining page).

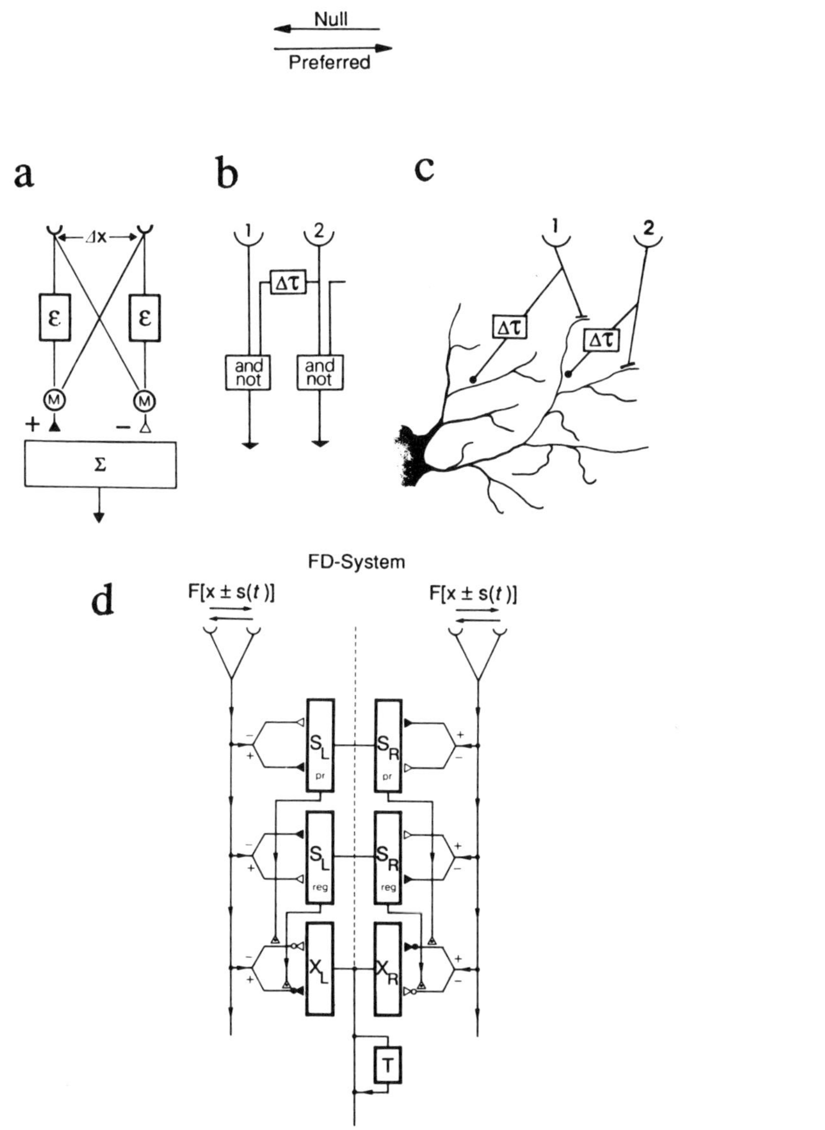
Null
Preferred
a
Δx
ε
ε
M
M
+
−
Σ
b
1
2
Δτ
and not
and not
c
1
2
Δτ
Δτ
d
FD-System
F[x ± s(t)]
F[x ± s(t)]
S L pr
S R pr
S L reg
S R reg
X L
X R
T

been confirmed experimentally. In humans, the psychophysical evidence favors a slight modification of the original correlation model (van Santen and Sperling, 1984; see also Gotz, 1965). Thus, the correlation model for motion detection is consistent with psychophysics in several different species. It is a realistic system level model.

Motion detection has also been explored at the cellular level. Barlow and Levick (1965) systematically studied the rabbit retina by recording extracellularly from the output cells of the retina. About 20% of all ganglion cells responded vigorously to the motion of both black and white spots in one direction but were silent when the spots moved in the opposite direction. Barlow and Levick's two principal conclusions were that inhibition is crucial for direction discrimination and that this discrimination must occur at many different sites within the receptive field. They proposed (figure 1b) that direction selectivity is based upon a scheme whereby the response in the null direction is vetoed or inhibited by appropriate neighboring inputs; directionality is achieved by a delay between the excitatory and inhibitory channels (extending from photoreceptors to ganglion cells). This proposal is a cellular-level version of Reichardt's correlation detector.

Techniques are being developed to study motion detection at the biophysical level. The key experimental finding is that lateral inhibition is induced in ganglion cells for all directions of movement except for the preferred direction (Wyatt and Daw, 1976). This inhibition could be generated by mechanisms presynaptic to the ganglion cell (Grzywacz and Koch, 1987). Werblin et al. (1988) have found evidence in the retina of the tiger salamander for movement-gated lateral inhibition generated by interactions among amacrine and bipolar cells. This mechanism has not yet been modeled at the cellular level. Another possible mechanism for motion detection is through postsynaptic interaction in the ganglion cell dendrite between an excitatory synapse and a delayed inhibitory synapse with a reversal potential close to the resting potential of the cell (shunting or silent inhibition), as shown in figure 1c (Rall, 1964; Torre and Poggio, 1978; Koch et al., 1982). The latter mechanism has been proven feasible with modeling studies, though a direct experimental test has not yet been performed in the retina.

Visual motion detection has many different uses. Reichardt and his group (Reichardt and Poggio, 1979; Reichardt, Poggio and

Hausen, 1983; Egelhaaf, 1985; Guo and Reichardt, 1986) have studied the use of motion information for pattern discrimination in the house fly, *Musca Domestica.* Pattern discrimination here is a special case of the more general figure-ground discrimination problem, where the figure is one pattern and the ground is the other. The fly can distinguish relative motion between two moving objects, even if both objects have a similar texture, such as random dot patterns. The cellular basis of this behavior was explored by using a combination of exacting behavioral studies, electrophysiology and systems analysis. The model circuitry (figure 1d) gives a satisfactory account of how the fly behaves under different conditions of relative motion (for example, a small figure oscillating on a background) and also explains the observed independence of the behavioral optomotor response from the spatial extent of motion – that is, the size of the moving object. Although the key neurons in the model (figure 1d) have been identified in the third optic ganglion of the fly (Egelhaaf, 1985), the proposed synaptic interactions are still under investigation.

What has been learned about visual processing from the fly? An engineer might have designed a motion system that first extracted pure velocity information. Motion detection is accomplished in the fly visual system using the same neurons that process local pattern information. The models of motion detection and motion processing have shown how populations of neurons with mixed pattern and motion signals can accomplish figure-ground segregation. We can begin to see how the fly uses time-varying visual information to control its behavior in ways that were not intuitively obvious. Models were used at the systems, network, cellular and biophysical levels to generate hypotheses and help guide experimental work.

Orientation selectivity in visual cortex Our second example is taken from the mammalian visual system. In an influential study, Hubel and Wiesel (1962) showed that most cells in cat striate cortex optimally respond to elongated bar stimuli, oriented in a specified direction. The receptive field of a simple cell is divided into several elongated subregions (figure 2a). Illumination of part or all of an excitatory region increases the cellular response, whereas such stimuli suppress the response if projected onto inhibitory regions. The input

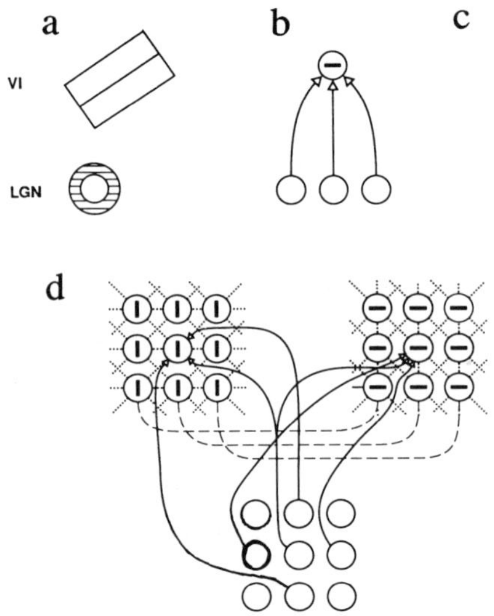

Figure 2: Wiring Underlying Orientation Selectivity. (a) The concentric center-surround receptive field of geniculate cells contrasts with the elongated receptive fields of their cortical target cells. (b) The excitatory-only model put forth by Hubel and Wiesel (1962). A row of appropriately located geniculate cells excites the oriented cortical cell. A number of geniculate cells whose receptive fields are located along a given axis monosynaptically excite the cortical cell (open arrows). The firing threshold of the simple cell determines the degree of orientation tuning. Here and in the following only the excitatory centers of the geniculate cells is drawn. (c) One instance of an inhibitory-only model (Bishop et al., 1971; Benevento et al., 1972; Sillito, 1975; Heggelund, 1981; Orban, 1984). Non-oriented cortical inhibitory (filled arrows) interneurons sculpture the orientation tuning of their target cell by suppressing its response at non-optimal orientations. For example, a horizontal bar will lead to activation of the interneuron, which will inhibit the vertically-oriented cortical cell. (d) "Eclectic" model combining features of all models (Ferster and Koch, 1987). An excitatory Hubel and Wiesel type of presynaptic arrangement is superimposed upon two inhibitory ones: reciprocal inhibition among similarly-oriented cortical cells with spatially nonoverlapping receptive fields (dotted lines) and cross-orientation inhibition among orthogonal oriented cells (not shown). Due to the massive feedback among the participating cortical neurons (the model can be formulated as an associative network (Hopfield, 1984)), each neuron acquires orientation selective via a collective computation.

to striate cortex, which arises primarily from cells in the lateral geniculate nucleus, also possess such excitatory and inhibitory subdivisions but are organized in a concentric fashion (figure 2a).

Hubel and Wiesel postulated that the orientation selectivity of simple cells was generated by a row of appropriately aligned excitatory geniculate cells converging on their cortical target neuron (figure 2b). The firing threshold of this cortical cell at its cell body is such that only simultaneous input from a large number of geniculate cells triggers a burst of action potentials. According to this hypothesis, the arrangement of the receptive fields of presynaptic inputs is sufficient to produce orientation selectivity (Hubel and Wiesel, 1962).

Blockage of cortical inhibition via pharmacological agents leads to a partial loss of orientation selectivity in some neurons (Sillito, 1975; Sillito et al. 1980). A different set of models for orientation selectivity postulates the use of intracortical inhibition (Bishop, Coombs and Henry, 1971; Benevento, Creutzfeldt and Kuhnt, 1972; Sillito, 1975; Heggelund, 1981; Orban, 1984). In the version shown in figure 2c, non-oriented cortical interneurons suppress the response of the oriented simple cell to non-optimal oriented stimuli via synaptically-mediated ihibition. However, the non-oriented inhibitory interneurons postulated by this model have not been found in the cat striate cortex. Alternatively, intracortical inhibition from oriented cells could suppress firing of cells to stimuli with orthogonal orientations (cross-orientation inhibition, Moronne et al. 1982).

More recently, an "eclectic" model has been proposed that accounts for most of the experimental data by conferring orientation selectivity upon all cells through massive feedback connections among cortical neurons (Koch, 1987; Ferster and Koch, 1987; see also Orban, 1984). It assumes a Hubel and Wiesel type of excitatory presynaptic arrangement superimposed upon two inhibitory systems: inhibition among similarly oriented cells but with spatially non-overlapping receptive fields and cross-orientation inhibition among cells with spatially overlapping receptive fields but differing orientation. Intracellular recordings of simple cells support the existence of at least the first two systems (Ferster, 1986).

It is a sign of the complexity of the brain that such a seemingly simple question – what is the circuitry underlying orientation selectivity – has not been satisfactorily answered 25 years after it was first posed. Realistic models of cortical processing could be used to test the strengths and weakness of the various proposals (Wehmeier, Dong, and Koch, 1988). A simulation of a realistic model including massive inhibitory cortical interactions is presented in figure 3. In addition to verifying the consistency of the model with experimental measurements, the simulation also makes interesting predictions for the gain control of responses to visual stimuli of varying contrasts (Wehmeier, Dong and Koch, 1988).

These models of orientation tuning for simple cells in visual cortex provide explanations for how the response properties of neurons are generated. The question remains what these properties contribute to visual processing. It is generally believed that because simple cells respond best to bars of light, they are used to detect the presence of an edge or a line at a particular orientation and position in the visual field. However, the same neurons also respond to spatial frequency gratings and textured patterns, so it is by no means clear how to infer their actual function. The problem of inferring function from response properties applies to neurons throughout the nervous system. One way to test a functional hypothesis is to construct a network model with processing units having the same response properties as those found in the nervous system and to show that the network does indeed perform the desired computation (Sejnowski, 1986). Although a working model can help generate hypotheses, and rule some out, it cannot prove that the brain necessarily solves the problem in the same way.

Using this approach, Lehky and Sejnowski (1988) have constructed a small layered feedforward network that takes as input a small patch of shaded image, as transformed by the retina and lateral geniculate nucleus, and computes the principal curvatures and orientation of the surface from which the image was derived. The resulting network can compute surface curvatures regardless of the direction of illumination, a computational problem in which the shape of a surface is recovered from the shading in the image (Ikeuchi and Horn, 1981; Pentland, 1984). The construction of the network model was made possible by recent advances in network learning algorithms that allow complex networks with many thousands of connections to be synthesized from an input-output level description of the problem (Sejnowski et al., 1986; Rumelhart et al., 1986). The network is trained by presenting images to the input layer and comparing the firing rates of the units on the output layer with the desired values. The error signals are used to correct the synaptic weights between the processing units. The learning algorithm was used solely as a technique to create the network and not as a model for learning in the brain, nor as a model for how orientation tuning actually develops[6].

The surprising finding was that most of the cells on the hidden layer, which received the geniculate input directly, had responses when probed with simulated spots and bars of light that were similar to those of the simple cells in visual cortex (figure 3b, see also figure 2a). These properties emerged through the learning procedure as the appropriate ones for extracting information about the principal curvatures and were not put into the network directly. Curvatures were represented on the output layer by units having the tuning curves shown in figure 3a. When probed with bars of light the output units had receptive field properties similar to complex cells in visual cortex that are end-stopped. (The response was reduced when the end of a bar of light was extended into an inhibitory region.[7])

A finding of general significance was that the function of the model neurons in the network could not be understood solely by examining their receptive field properties. For example, some cells with a simple receptive field were responsible not for signaling orientation to the output layer, but rather were providing information about the convexity or concavity of the surface (see figure 4). Examining the outputs of a simple cell, called its "projective" field by analogy with the input receptive field, was critical in uncovering its function in the network. Thus, the very same simple cells can be engaged in several functions – on one hand detecting edges, on the other hand participating in the computation of curvature from shading information – depending on the circuits to which the cell projects (Lehky and Sejnowski, 1988). Determining what was required to understand the function of cells in this network model has consequences for the more difficult task of determining the function of real neurons.

This simplifying network model of shape-from-shading has

---

6. For models of cortical plasticity see: Bear, Cooper and Ebner, 1987; Linsker, 1986; von der Malsburg and Willshaw, 1981, Finkel and Edelman, 1987.

7. Network models can also be constructed that extract the 1- D curvature of lines in a plane (Dobbins, Zucker and Cynader, 1987; Wilson and Richards, 1988).

Figure 3: Properties of model neurons in a network model that computes surface curvatures from shaded images of surfaces. (a) Representation of the tuning curve for one of the output cells in the network. Each cell is trained to respond jointly to one of the two principal curvatures and to the orientation of the long axis of the curved surface. The response of a single cell is ambiguous because of the broad tuning, but the pattern of responses in a population of 24 overlapping output cells uniquely represents the curvatures and orientation. When probed with simulated bars of light these cell show some properties of end-stopped complex cells in visual cortex. (b) Examples of three out of 12 cells in the middle layer of the network that receive geniculate inputs and project to the output units. The hexagonal region at the bottom represents an array of inputs to the unit from on-center geniculate cells, and the hexagonal region above it are inputs from off-center geniculate cells. The white squares represent excitatory influences and the black squares inhibitory influences and the area of the each square is proportional to the magnitude of the synaptic strength. The patterns in the synaptic weights accurately predict the response of a cell when stimulated with simulated patterns of light and can be interpreted as the receptive field of the cell. The rectangle at the top represents the values of the weights from the cell to the 24 output cells that represent the curvature. All the cells in each column have different preferred orientations, and each row is tuned to a different curvature, as indicated in (a) above. The isolated square in the upper right corner is a constant bias on the cell (negative of the threshold). The two cells on the left have oriented receptive fields and behave like simple cells when probed with bars of light (compare the wiring diagram in figure 2b). However, the middle cell provides information to the output layer not about orientation, but about the convexity of the surface, as determined by the pattern of projection to the cells on the output layer. The cell on the right has a circularly-symmetric receptive field and is sensitive to the ratio of the two principal curvatures (Figure is on adjoining page).

generated several hypotheses about cortical processing without directly simulating the detailed connectivity of the cortex, which in this case is not yet known. The prediction that end-stopped special complex cells should be selectively tuned for the surface curvature can be tested by using shaded images as stimuli along with bars of light. The model can also evolve toward a realistic model that incorporates lateral interactions between circuits at neighboring patches of the visual field and interactions with circuits that use other cues for computing shape.

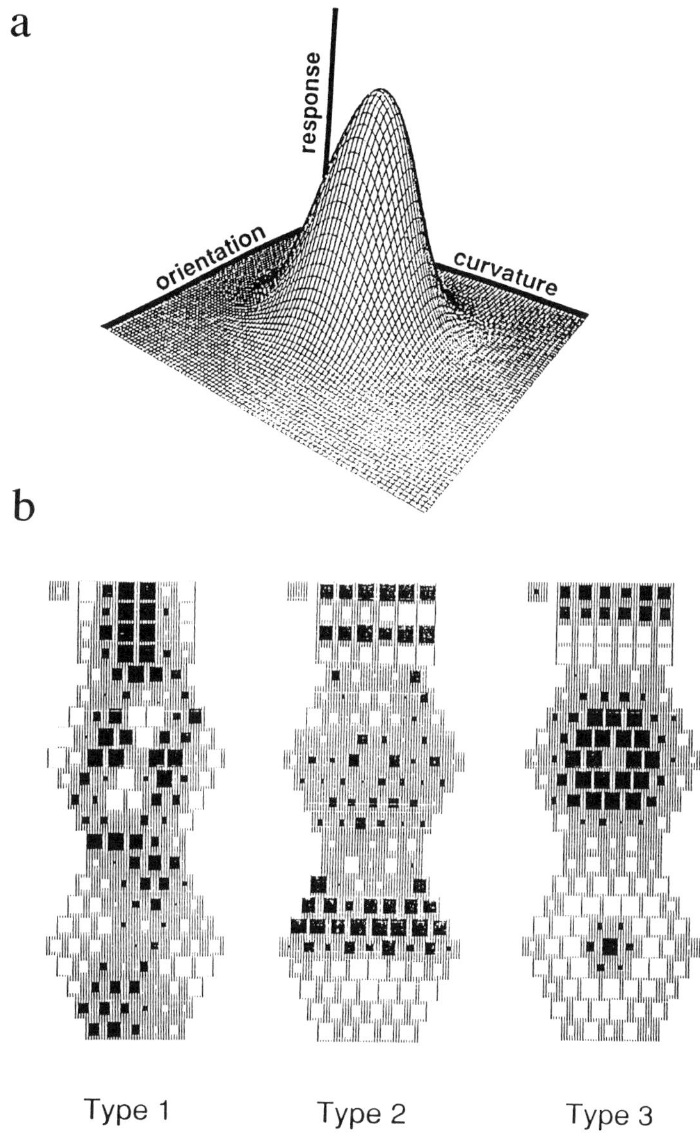
a
response
orientation
curvature
b
Type 1
Type 2
Type 3

Conclusions

A scientific field is defined primarily by its problem-space and its successful large-scale theories. Until there are such theories in computational neuroscience, the field is defined mostly by the problems it seeks to solve and the general methodology and specific techniques it hopes will yield successful theories. Models of brain function driven primarily by functional considerations can provide only the most general guidance about what might be happening in the brain; conversely, models driven primarily by signal measurements and anatomy can easily miss those aspects of the signals that are relevant for information processing. In this review we have presented several examples of models that attempt to combine insights from both the functional and the implementation levels of analysis. More examples can be found in other areas of neurobiology where models are being used to explore information processing by brain mechanisms at many different levels of structural organization.

Realistic and simplifying brain models have been distinguished to reveal their separate strengths and weakness. Neither type of model should be used uncritically. Realistic models require a substantial empirical database; it is all too easy to make a complex model fit a limited subset of the data. Simplifying models are essential but are also dangerously seductive; a model can become an end in itself and can lose touch with nature. Ideally these two types of models should complement each other. For example, the same mathematical tools and techniques that are developed for studying a simplifying model could well be applied to analyzing a realistic model, or even the brain itself. More accurately, the two types of models are really end points of a continuum, and any given model may have features of both. Thus, we expect future brain models to be intermediate types that combine the advantages of both realistic and simplifying models.

It may be premature to predict how computational models will develop within neuroscience over the next decade, but several general features are already emerging. First, in view of the many different structural levels of organization in the brain, and the realization that models rarely span more than two levels, we expect that many different types of models will be needed. It will be especially difficult to find a chain of models to cross the gap of at least three identified levels between the cellular and systems levels of investigation.

Second, a model of an intermediate level of organization will necessarily simplify with respect to the structural properties of lower level elements, though it ought to try to incorporate as many of that level's functional properties as actually figure in the higher level's computational tasks. Thus, a model of a large network of neurons will necessarily simplify the molecular level within a single neuron.

At this stage in our understanding of the brain, it may be fruitful to concentrate on models that suggest new and promising lines of experimentation, at all levels of organization. In this spirit, a model should be considered a provisional framework for organizing possible ways of thinking about the nervous system. The model may not be able to generate a full range of predictions owing to incompleteness, some assumptions may be unrealistic simplifications, and some details may even be demonstrably wrong[8]. Nevertheless, if the computational model is firmly based on the available experimental data, it can evolve along with the experimental program and help to guide future research directions.

Acknowledgements

A version of this chapter first appeared in Sejnowski, Koch and Churchland (1989). We are grateful to Drs. F. Crick, T. Poggio and W. Reichardt for helpful comments on this paper. TJS is supported by an NSF Presidential Young Investigator Award and grants from the Sloan Foundation, the Seaver Institute, the Lounsbury Foundation, the Air Force Office of Scientific Research, and the Office of Naval Research, and the General Electric Corporation; CK is supported by grants from the Office of Naval Research, the Advanced Engineering program at the National Science Foundation, the Sloan Foundation, and by the Powell Lee Foundation; PSC is supported by a grant from the National Science Foundation.

8. An example of an imperfect, but useful model from physics is the Bohr model of the hydrogen atom. This model served an essential role as a bridge between classical physics and quantum physics.

References

Abeles, M. (1982). Local Cortical Circuits. New York, Springer-Verlag.

Abu-Mostafa, Y.S., and Psaltis, D. (1987). Optical Neural Computers. Scientific American 256: 88-95.

Abu-Mostafa, Y.S. (1986). The complexity of information extraction. IEEE Transactions on Information Theory IT-32: 513-525.

Alkon, D.L. (1987). Memory Traces in the Brain. New York, Cambridge University Press.

Amari, S. (1983). Field theory of self-organizing neural nets. IEEE Transactions on Systems, Man, and Cybernetics SMC-13: 741-748.

Anastasio, T. J. and Robinson, D.A. (1989). Distributed parallel processing in the vestibulo-oculomoter system. Neural Computation, 1, 230-241.

Arbib, M. (1989). Brain theory: A historical perspective. In: Schwartz, E., Computational Neuroscience. Cambridge, MA: MIT Press.

Bachus, R., Mueller, E., Koenig, H., Braeckle, G., and Weber, H. (1987). Functional imaging using NMR. In: McCready, V.R.,Leach, M. and Ell, P.J., Functional Studies Using NMR. New York: Springer- Verlag, 43-60.

Barlow, H.B. (1986). Why have multiple cortical areas?. Vision Res. 26: 81-90.

Barlow, H.B., and Levick, W.R. (1965). The mechanism of directionally sensitive units in rabbit's retina. J. Physiol. 178: 477-504.

Bear, M., Cooper, L.N., and Ebner, F.F. (1987). A physiological basis for a theory of synapse modification. Science 237: 42-48.

Benevento, L.A., Creutzfeldt, O.D., and Kuhnt, H. (1972). Significance of intracortical inhibition in the visual cortex. Nature, New Biology 238: 124-126.

Bishop, P.O., Coombs, J.S., and Henry, G.H. (1971). Interaction effects of visual contours on the discharge frequency of simple striate neurones. J. Physiol. (Lond.) 219: 659-681.

Brodie, S. E., Knight, B. W. and Ratliff, F. (1978). The response of the *Limulus* retina to moving stimuli: a prediction by Fourier synthesis. J. Gen. Physiol., 72, 129-166.

Brown, T.H., Ganong, A.H., Kariss, E.W., Keenan, C.L., and Kelso, S.R. (1989). Long-term potentiation in two synaptic systems of the hippocampal brain slice. In: Byrne, J.H. and Berry, W.O., Neural Models of Plasticity. New York: Academic Press, in press.

Churchland, P.S., Koch, C., and Sejnowski, T.J. (1989). What is computational neuroscience?. In: Schwartz, E., Computational Neuroscience. Cambridge: MIT Press.

Churchland, P.S. and Sejnowski, T. J. (1989a), Perspectives on cognitive neuroscience, Science, 242: 741-745.

Churchland, P.S., and Sejnowski, T.J. (1989b). Neural representation and neural computation. In: Nadel, L., Neural Connections and Mental Computation. Cambridge, MA: MIT Press, 15-48.

Cowan, J.D., and Sharp, D.H. (1988). Neural nets and artificial intelligence. Daedalus 117, 85-122.

Dobbins, A., Zucker, S.W., and Cynader, M.S. (1987). Endstopped neurons in the visual cortex as as substrate for calculating curvature. Nature 329: 438-441.

Donegan, N.H., Gluck, M.A., and Thompson, R.F. (1988). Integrating behavioral and biological models of classical conditioning. In: G. H. Bower (Ed.), Psychology of Learning and Motivation : San Diego: Academic Press.

Egelhaaf, M. (1985). On the neuronal basis of figure-ground discrimination by relative motion in the visual system of the fly. II.: Figure-detection cells a new class of visual interneurons. Biol. Cybern. 52: 195-209.

Feldman, J.A., and Ballard, D.H. (1982). Connectionist models and their properties. Cognitive Science 6: 205-254.

Ferster, D. (1986). Orientation selectivity of synaptic potentials in neurons of cat primary visual cortex. J. Neurosci. 6: 1284-1301.

Ferster, D., and Koch, C. (1987). Neuronal connections underlying orientation selectivity in cat visual cortex. Trends in Neurosciences 10: 487-492.

Finkel, L.H., and Edelman, G.M. (1987). Population rules for synapses in networks. In: Edelman, G.M.,Gall, W.E. and Cowan, W.M., Synaptic Function. New York: John Wiley and Sons, 711-757.

Franceschini, N. (1985). Early processing of colour and motion in a mosaic visual system. Neurosci. Res. Suppl. 2: S17-S49.

Freeman, W.J. (1975). Mass Action in the Nervous System. New York, Academic Press.

Gerstein, G.L., and Aertsen, A.M.H.J. (1985). Representation of cooperative firing activity among simultaneously recorded neurons. Journal of Neurophysiology 54: 1513-1528.

Getting, P.A. (1983). Mechanisms of pattern generation underlying swimming in Tritonia. II. Network Reconstruction. Journal of Neurophysiology 49: 1017-1035.

Goldman-Rakic, P.S. (1988). Topography of cognition: Parallel distributed networks in primate association cortex. In: Cowan, W.M., Shooter, E.M.,Stevens, C.F. and Thompson, R.F., Ann. Rev. Neurosci.. Palo Alto: Annual Reviews Inc., 137-156.

Gotz, K.G. (1965). Optomotorische Untersuchung des visuellen Systems einiger Augenmutanten der Fruchtfliege Drosophila. Kybernetik 2: 215-221.

Grinvald, A. (1985). Real-time optical mapping of neuronal acitvity: from single growth cones to the intact mammalian brain. In: Cowan, W.M.,Shooter, E.M.,Stevens, C.F. and Thompson, R.F., Annual Review of Neurocience. Palo Alto: Annual Reviews, Inc., 263-305.

Grossberg, S. (1988). Nonlinear neural networks: Principles, mechanisms, and architectures. Neural Networks 1: 17-61.

Grossberg, S., and Kuperstein, M. (1986). Neural Dynamics of Adaptive Sensory-Motor Control. Amsterdam, North Holland.

Grzywacz, N.M., and Koch, C. (1987). Functional properties of models for direction selectivity in the retina. Synapse 1: 417-434.

Guo, A., and Reichardt, W. (1986). Figure-ground discrimination by relative movement in the visual system of the fly. Part III. Experiments and simulations of complex figure-ground situations with movement detectors input elements to the figure-ground circuit. Biol. Cybern. 53: 285-306.

Hassenstein, B., and Reichardt, W. (1956). Systemtheoretische Analyse der Zeit-, Reihenfolgen- und Vorzeichenauswertung bei der Bewegungsperzeption des Russelkafers Chlorophanus. Z. Naturforsch 11b: 513-524.

Hausen, K., and Hengstenberg, R. (1987). Multimodal convergence of sensory pathways on moto-neurons of flight muscles in the fly. Annual Meeting Soc. Neurosci. Abst. 17: 295.

Heggelund, P. (1981). Receptive field organization of simple cells in cat striate cortex. Exp. Brain Res. 42: 89-98.

Heiligenberg, W.J. (1987). Sensory processing of sensory information in electric fish. J. Comp. Physiol. A161: 621-631.

Hildreth, E., and Koch, C. (1987). The analysis of visual motion: From computational theory to neuronal mechanisms. Ann. Rev. Neurosci. 10: 477-533.

Hildreth, E.C., and Holerbach, J.M. (1987). The computational approach to vision and motor control. In: Plum, F., Handbook of Physiology. : Am. Physiol. Soc.

Hinton, G.E., and Anderson, J.A. (1981). Parallel Models of Associative Memory. Hillsdale, NJ, Lawrence Erlbaum Associates.

Hodgkin, A.K., and Huxley, A.F. (1952). Currents carried by sodium and potassium ions through the membrane of the giant axon of Loligo. Journal of Physiology 116: 449-472.

Hodgkin, A.L. (1976). Chance and design in electrophysiology: An informal account of certain experiments on nerve carried out between 1934 and 1952. J. Physiol. 263: 1-21.

Hopfield, J.J. (1982). Neural networks and physical systems with emergent collective computational abilities. Proc. Natl. Acad. Sci. USA 79: 2554-2558.

Hopfield, J.J. (1984). Neurons with graded response have collective computation abilities. In:, Proceedings of the National Academy of Sciences USA. :, 3088-3092.

Hopfield, J.J., and Tank, D.W. (1986). Computing with neural circuits: A model. Science 233: 625-633.

Houk, J.C., and Rymer, W.Z. (1981). Neural control of muscle length and tension. In: Brooks, V.B., Handbook of Physiology Section I: The Nervous System, Vol. II, Motor Control. Bethesda, MD: American Physiological Society, 257-323.

Hubel, D.H., and Wiesel, T.N. (1962). Receptive fields, binocular interaction and functional architecture in the cat's visual cortex. J. Physiol. (Lond.) 160: 106-154.

Hurlbert, A., and Poggio, T. (1988). Synthesizing a color algorithm from examples. Science 239: 482.

Hutchinson, J., Koch, C., Luo, J., and Mead, C. (1988). Computing motion using analog and binary resistive networks. IEEE Computer: 21, March No. 3, 52-64.

Ikeuchi, K., and Horn, B.K.ti 2em. (1981). Numerical shape from shading and occluding boundaries. Artificial Intelligence: 17: 141-184.

Jacob, F. (1982). The possible and the actual. Seattle, University of Washington Press.

Kandel, E.R., Klein, M., Hochner, B., Shuster, M., Siegelbaum, S.A., Hawkins, R.D., Glanzman, D.L., and Castellucci, V.F. (1987). Synaptic modulation and learning: New insights into synaptic transmission from the study of behavior. In: Edelman, G.M.,Gall, W.E. and Cowan, W.M., Synaptic Function. New York: John Wiley and Sons, 471-518.

Knudsen, E.I., du Lac, S., and Esterly, S.D. (1987). Computational maps in the brain. In: Cowan, W.M.,Shooter, E.M.,Stevens, C.F. and Thompson, R.F., Annual Review of Neuroscience. Palo Alto: Annual Reviews, Inc., 41-65.

Koch, C., and Segev, I. (1989) Methods in neuronal modeling. Cambridge, Mass.: MIT Press.

Koch, C., Poggio, T., and Torre, V. (1986). Retinal ganglion cells: A functional interpretation of dendritic morphology. Trends Neurosci. 9: 204-211.

Kohonen, T. (1984). Self-Organization and Associative Memory. New York, Springer-Verlag.

Kohonen, T. (1988). An introduction to neural computing. Neural Networks 1: 3-16.

Krauzlis, T. J. and Lisberger, S. G. (1989) A control systems model of smooth pursuit eye movements with realistic emergent properties. Neural Comp., 1, 116-122.

Lapdes, A., and Farber, R. (1988). How Neural Nets Work. In: Anderson, D.Z., Neural Information Processing Systems. New York: American Institute of Physics, 442-443.

Lee, C., Rohrer, W.H., and Sparks, D.L. (1988). Population coding of saccadic eye movements by neurons in the superior colliculus. Nature 332: 357-360.

Lehky, S.R., and Sejnowski, T.J. (1988). Network model of shape-from- shading: Neural function arises from both receptive and prejective fields. Nature 333: 452-454.

Linsker, R. (1986). Emergence of orientation columns. Proceedings

of the National Academy of Sciences USA 83: 8779-8783.
Lippmann, R. P. (1987). An introduction to computing with neural nets. IEEE Acoustics, Speech, and Signal Processing 4: 4-22.
Lisberger, S.G., Morris, E.J., and Tychsen, L. (1987). Visual motion processing and sensory-motor integration for smooth pursuit eye movements. Ann. Rev. Neurosci. 10: 97-129.
Llinas, R.R. (1985). Electronic transmission in the mammalian central nervous system. In: Bennett, M.E.L. and Spray, D.C., Gap Junctions. Cold Spring Harbor: Cold Spring Harbor Laboratory, 337-353.
Lynch, G. (1988). Structure-functions considerations in the study of memory in cortical networks. In: Gazzaniga, M.S., Perspectives in Memory Research and Training. Cambridge, MA: MIT Press, in press.
Marr, D. (1982). Vision. San Francisco, Freeman.
Marr, D., and Poggio, T. (1976). Cooperative computation of stereo disparity. Science 194: 283-287.
McNaughton, B.L., and Morris, R.G. (1987). Hippocampal synaptic enhancement and information storage within a distributed memory system. Trends in Neurosciences 10: 408-415.
Mead, C. (1989). Analog VLSI and Neural Systems. Reading, MA, Addison-Wesley.
Merzenich, M.M., and Kaas, J.H. (1982). Reorganization of mammalian somatosensory cortex following peripheral nerve injury. Trends in Neurosciences 5: 434-436.
Mitchison, G.J., and Durbin, R.M. (1988). The learning capacity of multi-layer networks. King's College Research Center, Cambridge, U.K. : Technical Report.
Morrone, N.C., Burr, D.C., and Massei, L. (1982). Proc. R. Soc. London Ser. B 216: 335.
Mountcastle, V.B. (1979). An organizing principle for cerebral function: The unit module and the distributed system. In: Adelman, G. and Smith, B.H., The Neurosciences. Cambridge, MA: MIT Press, 21-42.
Nakayama, K. (1985). Biological image motion processing: A review. Vision Res. 25: 625-660.
Orban, G.A. (1984). Neuronal Operations in the Visual Cortex. New York, Springer-Verlag.
Palm, G. (1982). Neural Assemblies An Alternative Approach to

Artificial Intelligance. Hiedelberg, Springer-Verlag.
Pellionisz, A., and Llinas, R. (1985). Tensor network theory of the metaorganization of functional geometries in the central nervous system. Neuroscience 16: 245-273.
Pentland, A. P. (1984). Local shading analysis. IEEE Transactions on Pattern Analysis and Machine Intelligence 6: 170-187.
Raichle, M.E. (1986). Neuroimaging. Trends in Neurosciences 9: 525- 529.
Rall, W. (1964). Theoretical significance of dendritic trees for neuronal input-output reltaions. In: Reiss, R.F., Neuronal Theory and Modeling. Palo Alto: Stanford University Press, 73-97.
Ratliff, F. (1962). Some interrelations among physics, physiology, and psychology in the study of vision. In: Koch, S., Psychology: A Study of a Science. New York: McGraw-Hill, 417-482.
Reichardt, W. (1957). Autokorrelations-Auswertung als Funktionsprinzip des Zentralnervensystems (bei der optischen Wahrnehmung eines Insektes). Z. Naturforsch 12b: 448-457.
Reichardt, W. (1961). Autocorrelation a principle for evaluation of sensory information by the central nervous system. In: Rosenbluth, W.A., Principles of Sensory Communications. New York: Wiley, 303-317.
Reichardt, W. E. and Guo, a. (1986) Elementary pattern discrimination (Behavioral experiments with the fly *Musca domestica*) Biol. Cyber. 53: 285-306.
Reichardt, W., and Poggio, T. (1979). Figure-ground discrimination by relative movement in the visual system of the fly. Part I. Experimental results. Biol. Cybern. 35: 81-100.
Reichardt, W., Poggio, T., and Hausen, K. (1983). Figure-ground discrimination by relative movement in the visual system of the fly. Part II. Towards the neural circuitry. Biol. Cybern. 46: Suppl. 46, 1-30.
Reichardt, W.E., and Poggio, T. (1981). Theoretical Approaches in Neurobiology. Cambridge, MA, MIT Press.
Reitboeck, H.J.P. (1983). A 19-channel matrix drive with individually controllable fiber microelectrodes for neurophysiological applications. IEEE Transactions on Systems, Man, and Cybernetics SMC-13: 676-683.
Robinson, D.A. (1981). The use of control systems analysis in the

neurophysiology of eye movements. Annual Review of Neuroscience 4: 463-503.

Rumelhart, D.E., and McClelland, J.L. (1986). Parallel Distributed Processing: Explorations in the Microstructure of Cognition. Cambridge, MA, MIT Press.

Schwartz, E. (1988). In: Schwartz, E., Computational Neuroscience. Cambridge, MA: MIT Press.

Sejnowski, T.J. (1987). Computing with Connections. J. Math. Psych. 31: 203-210.

Sejnowski, T.J., Koch, C., and Churchland, P.S. (1988) Computational neuroscience. Science, 241, 1299-1306.

Sejnowski, T.J., and Churchland, P.S. (1989). Brain and Cognition. In: Posner, M.I., Foundations of Cognitive Science. Cambridge, MA: MIT Press, in press.

Sejnowski, T.J. (1986). Open Questions About Computation in Cerebral Cortex. In: McClelland, J.L., and Rumelhart, D.E., Parallel Distributed Processing, Volume 2. Cambridge, MA: MIT Press, 327-389.

Sejnowski, T.J., Kienker, P.K., and Hinton, G.E. (1986). Learning symmetry groups with hidden units: Beyond the perceptron. Physica 22D: 260-275.

Sejnowski, T.J., and Rosenberg, C.R. (1987). Parallel networks that learn to pronounce English text. Complex Systems 1: 145-168.

Sejnowski, T.J., and Tesauro, G. (1989). Building network learning algorithms from Hebbian synapses. In: McGaugh, J.L.,Weinberger, N.M. and Lynch, G., Brain Organization and Memory. New York: Oxford University Press, in press.

Selverston, A.I., Miller, J.P., and Wadepuhl, M. (1983). Neural mechanisms for the production of cyclic motor patterns. IEEE Transactions on Systems, Man and Cybernetics SMC-13: 749-757.

Shriver, B.D. (1988). Special Issue: Artificial Neural Systems. Computer, 21, no. 3.

Sillito, A.M. (1975). The contribution of inhibitory mechanisms to the receptive field properties of neurons in cat striate cortex. J. Physiol. (Lond.) 250: 305-329.

Sillito, A.M., Kemp, J.A., Milson, J.A., and Berardi, N. (1980). The re-evaluation of the mechanisms underlying simple cell orientation selectivity. Brain Res. 194: 517-520.

Sivilotti, M., Mahowald, M., and Mead, C. (1987). Real-time visual computation using analog CMOS processing arrays. In: Losleben, P., Stanford VLSI Conference. :, 295-312.

Sokoloff, L. (1984). Metabolic Probes of Central Nervous System Activity in Experimental Animals and Man. Sunderland, MA, Sinauer Associates.

Tanner, J., and Mead, C. (1986). In: Kung, S.Y.,Owen, R. and Nash, G., VLSI Signal Processing II. New York: IEEE Press, 59-76.

Tesauro, G., and Janssens, B. (1988). Scaling relationships in backpropagation learning. Complex Systems 2: 39-44.

Thompson, R.F. (1986). The neurobiology of learning and memory. Science 233: 941-947.

Torre, V., and Poggio, T. (1978). A synaptic mechanism possibly underlying directional selectivity to motion. Proc. R. Soc. Lond. B 202: 409-416.

Traub, R.D., Miles, R., and Wong, R.K.S. (1987). Models of synchronized hippocampal bursts in the presence of inhibition. I. Single population events. Journal of Neurophysiology 58: 752-764.

Tsien, R.Y., and Poenie, M. (1986). Fluorescence ratio imaging: a new window into intracellular ionic signaling. Trends in Biochemical Sciences 11: 450-455.

Ullman, S. (1986). Artificial intelligence: Computational studies of the visual system. Ann. Rev. Neurosci. 9: 1-26.

van Santen, J.P.H., and Sperling, G. (1984). A temporal covariance model of motion perception. J. Opt. Soc. Am. A1: 451-473.

Varju, D., and Reichardt, W. (1967). Ubertragungseigenschaften im auswertesystem fur das bewegungssehen. Z. Naturforsch 22b: 1343- 1351.

von der Malsburg, C., and Willshaw, D. (1981). Co-operativity and brain organization. Trends in Neurosciences 4: 80-83.

Wehmeier, U., Dong, D., Koch, C., and Van Essen, D. (1988). Modeling the mammalian visual system. In: Koch, C. and Segev, I., Methods in Neuronal Modeling. Cambridge, MA: MIT Press, 335-360.

Werblin, F., Maguire, G., Lukasiewicz, P., Eliasof, S., and Wu, S. (1988). Neural interactions mediating the detection of motion in the retina of the tiger salamander. Visual Neuroscience 1:

317-329.

Williamson, S.J., Romani, G.L., Kaufman, L., and Modena, I. (1983). Biomagnetism: An Interdisciplinary Approach. New York, Plenum Press.

Wilson, H.R., and Richards, W.A. (in press). Mechanisms of contour curvature discrimination. J. Opt. Soc. Am. :.

Wilson, M.A., and Bower, J.M. (1989). The simulation of large-scale neural networks, In: Koch, C. and Segev, I., Methods in Neuronal Modeling. Cambridge MA: MIT Press, 291-334.

Wyatt, H.J., and Daw, N.W. (1976). Specific effects of neurotransmitter antagonists on ganglion cells in rabbit retina. Science 191: 204-205.

Zipser, D., and Andersen, R. (1988). Back propagation learning simulates response properties of a subset of posterior parietal neurons. Nature 332: 679-684.

CHAPTER 3

The Behavioral Analysis of Associative Learning in the Terrestrial Mollusc *Limax maximus*: The Importance of Inter-Event Relationships

C. L. Sahley

Introduction

Associative learning is a unique ability that gives an animal the capability to assess regularities in its environment; that is, to extract causal relations between stimuli in the environment and then modify its behavior accordingly. Thus any organism that can demonstrate associative learning has gained a great selective advantage for survival. Animals from many species, both vertebrate and invertebrate, have been shown to be capable of associative learning, and tremendous progress has been made toward the delineation of the cellular and molecular events underlying this behavioral phenomenon (see Byrne, 1987; Carew and Sahley, 1986; Hawkins, Clark, and Kandel, 1987; Mpitsos and Lukowiak, 1985; Farley and Alkon, 1985 for reviews).

Moreover, in recent years a tremendous interest has emerged in the development of models of learning. Before we can have comprehensive models of the associative process, whether dealing with cellular or computer models, it is important to understand the environmental conditions and temporal parameters that produce learning. In experimental terms, associative learning is a theoretical construct, an inference made on the basis of observed relationships between an organism's behavior and its known past experiences. To conclude that associative learning has occurred is to say that the organism's behavior is a product of its experience with a relationship between at least two stimulus events. It is critical to demonstrate that it is the organism's experience with the inter-event relationship per se and not some other variable confounded with that relationship that produced the observed change in behavior. In the Pavlovian conditioning experiment, the experimenter is required to assess the effects of the conditioned stimulus-unconditioned stimulus (CS-US) pairing operation against the effects of a host of control procedures–for example, CS alone, US alone, random presentations of the CS and US–before concluding that associative learning processes were responsible for the ability of the CS to evoke conditioned responding. Once this phase is complete it is essential to characterize the important operating properties of the associative process involved by systematically varying the inter-event relation that was originally observed to produce the learned behavior changes. From this point of view, the study of associative learning is the experimental analysis of the kinds of inter-event relations that influence the organism's behavior.

Several inter-event relationships have been proved to be critical in the determination of the kind of learning and thus the behavioral outcome. These include: (1) temporal pairing; (2) contingency, forward, and backward relationships of the stimuli; and (3) second-order relationships. The goal of this chapter is to discuss these relationships and their behavioral outcomes in such a way that they can serve as a basis for the development of models, at both the level of the computer and the neural network. The examples used will be from experiments conducted over the past several years on the terrestrial mollusc *Limax maximus*, in collaboration with Alan Gelperin, Jerry Rudy, and Kathleen Martin.

The Behavioral Preparation

The animal used in these experiments is the gastropod mollusc, *Limax maximus*. *Limax* is but one of the many invertebrates that have been used to study brain-behavior relationships, including Apis mellifera (Bitterman, 1982; Couvillon and Bitterman, 1980; Erber, 1976; 1982; Erber and Schildberger, 1980; Kuwabara, 1957; Menzel, 1967, 1968; Menzel and Erber, 1972; Grossman, 1970), Aplysia californica (Cook and Carew, 1985; Carew, Walters, and Kandel, 1981; Colwill, et. al., 1988; 1983; Lukowiak and Sahley, 1981; Lukowiak, 1982; Susswein and Schwarz, 1983; Susswein, Schwarz and Feldman, 1986; Walters and Byrne, 1983, Walters, Carew, and Kandel, 1979), Drosophila melanogaster (Aceves-Pina and Quinn, 1979; Booker and Quinn, 1981; Dudai, 1977; Menne and Spatz, 1981; Quinn, Harris, and Benzer,1974; Tempel et al., 1983), Pleurobranchaea californica (Mpitsos and Cohan, 1986a, Mpitsos and Cohan, 1986b; Mpitsos and Collins, 1975; Mpitsos and Davis, 1973), Lymnaea stagnalis (Alexander, Audesirk, and Audesirk, 1984; Audesirk et al.,1982) Achatina fulica (Croll and Chase, 1977; 1980) Hirudo medicinalis (Henderson and Strong, 1972; Sahley and Ready, 1988; Sahley, 1988), Schistocerca gregaria (Forman and Hoyle, 1978; Hoyle, 1980; Hoyle and Field, 1983; Horridge, 1962), Hermissenda crassicornis (Alkon, 1983; Crow, 1983; 1988; Crow and Alkon, 1978; Crow and Offenbach, 1983). Invertebrates have proved fruitful as powerful model systems in which to explore the cellular and molecular mechanisms of many behaviors, including learning and memory. This is largely due to the fact that these animals have simpler behaviors and simpler nervous systems than those found in vertebrates; thus, it is possible to identify the neural circuit mediating a particular behavior. Once a behavioral circuit is identified, then it is possible to explore the synaptic interactions between cells within that circuit.

In invertebrates, neurons can be uniquely identified by way of size, location, pigment (in some invertebrates), and input-output characteristics. These elements remain constant from one preparation to another. In addition, the neurons are large making them amenable for electrophysiological (standard intra and extracellular recording, voltage clamp, and patch clamp), biochemical, and molecular analyses.

Another important advantage provided by invertebrates is the ease with which semi-intact preparations can be developed to study brain-behavior relationships. Semi-intact preparations provide the opportunity to monitor behavior and cellular activity simultaneously. In this manner it is possible to monitor neural activity as the behavior occurs and thus determine the causal cellular events mediating behavior and, in the case of learning, the causal cellular changes that result in the learned change in behavior.

In a more general sense, the use of invertebrates reflects a reductionistic and an evolutionary strategy. Associative learning is the way an animal extracts causal relations from the environment. In order for an association to be formed, learning psychologists have determined that the stimuli in the environment must occur in a precise temporal relationship. The inter-event relations between the cells are the important variables in the associative process. The temporal relationship, the predictive relationship, and the order relationship of the stimulus events are critical determinants of whether an association is formed and what kind of association is formed.

To date, the rules that influence associative learning across many species both, vertebrate and invertebrate, remain conserved. In retrospect this is not surprising, since all species must survive in the same general environment. Although one cannot make assumptions of similarities in cellular functioning from similarities in behavior, it is intriguing to think that some mechanistic similarities may exist. There is no doubt that differences between vertebrates and invertebrates exist; however, the use of invertebrates may reveal cellular processes and molecular processes mediating learning which are the fundamental building blocks of the associative process which will be shared between vertebrates and invertebrates.

Limax is a generalized herbivore whose feeding behavior is dominated by two classes of stimuli, olfactory and gustatory. The slug is attracted to potential food sources by their odors, which it detects with olfactory receptors located on its superior tentacles. When a potential food source is located, *Limax* everts its lips so that the gustatory receptors contact the food surface. If a bitter taste is encountered, the food is rejected.

Following the lead of Gelperin (1975) our goal was to exploit in the laboratory the natural odor-taste sequence encountered by the foraging slug. If, for example *Limax* could associate the independently sensed odor and taste properties of various potential food sources, it might forage more economically by avoiding odors that had learned signal aversive taste.

To study odor-taste associations we developed a simple behavioral preparation. Slugs were exposed to normally preferred food odors (either potato or carrot odor) for two minutes, and then were exposed to a bitter taste provided by a saturated solution of quinidine sulfate applied to their lips (in molluscs chemoreceptors are located directly on the lip surface).

After slugs were exposed to an odor-quinidine pairing, their preference for the odor was assessed (usually 24 hours following training) in a two-choice test procedure. The slug was placed in the test apparatus where it was presented with two food odors, one (carrot or potato odor) that had been paired with the quinidine, and another, generated by laboratory rat chow, that had been its normal food source. From a series of three 60-sec tests we calculated the percentage of total test time (180 sec) that the slug spent over the odor paired with the quinidine (see Sahley, Gelperin and Rudy, 1981; Sahley, Rudy and Gelperin, 1981 for further details).

Pairing

The original idea regarding associative learning was that the the critical variable determining the formation of associations between two stimuli was the temporal relation of the CS and US, that is, the pairing of these two stimuli (see Mackintosh, 1975). For all animals, regardless of species, the time interval separating the CS and US governs the probability that CS will become associated with a US. Indeed, depending on the species and the response measure, the time interval which will support conditioning varies from milliseconds (rabbit nictitating membrane response; Gormezano, 1972) to hours (taste aversion learning in rats; Revusky and Garcia, 1970). Nevertheless, as this interval increases, conditioning decreases. In the basic experiment slugs were exposed to normally preferred food odors (either potato or carrot odor) (CS) for two minutes. They were then presented immediately with a bitter taste, a saturated solution of quinidine sulfate (US). Following training, the effect of the pairings was assessed on the slug's odor preference behavior in a two choice test procedure.

A schematic representation of the training protocol for the pairing experiment is presented in figure 1. Four groups of slugs were used in this experiment, two experimental groups, CQ and PQ, and two control groups CS and PS. Slugs in each of the experimental groups experienced both food odors, however only one food odor in each group was paired with the US (quinidine). To control for the effect of the non-aversive qualities of the US, the food not paired with the US was paired with a dilute salt solution (saline). If the training procedures produced associative learning the learning should be specific to the odor paired with the US and the slug's behavior toward the unpaired odor should remain unchanged. Slugs in each of the control groups experienced the same conditions as slugs in the experimental groups except that both odors were paired with saline. We included these two additional groups to control for the nonassociative effects that repeated presentations of the odors might have on the slugs' preference behavior. In subsequent testing, the slug was presented with two odors, one (carrot or potato) that had been paired with quinidine, and another, provided by laboratory rat chow (the normal slug diet). Typically, slugs prefer carrot or potato odor when given a choice between carrot versus rat chow or potato versus rat chow odor.

Pairing also has been shown to be critical in the formation of associations in the slug. But following the CS-US pairings the slugs

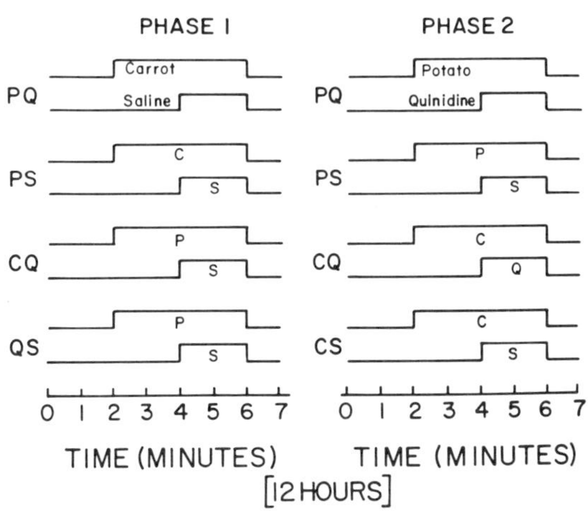

Figure 1: Training Procedure for Experiment 1. Schematic representation of the training protocol for the four training conditions. Slugs in the PQ group experienced paired presentations of potato odor and quinidine in phase 2 and carrot and saline in phase 1. Slugs in the PS and CS groups experienced both potato and carrot odor paired with saline. The difference between the two experimental groups and the two control groups is simply the order of the odor presentation. Slugs in group CQ experienced the carrot odor paired with quinidine in phase 2 and the potato odor paired with saline presentation in phase 1.

showed a dramatic reduction in the amount of time spent over the odor paired with the quinidine. Neither the odor experience nor the quinidine experience alone or in an unpaired relationship was sufficient to reduce the normal odor preference to potato or carrot odor (see Figure 2). That is, the learned changes in *Limax's* food odor preferences were selective. Conditioned reductions in *Limax's* preference for potato odor did not generalize to carrot odor and vice versa.

After demonstrating the slug's ability to learn to avoid an odor paired with a bitter taste we asked the question: Can a odor preference be established as a result of appetitive conditioning? Only a few instances of learning using appetitive reinforcers have been reported (Alexander, Audesirk, and Audesirk, 1984; Audesirk, et al., 1982; Croll and Chase, 1977; 1980; Sahley, et al. ,1982). It is important to establish appetitive conditioning in *Limax* for a couple of reasons. First, we had previously demonstrated that *Limax* can learn to suppress an odor preference as a result of associative conditioning and if we could demonstrate that *Limax* could establish a new odor preference, also as a result of associative learning, then in a single preparation we can compare and contrast the cellular and molecular mechanisms of both of these types of associative learning. Second, a synaptic model of the associative learning mechanism used by *Limax* to alter its responses to odor and taste inputs (called LIMAX) has been developed (Gelperin, Hopfield, and Tank, 1985; Gelperin, Tank, and Tesauro, 1988) and is presented in by Tesauro in Chapter 4. The model provides for the storage of odor and taste inputs and predicts precise behavioral outcomes as a result of learning dependent changes in the representations of the odor and taste inputs. Since the model contains inputs for both aversive and appetitive USs, it is important to have a full range of behavioral data on appetitive conditioning as well as aversive conditioning to test the accuracy of the LIMAX learning model.

Did the specific appetitive experiments assess the slug's ability to learn to approach an odor paired with the opportunity to feed? To answer this question, two chemicals were used as food odor sources for the CS's: amyl acetate (AA) and methylcyclohexanol (MCH). When each of these odors are tested in comparison to rat chow odor, slugs tend to spend more time over the rat chow odor. The US was provided by a powdered rat chow solution. For half of the slugs AA was the CS+ and was paired with the US. For these animals MCH odor was presented in an unpaired relationship with the food. The remaining slugs experienced MCH odor paired with the food and the AA odor in an unpaired relationship with the US. Training consisted of multiple trials per day, five days a week, for two weeks. Twenty-four hours following training the effect of this regimen on AA and MCH preferences versus rat chow were assessed in the two-odor choice testing apparatus. The results are presented in figure 3.

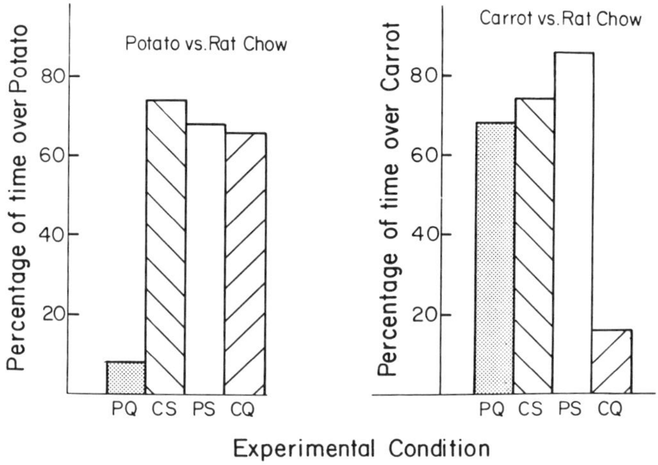

Figure 2: Results for Experiment 1. Left histogram: Mean percentage of time spent over potato odor in a potato versus rat chow choice test for the slugs in the four training groups . Only those slugs with prior experience of the potato - quinidine pairings (Group PQ) showed a decrease in the time spent over the potato odor. Right histogram: The mean percentage of time spent over carrot odor in a carrot versus rat chow choice test for the same slugs. Only those slugs with prior experience of carrot-quinidine pairings (Group CQ) showed a reduction in the time spent over carrot odor. PQ, CS, PS, CQ: training groups defined in Figure 1.

As seen in the figure the learning was both robust and specific. That is, slugs showed an increase in preference only for the odor that was directly paired with the US. The slugs' performance toward the unpaired odor did not change. More recent experiments have indicated that the same pattern of results is obtained when food odors such as garlic and onion are substituted for the chemical odors (Martin and Sahley, 1986; Sahley, Martin, and Gelperin, 1989).

Contingency

The importance of contingency or predictability for associative learning was first recognized by Rescorla (1968) and Kamin (1969). To assess the relative importance of pairing number and predictability Rescorla (1968) held the pairing number constant, while introducing unpaired US presentations during the inter-CS intervals. The importance of CS-US contingency emerged from these studies—

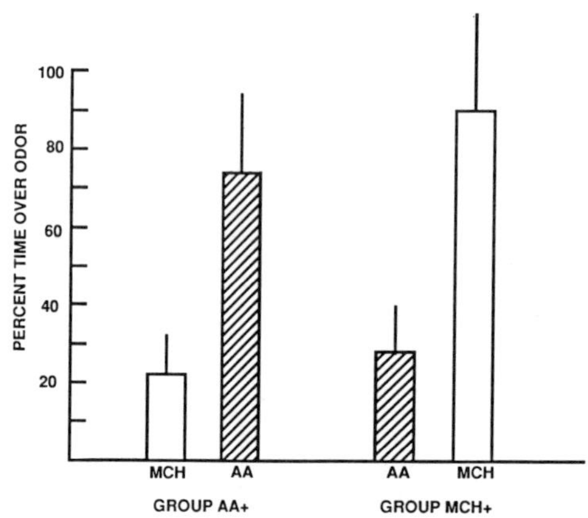

Figure 3: Results for Experiment 2 (appetitive conditioning). Mean percent of time spent over the CS+ and CS- odors in CS versus rat chow tests for slugs in the two training groups. Half of the slugs experienced AA as the CS+ and MCH as the CS-. The remaining slugs experienced MCH as the CS+ and AA as the CS-. Slugs which experienced the AA paired with the US showed an increased preference to AA. Slugs which experienced MCH paired with the US showed an increased preference for MCH. (AA, amyl acetate; MCH, methylcyclohexanol).

excitatory conditioning to a CS occurs only when the occurrence of the US is more probable in the presence of the CS than in its absence. The inter-event relations which support conditioned inhibition are the opposite; the occurrence of the US must be less probable following the CS. Conditioning to a stimulus correlated with the US will occur only if no other available stimulus is a more accurate predictor of the US (Rescorla, 1975).

In Kamin's initial experiments he conditioned rats to display a fear reaction to a noise (S1) stimulus by exposing the animals to pairings of noise and electric shock US. In a second phase, these animals were exposed to pairings of a compound stimulus, consisting of the noise (S1), light (S2), and shock. When tested to the light (S2) alone, he found that the rats failed to display a conditioned fear reaction. In comparison, animals that were conditioned only to the noise and the light compound displayed a fear reaction (freezing) to the light as well as to the noise. Prior conditioning to one element of the compound thus blocked or prevented conditioning to the other element. This phenomenon has been termed blocking. Blocking has been observed with a variety of conditioning tasks (Cheatle and Rudy, 1978; Gillan and Domjan, 1977; Holland, 1977; Kamin, 1969, Marchant and Moore, 1973, Tennent and Bitterman, 1975; vom Saal and Jenkins, 1970).

To assess the role of predictability in the slug, we used a modified Kamin protocol which is presented in figure 4 (Sahley, Rudy, and Gelperin, 1981). Slugs in the blocking condition, group B, first received three pairings of carrot odor and quinidine (S1-US pairings). In phase 2 they received three pairings of a compound stimulus (carrot and potato odor) and quinidine (S1 S2-US pairings). Since the question was–would prior conditioning to carrot odor (S1) block conditioning to the potato odor (S2)–we needed to demonstrate that both the phase 1 S1-US pairings and the presence of S1 on the phase 2 conditioning trials are necessary for blocking to occur. To accomplish this, we used several control groups. To assess the importance of the phase 1 S1-US pairings we included two groups, Group C1, that only received the phase 2 compound conditioning trials and group C2, that also received the phase 2 trials but in phase 1 received backward pairings of quinidine and carrot odor (S1). To assess the importance of the presence of preconditioned S1 during phase 2 trials, slugs in group C3 experienced the phase 1 carrot odor-quinidine pairings of phase 1, but in phase 2 these slugs only experienced potato odor-quinidine pairings; carrot odor (S1) was not presented. The results of the potato odor (S2) versus rat chow odor preference test displayed in figure 5 it reveals a powerful blocking effect. Slugs in group B manifested no aversion to potato odor, in contrast to slugs in the three control groups, even though all groups experienced the same number of potato odor-quinidine pairings. We conclude that the temporal pairing of potato odor and quinidine was not a sufficient condition for producing associative learning. What was important was the predictive value of potato odor. When its predictive value was reduced by the presence of the preconditioned carrot odor, it was not associated with the quinidine. Just as for vertebrates excitatory conditioning in the slug then occurs when the CS predicts the occurrence of the US.

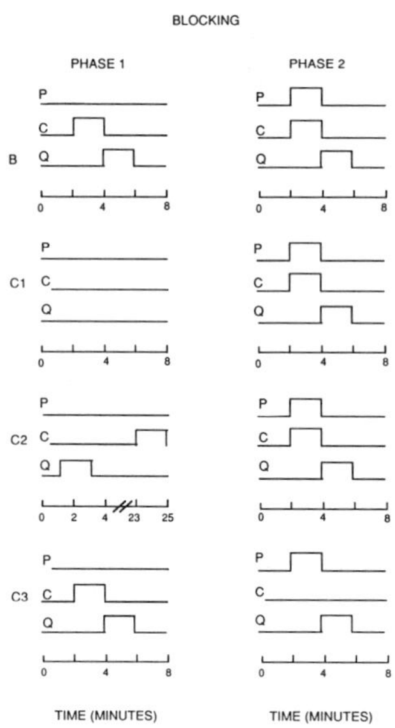

Figure 4: Training protocols for the blocking experiment (Experiment 3). B: Blocking group. C1, C2, C3: Control Groups. P, C, Q: Potato Odor, Carrot odor, Quinidine. For each group the sequence of events depicted under Phase 1 was repeated 3 times; then the sequence depicted under Phase 2 was repeated 3 times. Testing occurred 24 hours following phase 2.

All of the above examples are instances of excitatory associative learning in which the CS predicts the presence of the US. In contrast to excitatory conditioning, inhibitory learning is characterized by a negative relationship between an environmental stimulus, the CS, and a biologically significant US. That is, the environmental CS is correlated with the absence of the US. In order for a CS to become inhibitory it must be nonreinforced in an excitatory context in which the nonpresence of the US is surprising (Wagner and Rescorla, 1972). Backward conditioning procedures fulfill this requirement. The amount of inhibitory conditioning produced by backward conditioning procedures is related to the duration of CS and the time of onset in relation to US termination (Larew, 1986). The procedure we designed

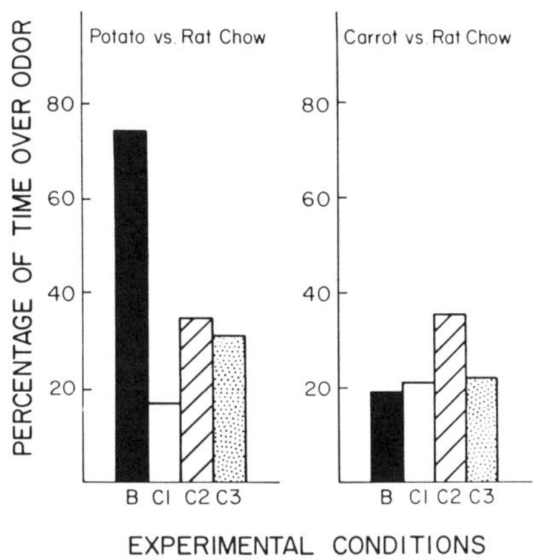

Figure 5: Results for Experiment 3. Mean percentage of time spent over potato-odor (left panel) or carrot-odor (right panel) when each was tested in in a choice with rat chow. Slugs in the blocking group (B) failed to condition to the potato-odor (S2) component of the compound stimulus as compared to slugs in the three control groups. Slugs in all groups showed conditioning to the carrot odor (S1).

to investigate excitation and inhibition to a single US combined the short-delayed procedure which we have used previously to study first-order conditioning with backward pairings of an odor with the absence of the US (Martin and Sahley, 1986). In these experiments food odors provided by garlic and onion were used as the stimuli. Garlic odor (CS+) was paired with the onset of the US and potato odor (CS-) was paired with the termination of the US. A schematic representation of the training procedure is presented in figure 6. A group of slugs that experienced random presentations of the three stimuli was also evaluated. The effect of these training procedures was evaluated several times during training to determine the rate of acquisition. Retention of excitatory and inhibitory learning was measured on days five and six following training. Excitatory learning was evaluated in a garlic versus rat chow odor test and inhibitory learning was evaluated in a potato versus rat chow test. The results of this experiment are presented in figure 7. In pretraining tests, the experimental and control groups showed similar odor preferences. They spent comparatively little time (approximately 20%) over garlic when it was paired with rat chow, and spent considerably more time

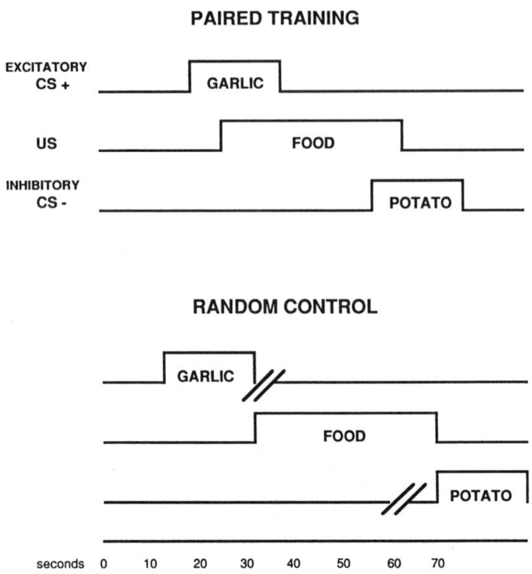

Figure 6: Training protocol for the Inhibitory Conditioning Experiment (Experiment 4). Experimental (top panel) and control group (bottom panel). The control group differed from the experimental group in that relative timing of Garlic Odor, Potato Odor and Food was random.

(approximately 60%) over potato when the latter was paired with rat chow (figure 7 A,B). As expected, the training procedure produced appetitive conditioning. After 18 training trials, slugs in the paired group showed a significant increase in the time spent over the garlic odor as compared to slugs in the unpaired control group. More significantly, there was also evidence of inhibitory conditioning. Following 30 training trials, slugs in the paired group showed a significant decrease in the amount of time spent over the potato odor. No such change in preference was exhibited by slugs in the control group. This learning (both excitatory or inhibitory) was retained for at least five days. This then is the first demonstration of inhibitory learning in a mollusc and only the second

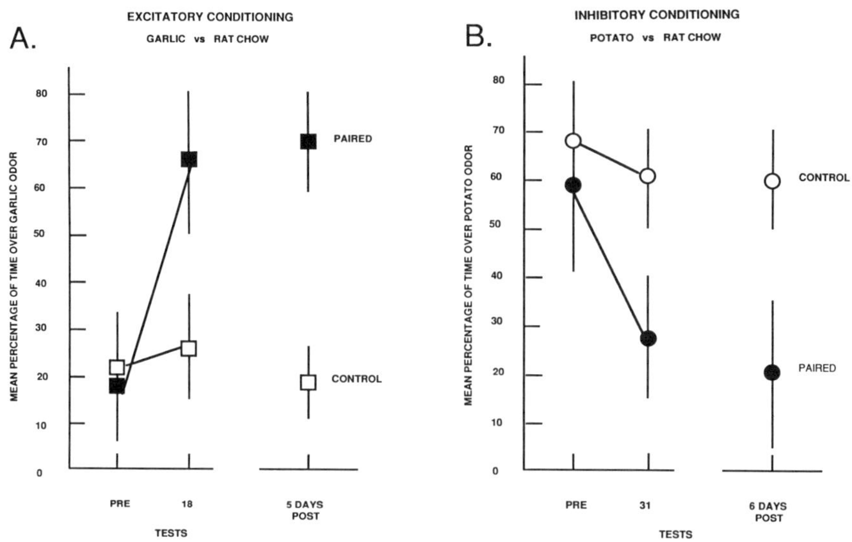

Figure 7: The results from the excitatory conditioning portion of inhibitory learning experiment (garlic-odor vs. rat chow-odor test) are presented in panel A. Slugs that received forward paired presentations of the CS and US showed a significant increase in garlic-odor preference following 18 training trials. Slugs that received random presentations of the CS and US showed no increase in garlic-odor preference. Inhibitory conditioning results (potato-odor vs. garlic odor) for the same slugs are presented in panel B. Slugs that received backward paired presentations of the CS and US showed a decrease in preference for potato-odor. Slugs in the random control showed no significant decrease in potato-odor preference.

demonstration in any invertebrate (Bitterman et al., 1983).

Higher Order Relationships

The defining feature of second-order conditioning as first described by Pavlov (1927), is that a CS can acquire the ability to elicit conditioned responding without ever being directly paired with the US. In a two phase experiment, first-order conditioned responding is established to one conditioned stimulus, S1, by pairing it with the US (S1-US pairings). In phase 2 another CS, S2, is introduced and paired with S1 (S2-S1 pairings). The result of this two-phase training procedure is that S2 acquires the ability to evoke conditioned responding even though it was never paired with the US. Second-order conditioning has now been demonstrated in a wide variety of vertebrates (Amiro and Bitterman, 1980; Cheatle and Rudy, 1979; Holland and Rescorla, 1975; Rescorla, 1980).

Using experimental designs similar to those employed in vertebrates, we (Sahley, Rudy, and Gelperin, 1981) exposed slugs to two phases of training. In phase 1, slugs in the second-order

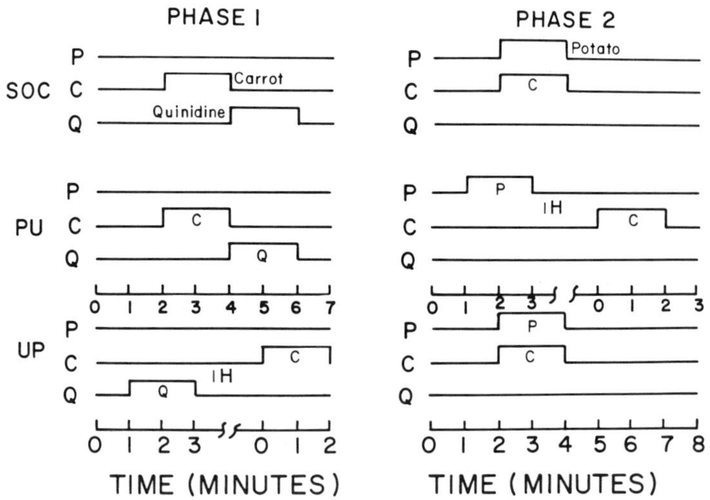

Figure 8: Training protocol for the Second-Order Conditioning Experiment (Experiment 5). SOC: Second-Order Group. PU, UP: control groups. For each group the sequence of events depicted under Phase 1 was repeated 3 times; then the sequence depicted under Phase 2 was repeated 3 times. Testing followed.

conditioning group (SOC) experienced three pairings of carrot odor and quinidine (S1-US pairings) and in phase 2 they received three pairings of potato odor and carrot odor (S2-S1 pairings). A schematic representation of the training protocol is presented in figure 8. It is important to note that in phase 2 of this experiment, potato and carrot odor were were presented to the slug simultaneously. The question was, would this training reduce the slug's preference for potato odor (S2), even though it was never directly paired with the quinidine? To conclude that such a result is truly second-order conditioning requires that the treatment given group SOC be evaluated against two control groups: one that receives the phase S1-US pairings but not the phase 2 S2-S1 pairing, and another that receives the phase 2 but not the phase 1 pairings. We included two such control groups in our experiment. Group PU received the phase 1 carrot odor-quinidine pairings but in phase 2 the carrot and potato were unpaired. Group UP received unpaired presentations of carrot odor and quinidine in phase 1 and paired presentations of potato an carrot odor in phase 2. Following training, all slugs were tested for first-odor conditioning to carrot odor (S1) and second-order conditioning to potato odor (S2). The results of this experiment are presented in figure 9. First-order conditioning occurred, as demonstrated by data in the right panel: groups experiencing paired presentations of carrot and quinidine, (SOC and PU), displayed a reduced preference for carrot odor in comparison to slugs experiencing unpaired presentations, group (UP).

Furthermore, second-order conditioning occurred, as shown in the left panel: slugs in group SOC displayed a reduced preference for potato odor in comparison to the slugs in the two control groups. Thus as for pairing and contingency, the behavior of *Limax* parallels closely the behavior of vertebrates.

Internal Representations in Second-Order Conditioning Many psychologists have begun to explore the nature of the learned association in second-order conditioning. For example, Rescorla and his colleagues (Holland and Rescorla, 1975; Rescorla, 1973, 1980; Rizley and Rescorla, 1972) have examined the associations mediating second-order conditioning in vertebrates. During phase 2 of a second-order conditioning experiment (S2-S1 pairings), there are several associations that the subject can learn that could mediate a conditioned reaction to S2. One possibility is an S2-S1 association. In this case it is assumed that the subject associates representations of the two CS events. For example, this would imply that the slugs associated representations activated by the potato odor (S2) and carrot odor (S1) experience. According to this view, the second-order conditioned response to potato odor (S2) would be mediated by a two-link chain of associations. First, potato odor (S2) would activate the slug's representation of carrot odor (S1) which in turn due to its first-order association with quinidine would evoke the

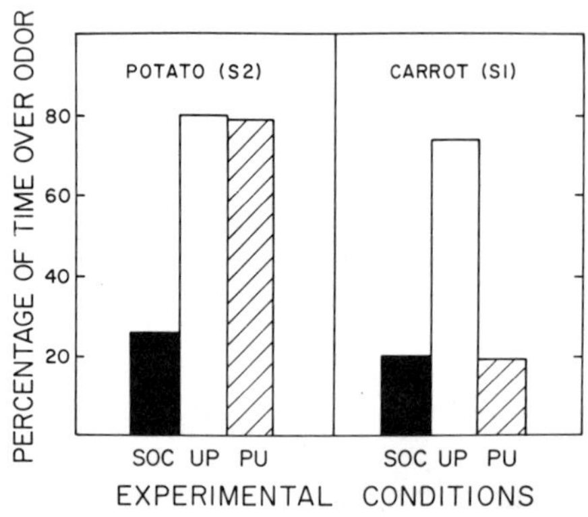

Figure 9: Results for Experiment 5. Mean percentage of time spent over potato - odor (left panel) or carrot - odor (right panel); SOC, second-order conditioning group; UP, unpaired - paired group; PU, paired - unpaired group. Slugs in the SOC group spent significantly less time over potato (S2) odor than slugs in either of the control groups, PU and UP. Slugs receiving carrot - quinidine pairings in phase 1 (Groups SOC and PU) demonstrated first-order conditioning, whereas slugs receiving unpaired presentations of carrot and quinidine (Group UP) showed no first-order conditioning.

conditioned response. A second alternative is the S2-CR association. This account assumes that during the phase 2 (S2-S1) pairings, S1, due to its previous pairings with the quinidine US, evokes a conditioned response (CR). This allows for the possibility that the subject will associate S2 with the CR activated by the S1. For example, this would mean that the slugs associated potato-odor (S2) with the conditioned response that the carrot odor (S1) activated. The second-order response to potato odor (S2) during testing would then be directly evoked by potato odor and not indirectly through carrot odor.

Several experiments have been designed which make it possible to differentiate between these two options. The distinguishing experiments make use of what Rescorla (1975) terms a postconditioning treatment strategy. The S2-S1 association account

	Phase 1	Phase 2	Phase 3	Test
SOC	S1-US	S2-S1	---------	S1,S2
SOC-E	S1-US	S2-S1	S1	S1,S2
PU	S1-US	S2/S1	---------	S1,S2
UP	S1/US	S2-S1	---------	S1,S2

Figure 10: Training protocol for the S1 extinction treatment experiment (Experiment 6). Groups SOC, PU and UP were treated exactly like the similarly designated groups of the previous experiment. Slugs in Group SOC-E were treated exactly like Group SOC in the previous experiment, but prior to testing these slugs received a postconditioning extinction treatment .

implies that conditioned responding to S2 is dependent upon the ability of S1 to evoke a conditioned response. In contrast, the S2-CR association view implies no such dependency. The postconditioning strategy inserts an extinction procedure following the two training phases but prior to testing (S1 extinction treatment). The S1-S2 association view would predict that the S1 extinction treatment should eliminate S2's ability to evoke a second-order conditioned response, because S2's ability depends on the integrity of S1's first-order conditioning properties. In contrast, the S2-CR association view predicts that the S1 extinction treatment would not influence S2's ability to evoke a conditioned response, because S2 is assumed to be able to directly evoke conditioned responding.

In the case of vertebrates, recent experiments have identified some of the variables that determine whether vertebrates learn S2-S1 associations or S2-CR associations are learned (see Cheatle and Rudy, 1978, 1979; Rescorla, 1982). What appears to be especially important is the manner in which S2 and S1 are presented during phase 2. When they are presented simultaneously (S2 + S1), the acquisition of S2-S1 associations is favored. When they are presented sequentially, (S2, S1), acquisition of S2-CR associations is favored.

We have recently completed experiments designed to determine whether the same rule applies to *Limax*. First we asked what associations are learned by slugs exposed to simultaneous phase 2 pairings of potato (S2) and carrot (S1) odor. Three of the groups, SOC, PU, and UP, were treated just like the similarly designated groups of the previous experiment. A fourth group SOC-E, was treated exactly like group SOC, but prior to testing these slugs received a post-conditioning S1 extinction treatment. They were repeatedly exposed to nonreinforced presentations of carrot odor (S1). A schematic representation of the training protocol is presented in figure 10. The results of this experiment are presented in figure 11. First, it can be seen (left panel) that the S1 extinction treatment administered to slugs in group SOC-E eliminated their first-order conditioned reaction to carrot odor. Compared to slugs in both groups, PU and SOC, slugs in SOC-E spent more time over carrot odor. Slugs in groups SOC-E did not differ significantly from those in group UP that never received the carrot odor-quinidine pairings.

Furthermore (right panel), the S1 extinction treatment also eliminated the slug's second-order reaction. The aversion for potato odor that was markedly evident in group SOC was entirely lacking in group SOC-E, as in controls. This outcome strongly suggests that the second-order reaction produced by simultaneous S2-S1 pairings was mediated by an association between the representations of the two odors. Our second experiment was designed to assess what associations are learned when the slugs are exposed to sequential pairings of potato (S2) and carrot (S1) odor. It was similar to the previous experiment in all regards, except that during phase 2 the slugs in groups UP, SOC, and SOC-E received presentations of potato odor (S2) by carrot-odor (S1) in immediate sequence rather than simultaneously. The results of this experiment are presented in figure 12. As was the case in the previous experiment, the S1 extinction treatment reduced the first-order reaction to carrot-odor (S1). Slugs in group SOC-E spent significantly more time over carrot odor than did slugs in either group SOC or PU. The S1 extinction treatment, however, did not have a detectable effect on the slugs' second-order

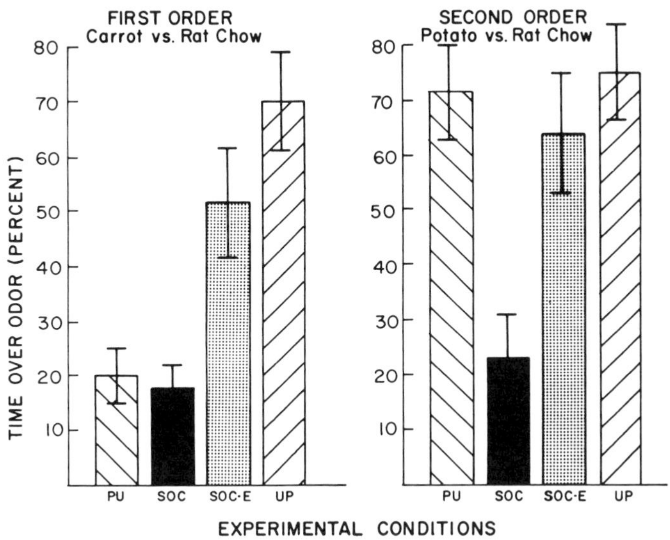

Figure 11: Results for Experiment 6 - Simultaneous S1 and S2 presentations. Mean percentage of time spent over carrot odor (the first- order stimulus; left panel) and potato odor (the second-order stimulus; right panel) for slugs experiencing simultaneous presentations of the S1 and S2 during second-order conditioning. As seen in the left panel, the slugs in the SOC-E group showed a reduction in first-order conditioning as compared to slugs in group SOC which did not experience the extinction treatment. As seen in the left panel, the slugs in Group SOC-E showed a reduction in their second-order conditioning as well as a result of the extinction of S1.

reaction to potato odor (S2). Slugs in group SOC-E, like those in group SOC displayed a reduced preference for potato odor, in comparison to controls. This result provides evidence in support of the so-called S2-CR association view. Thus slugs evidently associated potato odor with some component of the conditioned response complex evoked by carrot odor (S1) during the phase 2 potato-carrot odor pairings.

These experiments lead to conclusions remarkably similar to those established in the vertebrate literature. In particular, (1) the second-order conditioned change in *Limax's* odor preference behavior evidently can be mediated by alternative associative structures; and (2) simultaneous S2-S1 presentations favor the contribution of an S2-S1 association, while sequential presentations favor the contribution

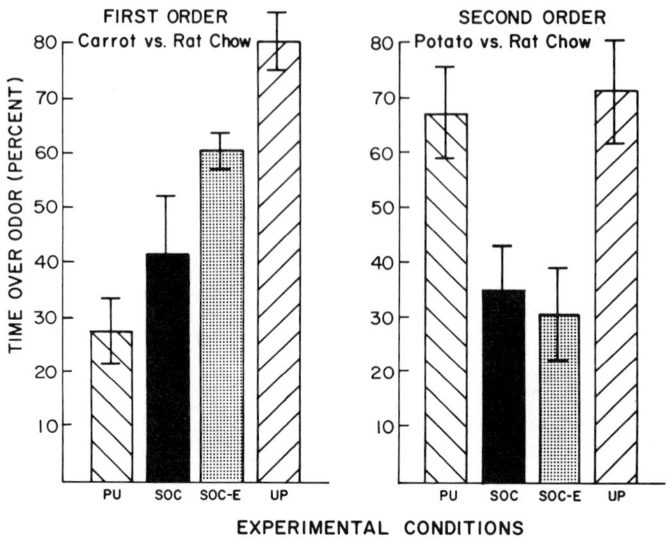

Figure 12: Results for Experiment 6 - Sequential S1 and S2 presentations. Mean percentage of time spent over carrot odor (the first-order stimuli; left panel) and potato odor (the second-order stimulus; right panel) for slugs experiencing sequential presentations of the S1 and S2 during second-order conditioning. As seen in the left panel, the slugs in group SOC-E showed a reduction in first-order conditioning as compared to slugs in Group SOC which did not experience the extinction treatment. As seen in the left panel, in contrast to the simultaneous conditioning, slugs in Group SOC-E showed no reduction in their second-order conditioning.

of an S2-CR association. In more general terms, the results of the behavioral experiments presented so far in this chapter underscore the multifaceted nature of the associative process. Learning, for invertebrates as well as for vertebrates is critically dependent on the interevent relationships between stimuli. Thus what might seem to be subtle differences in inter-event relationships between stimuli dramatically alter what is learned.

Learning in the semi-intact preparation Since we feel that the interevent relationships during the training procedure, and the consequent cellular events during training, are the key to the associative process, any attempt to study the cellular and molecular processes mediating associative learning require a semi-intact preparation in which behavioral and neural responses can be monitored during training. In this manner it is possible to determine the causal events mediating the associative process as it occurs, rather than measuring the differences in neural activity prior to and following training. It is likely that the important cellular events underlying the formation of the association are observable only at the time of the actual association and the cellular events measured following learning are more related to a storage or consolidation mechanism. Such a semi-intact *Limax* preparation has been developed by Prior and Gelperin (1977; Gelperin, Chang, and Reingold, 1978). Further, this preparation shows robust associative learning (Chang and Gelperin, 1980). The preparation consists of the cerebral and buccal ganglia, the gustatory input pathway, the lips and the three lip nerves, and the feeding motor output pathway, with the nerve roots, all maintained in a petri dish filled with a Limax Ringer's solution. A schematic representation of this preparation is presented in figure 13. When a taste, such as potato or carrot extract, is applied to the lips of the preparation a stereotypic pattern of bursting activity, termed feeding motor program (FMP), is recorded from the buccal nerve roots.

FMP has been shown to be the neural activity underlying the protraction and retraction of the radula scraping over a food source in the intact animal (Gelperin et al., 1978; Reingold and Gelperin, 1978). The pairing of the taste of a food extract on the lips with the application of the bitter taste of quinidine on the lips results in a cessation of feeding to the specific food taste paired with the quinidine when this taste is presented in a test trial several minutes following training. Food odors presented in an unpaired manner with the quinidine show no decrease in their ability to elicit FMP, thus eliminating non-associative effects. The neurons mediating this learning appear to be centrally rather than peripherally located since learning can be obtained when training is done on one lip and testing is done on the other (Culligan and Gelperin, 1983). As discussed above, this preparation provides a unique opportunity to allow simultaneous measurement of the change in behavior as a result of learning and the cellular events mediating this change. The efficacy of learned odor-taste associations on the initiation or maintenance of FMP has not been examined. However, it seems likely that

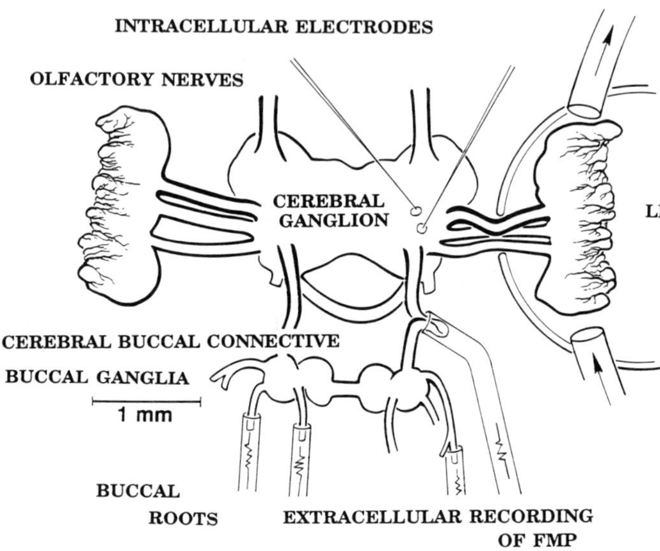

Figure 13: Semi-intact preparation used for in vitro experiments. The cerebral and buccal ganglia are dissected with the lips remaining connected via the three lip nerves. The preparation is maintained in a petri dish and continuously bathed with a solution of Limax Ringers. The lips are isolated from the bath in separate small plexiglass chambers thus allowing the application of taste stimuli only to the lips. Olfactory input from the tentacles enters the cerebral ganglion via the olfactory nerve (only partially shown here) and presumably converges with gustatory information entering the same ganglion via the lip nerves. The cerebral ganglion is connected to the buccal ganglion by the cerebral-buccal-connectives (CBCs). Axons from the feeding motor neurons leave the buccal ganglion through the motor roots to innervate the feeding musculature.

odor-taste associations that influence feeding in the semi-intact preparation will be possible.

Interaction between preference behavior and feeding Although associative learning has been observed in both the intact slug and in the semi-intact preparation, there are several differences between the two preparations. One, is that, in the intact slug, odor-taste associations modify an appetitive behavior (locomotor behavior) and, in contrast, in the semi-intact preparation, a taste-taste association modifies consummatory behavior (feeding behavior). Recent behavioral experiments however, have suggested that these two separate behaviors are in fact intimately intertwined and moreover, several experiments suggest that both the whole animal and the isolated brain learning involve similar neural events. Gelperin and Culligan (1982) and, subsequently Culligan and Gelperin (1983) reported that following food taste-quinidine pairings in the whole animal, which resulted in odor-avoidance, selective suppression of the feeding motor program was evident in the isolated brain. In addition, we have recently found that in hungry slugs feeding to an agar pellet can be initiated in the presence of an attractive odor. Non-hungry slugs will not feed to agar in the presence of odor, and hungry slugs will not feed to the agar pellet in the absence of the odor. Thus odor can modulate feeding. Also, we have found that following appetitive associative training, in addition to the observed change in preference behavior, the odor paired with the US can initiate feeding when the slug is presented with a tactile stimulus to the lips, such as an agar pellet. Odors not paired with US remain incapable of initiating feeding (Sahley et al., 1982). Consequently, a single association is capable of modulating both the appetitive and the consummatory phases of feeding. Moreover, the sites and mechanisms of cellular plasticity revealed in the lip- brain preparation will relate directly to the behavioral plasticity uncovered in the intact animal.

General Discussion

From this brief review of the behavioral work on the garden slug, it can be seen easily that *Limax* is capable of a broad array of associative learning phenomena. More important, the same variables that influence leaning in vertebrates influence learning in at least one invertebrate species. This keeps alive the hope that similar behavioral phenomena may indeed translate into similar cellular and subcellular mechanisms.

For the most part, the analysis of learning in Limax has been a behavioral one. Our goal was to understand the behavior as completely as possible and then let the behavior guide the cellular analysis. This is not the usual tack taken by neurobiologists, but we feel that it is an important one. The phenomenon of associative learning is not simple but, rather, is rich and complex. All aspects of the events to be associated are important to the final outcome and the expression of learning. Subtle changes in the temporal parameters of the stimuli in relationship to each other, for example, can dramatically alter the sign of the learning (excitatory versus inhibitory) or the nature of the learned association (SS versus SR). It is important to keep the total learning capability of an animal in perspective.

In terms of *Limax*, the behavioral approach has provided us with a framework with which to approach the delineation and characterization of the cellular mechanisms mediating the behavioral plasticity. The results of the interaction studies discussed above, for example, force several predictions regarding the neural machinery mediating these behaviors. First, the circuits mediating these behaviors (locomotion and feeding) must have overlap. The behavioral interaction necessitates the existence of physiological and anatomical sites of convergence between the two circuits. Second, the behavioral results predict that the circuits can function independently. Third, each individual circuit can modulate the other in an excitatory or inhibitory manner and thus can cause the initiation or cessation of the behavior or change the vigor of the behavior; finally, the effectiveness of chemostimuli on the behavioral outcome is dependent on the motivational level of the organism. A summary diagram of the odor-taste influences on feeding behavior and of the odor-taste influences on preference behavior compiled from the behavioral experiments are presented in figure 14. It is apparent from the figure that the final behavioral outcome is dependent on the interaction of the stimulus inputs and the "state of the system." That is to say, a given stimulus input, such as potato odor, can set up the system to feed even though the final decision to feed and the vigor of feeding is dependent on other sensory inputs such as the taste of the potato and any past experience (learning) with the stimulus.

More specifically, the behavioral results predict the existence of several categories of cells. In addition to sensory and motor neurons, interneurons responsible for the initiation, maintenance and modulation of the system must exist. Although the delineation of the neural circuits mediating preference behavior and feeding is far from complete these categories of cells have been defined and examples of each of the categories has been identified. In addition, given that

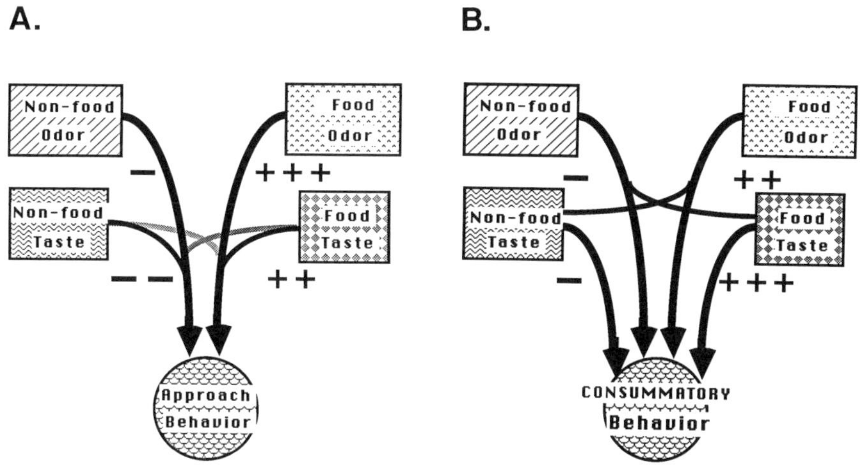

Figure 14: Summary diagram of the olfactory and gustatory influences on feeding behavior. Panel A depicts the odor taste influences on the approach to food and Panel B depicts the odor-taste modulation of consummatory behavior.

the semi-intact preparation is capable of associative learning, the necessary and sufficient circuitry for learning must be present within the cerebral and buccal ganglia, with the inputs coming from the lips via the lip nerves. A schematic representation of the circuits mediating feeding and locomotion and their interactions is presented in figure 15. This diagram was compiled from actual cellular and anatomical data discussed below as well as from predictions based on the behavioral results discussed above.

Putative gustatory and olfactory sensory neurons have been identified using backfill techniques within the cerebral ganglion (K.

Delaney, personal communication; C. Sahley unpublished observations), although these remain to be characterized. FMP motor neurons have been identified in the buccal ganglion using retrograde dye fills and electrophysiological techniques. These motor neurons, which innervate the protractor and retractor muscles controlling the radula, are activated as a result of food extracts applied to the lips of the preparation (Gelperin and Forsythe, 1976; Gelperin, 1981). Salivary neurons within the cerebral ganglion have also been identified and appear to have a role in the sequence of actions associated with the ingestion of food (Prior and Gelperin, 1977; Copeland and Gelperin, 1983). No physiological data is available regarding olfactory inputs to the feeding motor neurons or to the salivary neurons; however, backfills have revealed processes from the olfactory nerve which extend into the buccal ganglion (Delaney and Gelperin, personal communication; Sahley, unpublished data). These could be the source of the odor modulated feeding. That is, the behavior would predict that odors make contact with the feeding system via modulation of the command-like neurons in the cerebral ganglion or directly to the motor neurons in the buccal ganglion. In addition, we know from the behavioral results that tactile inputs through the lip appear to be important in feeding. This has been confirmed in the semi-intact preparation (Delaney, personal communication). A non-food, tactile stimulus applied to the lips can initiate FMP. Perhaps the olfactory inputs make the command neurons or the motor neurons themselves more excitable so that when the tactile input comes in from the lips feeding is initiated. This may indeed be a function of the metacerebral giant interneuron discussed below, since odors applied to the tentacles odor change the firing characteristics of this neuron (Egan and Gelperin, 1981).

Two classes of interneurons have been implicated in feeding, the metacerebral giant cell (MCG) and the FMP initiating cerebral to buccal neurons (CBs). The MCGs first identified by Gelperin (1981) are a bilateral pair of large serotonin containing interneurons. The MCGs respond to both gustatory and olfactory chemostimuli and are capable of triggering FMP (Delaney and Gelperin, 1989), they are in a key position to play an important role in the initiation and modulation of feeding. Activated by food application to the lips, they show increased spiking just before the onset of FMP and during the initial phases of the program. The MCG also responds by an increase in spiking to the application of an odor stimulus to the exposed sensory epithelium of the tentacle (Egan and Gelperin, 1981). The precise synaptic input-output characteristics of the MCG have not been specified; however, Lucifer yellow fills have revealed an extensive branching pattern extending into the buccal ganglion, suggesting that the MCG could influence feeding motor neurons. Interestingly, the MCGs are capable of increasing the effectiveness of the cerebral to buccal interneurons (CBs) as discussed below. Moreover, the exact function of this cell in regard to feeding is not known. Preventing the cell from firing during food application does not prevent the commencement of FMP. The existing anatomical and physiological data strongly suggest that the MCG is a site of convergence of olfactory and gustatory inputs for the feeding system. Thus this cell could turn out to be extremely important for the expression of learning dependent changes in feeding.

The cerebral to buccal interneurons (CBs) are important for the initiation and modulation of feeding (Delaney and Gelperin, 1986;1987; 1989 a,b,c). They are a group of 14–18 neurons located in the cerebral ganglion in small clusters in the cerebral ganglion near the insertion of the cerebral-buccal connective. (CG) and projecting to the buccal ganglia (BG) via the cerebral–buccal connectives (CBC) (Delaney and Gelperin, 1989a). They are capable of initiating FMP and are modulated by taste inputs. Chemostimuli such as fructose or food-plant extracts that stimulate feeding in intact slugs excite CBs whereas chemostimuli aversive to intact slugs inhibit CBs. It is not known whether they are modulated by odor. From the behavioral results it seems likely, since we know that olfactory stimuli can initiate feeding in hungry slugs. Moreover, olfactory neuron axons terminate in the area of the command-neurons. However, physiological experiments have not been conducted to determine whether olfactory stimuli influence the activity of the command neurons or whether olfactory inputs synapse directly (monosynaptic) or indirectly (poly–synaptic) onto the command neurons. Interestingly, the CB population is not homogeneous. Some CBs receive monosynaptic inputs from the taste-sensory afferents, while others recieve taste inputs via polysynaptic pathways (Delaney and Gelperin, 1989c). It is tantalizing to think that these polysynaptic pathways would be the loci for olfactory and modulatory inputs.

In terms of modeling, though, a major implication of this body of work is the critical importance of the inter-event relationships between stimuli in the environment for associative learning. For example, CS-US pairing is important in associative learning, but so also is the degree of contingency or predictability that is set up between the stimuli as demonstrated by the blocking experiment. Moreover, the placement of the CS either just preceding the US or just following the US determines the sign of the association, excitatory or inhibitory

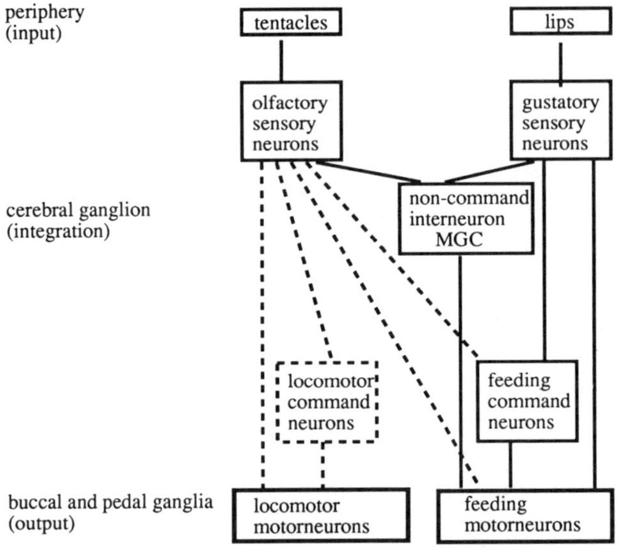

Figure 15: Schematic flow chart of the neural circuits mediating the approach and consummatory components of Limax's feeding behavior and their interactions. This diagram is compiled from implications drawn solely from the behavioral results (dashed lines) and from a combination of behavioral, anatomical and physiological data (solid lines). Monosynaptic or polysynaptic connections are not differentiated. The diagram represents the stimulus inputs from the periphery, the sites of convergence and integration most likely located in the cerebral ganglion and the motor neurons mediating the behavioral output in the pedal and buccal ganglia. It is clear from this chart that considerable convergence exits between the olfactory and gustatory systems thus providing considerable opportunity for interaction and modulation.

respectively as demonstrated by the inhibitory learning experiments. In addition, these same experiments illustrate the interaction between associations in the development of learning. That is, the emergence of inhibitory learning is dependent on the establishment of an excitatory context, in these experiments provided by the forward CS-US pairings experienced simultaneously with the backward pairings. Finally, as illustrated by the second-order conditioning experiments, the nature of the stored association is critically determined by the temporal relations between the stimuli used in the training phase of the experiment. Stimuli presented simultaneously in time favor the establishment of associations between the sensory aspects of the two stimuli; thus the second-order association is dependent on the integrity of the first-order conditioning to the S1. In contrast, stimuli presented sequentially in time favor the establishment of associations between the S2 and the US or UR qualities of the S1 and this association is not dependent on the integrity of the first-order conditioning to the S1. A take-home message to modelers then, is that in order for complete models to be built, it is essential to think in global terms and to incorporate into the models the many inter-event relationships important to the associative process.

Acknowledgement

I would like to thank the editors for their patience, indulgence, good humor and help in completing this chapter. I would also like to thank Dr. Alan Gelperin for supplying unpublished manuscripts and figures. Finally, I would like to thank Dr. Don Ready for his help with the preparation of this chapter.

References

Alexander, J., Audesirk, T. and Audesirk, G. (1984) One-trial reward learning in the snail Lymneae stagnalis. J. Neurobiology, 15, 67-72.

Aceves-Pina, E. O. and Quinn, W. G. (1979) Learning in normal and mutant Drosophila larvae. Science, 206, 93-95.

Alkon, D. L. (1983) Learning in a marine snail. Sci Amer, 249, 70-84.

Amiro, T. W. and Bitterman, M. E. (1980) Second-order conditioning in goldfish. J. Exp. Psychol. (Anim. Behav.) 6, 41-48.

Audesirk, T. E., Alexander, J. Jr., Audesirk, G. J. and Moyer, C. M. (1982) Rapid nonaversive conditioning in a freshwater gastropod. I. Effects of age and motivation. Behavioral and Neural Biology, 36, 379-390.

Bitterman, M. E., Menzel, R., Fietz, A. and Schader, S. (1983) Classical conditioning of proboscis-extension in honeybees. J. Comp. Physiol. Psychol., 97, 107-119.

Booker, R. and Quinn, W. G. (1981) Conditioning of leg position in normal and mutant Drosophila. Proc. Natl. Acad. Sci. USA 78, 3940-3944.

Byrne, J. H. (1987) Cellular analysis of associative learning. Physiol. Review, 9, 435-487.

Carew, T. J. and Sahley, C. L., (1986) Invertebrate learning and memory: From behavior to molecules. Ann. Rev. Neuroscience, 9, 435-487.

Carew, T. J., Walters, E. T., and Kandel, E. R. (1981) Classical conditioning in a simple withdrawal reflex in Aplysia californica. J. Neuroscience, 1, 1426-1437.

Chang, J. J. and Gelperin, A. (1980) Rapid taste-aversion learning by an isolated molluscan central nervous system. Proc. Natl. Acad. Sci. USA, 77, 6204-6206.

Cheatle, M. D. and Rudy, J. W. (1978) Analysis of second-order order-aversion conditioning in neonatal rats: Implications for Kamin's blocking effect. J. Exp. Psychology (Anim. Behav.), 4, 237-249.

Cheatle, M. D. and Rudy, J. W. (1979) Ontogeny of second-order order-aversion conditioning in neonatal rats. J. Exp. Psychol (Anim. Behav.), 5, 142-151.

Colwill, R. M., Absher, T. A. and Roberts, M. L. Context-US, Learning in Aplysia Californica. J. of Neuroscience, 1988, 12,4434-4437.

Cook, D. G. and Carew, T. J. (1985) Operant conditioning of head-waving in Aplysia. Proc. Natl. Acad Sci., USA, 83, 1120-1124.

Cooke, I., Delaney, K. and Gelperin, A. (1985) Complex computation in a small neural network. In: Memory Systems of the Brain, eds., N. M. Weinberger, J. L. McGaugh and G. Lynch., New York, London, The Guilford Press., 173-(192.

Copeland, J. and Gelperin, A. (1983) Feeding and a serotonergic interneuron activate an identified autoactive salivary neuron in Limax maximus. Comp. Biochem. Physiol., 94, 878-885.

Couvillon, P. A. and Bitterman, M. E. (1980) some phenomenon of associative learning in honeybees. J. Comp. Physiol. Psychol., 94, 878-885.

Couvillon, P. A. and Bitterman, M. E. (1982) Compound conditioning in honeybees. J. Comp. Physiol. Psychol., 96, (192-99.

Croll, R. and Chase, R. (1977) A long-term memory for food odors in the land snail, Achatina fulica. Behav. Biol., (19, 261-268.

Croll, R. and Chase, R. (1980) Plasticity of olfactory orientation to foods in the snail Achatina fulica. J. Comp. Physiol., A 136, 267-277.

Crow, T. J. (1983) Conditioned modification of locomotion in Hermissenda crassicornis: Analysis of time-dependent associative and nonassociative components. J. Neuroscience, 5, 209-214.

Crow, T. J. (1988) Cellular and molecular analysis of associative learning and memory in Hermissenda. Trends in Neurosciences 11, 136-142.

Crow, T. J. and Alkon, D. L. (1978) Retention of an associative behavioral change in Hermissenda. Science, 201, 1239-1241.

Crow, T. J. and Offenbach, N. (1983) Modification of the initiation of locomotion in Hermissenda. Brain Research, 271, 301-310.

Culligan, N. and Gelperin, A. (1983) One-trial associative learning by an isolated molluscan CNS: Use of different chemoreceptors for training and testing. Brain Research, 266, 319-27.

Delaney, K. and Gelperin, A. (1985) Cerebral to buccal interneurons which initiate fictive feeding in Limax maximus. Soc. Neurosci. Abstr., 11, 509.

Delaney, K. and Gelperin, A. (1986) Cerebral to buccal interneurons controlling initiation and modulation of feeding motor program in Limax maximus. Soc. Neurosci. Abstr., 12, 39.

Delaney, K. and Gelperin, A. (1987) Central correlates of taste discrimination in the slug Limax maximus. Soc. Neurosci. Abstr., 13, 388.

Delaney, K. and Gelperin, A. (1989a) Cerebral interneurons controlling fictive feeding in Limax maximus I: Anatomy and criteria for re-identification. J. Comp. Physiol., A., in press.

Delaney, K. and Gelperin, A. (1989b) Cerebral interneurons controlling fictive feeding in Limax maximus II: Initiation and modulation of fictive feeding. J. Comp. Physiol. A., in press.

Delaney, K. and Gelperin, A. (1989c) Cerebral interneurons controlling fictive feeding in Limax maximus III: Integration of sensory inputs. J. Comp. Physiol., A., in press.

Dudai, Y. (1977) Properties of learning and memory in Drosophila melanogaster. J. Comp. Physiol, 130, 271-285.

Egan, M. and Gelperin, A. (1981) Olfactory inputs to a bursting serotonergic interneuron in a terrestrial mollusc. J. Molluscan Studies, 47, 80-88.

Erber, J. (1976) Retrograde amnesia in honeybees (Apis mellifera carnica). J. Comp. Physiol. Psychol., 90, 41-46.

Erber, J. (1982) Movement learning of free flying honeybees. J. Comp. Physiol., 146, 273-282.

Erber, J. and Schildberger, K. (1980) Conditioning of an antennal reflex to visual stimuli in bees (Apis mellifera L.). J. Comp. Physiol., 135, 217-225.

Farley, J. and Alkon D. L. (1985) Cellular mechanisms of learning, memory and information storage. Annual Review Psychology, 36, 419-494.

Forman, R. and Hoyle, G. (1978) Position learning in behaviorally appropriate situations. Soc. Neurosci. Abstr., 4, 193.

Gelperin, A. (1975) Rapid food-aversion learning by a terrestrial mollusc. Science, 189, 567-70.

Gelperin, A. (1981) Synaptic modulation by identified serotonin neurons. In: Serotonin Neurotransmission and Behavior, eds. B. Jacobs and A. Gelperin, Cambridge: MIT Press, 288-304.

Gelperin, A. and Culligan, N. (1982) In vitro expression of in vivo learning by the cerebral ganglion of the terrestrial mollusc, Limax maximus. Soc. Neurosc. Abstr., 8: 823.

Gelperin, A., Chang, J. J. and Reingold, S. C. (1978) Feeding motor program in Limax, I. Neuromuscular correlates and control by chemosensory input. J. Neurobiology, 9, 285-300.

Gelperin, A. and Forsythe, D. (1976) Neuroethological studies of learning in mollusks. In: J. C. Fentress (ed.), Simpler Systems: An Approach to Patterned Behavior and Its Foundations, Sunderland, Mass. Sinauer Associates, 239-46.

Gelperin, A., Hopfield, J. J. and Tank, D. W. (1984) The logic of

Limax learning. In: Model Neural Networks and Behavior, ed., A. I. Selverston, Plenum Press, New York, 237-261.

Gelperin, A., Tank, D. W. and Tesauro, J. (1989) Olfactory processing and associative learning: Cellular and modeling studies. In: Neural Models of Plasticity, eds., J. H. Byrne, and W. Berry, W. , Academic Press, Orlando, in press.

Gillan, D. J. and Domjan, M. (1977) Taste aversion conditioning with expected versus unexpected drug treatment. J. Exp. Psychol (Anim. Behav.), 3, 297-309.

Gormezano, I. (1972) Investigations of defense and reward conditioning in the rabbit. In: Classical Conditioning, vol 2, eds., A. H. Black and W. F. Prokasy, New York: Appleton-Century-Crofts, 151-181.

Grossman, K. E. (1970) Erlernen von farbreizen an der futterstelle durch honigbienen wahrend des anflugs und wahrend des saugens. Z. Tierpsychol. 27, 553-562.

Hawkins, R. D., Carew, T. J. and Kandel, E. R. (1986) Effects of interstimulus interval and contingency on classical conditioning in Aplysia. J. Neuroscience, 6, 695-1701.

Hawkins, R. D., Clark, G. A. and Kandel, E. R. (1987) Cell biological studies of learning in simple invertebrate and vertebrate systems. In: V. Mountcastle and F. Plum (Eds.) Handbook of Physiology, Section 1: The Nervous system, American Physiological Society.

Henderson, T. B. and Strong, P. N. (1972) Classical conditioning in the leech, Macarabdella ditetra, as a function of CS and UCS intensity. Conditioned Reflex, 7, 210-215.

Holland, P. C. and Rescorla, R. A. (1975) Second-order conditioning with food unconditioned stimulus. J. Comp. Psychol., 88, 459P467.

Holland, P. C. (1977) Conditioned stimulus as a determinant of the form of the Pavlovian conditioned response. J. Exp. Psychol., 88, 459-467.

Horridge, D. (1962) Learning of leg position by headless insects. Nature, (193, 697-698.

Hoyle, G. (1980) Learning using natural reinforcements in insect preparations that permit cellular neuronal analysis. J. Neurobiology, 11, 323-354.

Hoyle, G. and Field, L. H. (1983) Elicitation and abrupt termination of behaviorally significant catchlike tension in a primitive insect. J. Neurobiology, 14, 299-312.

Kamin, L. J. (1969) Predictability, surprise, attention and conditioning. In: R. Churck and B. A. Campbell (Eds.)_, Punishment and Aversive Behavior, New York: Appleton-Century-Crofts, 279-296.

Kuwabara, (1957) Bildung des bedingten reflexes, vom Pavlov typus bei der honigbiene, J. Sac. Sci. Hokkaido Univ. ,13, 456-64.

Larew, M. (1986) Inhibitory learning in one-trial backward conditioning procedures. Unpublished doctoral dissertation, Yale University.

Lukowiak, K. and Sahley, C. L. (1981) The in vitro classical conditioning of the gill withdrawal reflex of Aplysia californica. Science, 212, 1516-1518.

Lukowiak, K. (1982) Associative learning in the isolated siphon, mantle, gill and abdominal ganglion preparation of Aplysia: A new paradigm. Soc. Neurosci. Abstr., 8, 385.

Mackintosh, N. J. (1975) The Psychology of Animal Learning. New York, Academic Press.

Marchant, H. G. and Moore, J. W. (1973) Blocking of the rabbit's conditioned nictitating membrane response in Kamin's two-stage paradigm. J. Exp. Psychol., 101, 155-158.

Martin, K. A. and Sahley, C. L., (1986) Excitatory and inhibitory appetitive conditioning in the terrestrial mollusc. Soc. Neurosci. Abstr., 12, 39.

Menne, D. and Spatz, H. C. (1981) Color learning in Drosophila. J. Comp. Physiology, 114, 301-312.

Menzel, R. (1967) Untersuchungen zum erlernen von spektralfarben durch die honigbeine. Z. Verlag. Physiol. 56, 22-62.

Menzel, R. (1968) Das gedachtnis der honigbienne fur spektralfarben. I. Kurzzeitiges und langzeitiges behalten. Z. Verlag Physiol., 60, 82-102.

Menzel, R. (1979) Behavioral access to short-term memory in bees. Nature, 281, 368-369.

Menzel, R. and Erber, J. (1972) Influence of the quality of reward on the learning performance in honeybees. Behavior, 41, 27-42.

Mpitsos, G. J. and Lukowiak, K. (1985) Learning in gastropod molluscs. In: The Mollusca, ed., A. O. D. Willows, New York, Academic Press, 95-267.

Mpitsos, G. J. and Cohan, C. S. (1986a) Discriminative behavior and Pavlovian conditioning in the mollusc Pleurobranchaea. J. Neurobiology, 17, 469-486.
Mpitsos, G. J. and Cohan, C. S. (1986b) Differential Pavlovian conditioning in the mollusc Pleurobranchaea. J. Neurobiology, 17, 487-497.
Mpitsos, G. J. and Collins, S. D. (1975) Learning: Rapid aversive conditioning in the gastropod mollusc Pleurobranchaea. Science 108, 954-957.
Mpitsos, G. J. and Davis, W. J. (1973) Learning: Classical and avoidance conditioning in the mollusk Pleurobranchaea. Science, 180, 317-320.
Pavlov, I. P. (1927) Conditioned Reflexes. London: Oxford University Press.
Prior, D. and Gelperin, A. (1977) Autoactive molluscan neuron: Reflex function and synaptic modulation during feeding in the terrestrial slug Limax maximus. J. Comp. Physiol 114, 217-32.
Quinn, W. G., Harris, W. A. and Benzer, S. (1974) Conditioned behavior in Drosophila melanogaster. Proc. Natl. Acad. Sci. USA, 71, 708-712.
Reingold, S. C. and Gelperin, A. (1980) Feeding motor program in Limax II. Modulation by sensory inputs in intact animals and isolated central nervous system. J. Exp. Biology, 85, 1-(19.
Rescorla, R. A. (1968) Probability of shock in the presence and absence of the CS in fear conditioning. J. Comp. Physiol. Psychol., 66, 1-5.
Rescorla, R. A. (1973) Second-order conditioning: Implications for theories of learning. In: Contemporary Approaches to Conditioning and Learning, eds., F. J. McGuigan and D. B. Lamsden, Washington, D. C., Winston, 7-33.
Rescorla, R. A. (1975) Pavlovian excitatory and inhibitory conditioning. In: W. K. Estes (Ed.), Handbook of learning and cognitive processes, vol 2., Hillsdale, N. J., Erlbaum, 7-35.
Rescorla, R. A. (1980) Pavlovian Second-order conditioning Studies in Associative Learning. Hillsdale, N. J., Erlbaum.
Rescorla, R. A. (1982) Simultaneous second-order conditioning produces S-S learning in conditioned suppression. J. Exp. Psychol., (Anim. Behav.), 8, 23-32.
Rescorla, R. A. (1988) Behavioral studies of Pavlovian conditioning.

Ann. Rev. Neuroscience, 11, 329-352.

Revusky, S. and Garcia, J. (1970) Learned associations over long delays. In: The Psychology of Learning and Motivation, vol 4, ed., G. H. Bower, New York, Academic Press, pp 1-84.

Rizley, R. C. and Rescorla, R. A. (1972) Associations in second-order conditioning and sensory preconditioning. J. Comp. Physiol. Psychol., 81, 1-11.

Sahley, C. L. and Ready, D. F. (1988) Associative learning modifies two behaviors in the leech, Hirudo medicinalis. J. Neuroscience, 8, 4612-4620.

Sahley, C. L. (1988) Associative learning in the leech: Behavioral and cellular effects of predictability. Soc. Neurosci. Abstr., 14, 838.

Sahley, C. L., Gelperin, A. and Rudy, J. W. (1981) One-trial associative learning in the terrestrial mollusc. Proc. Nat. Acad. Sci, 78, 640-642.

Sahley, C. L., Martin, K. A. and Gelperin, A., (1989) An analysis of associative learning in the terrestrial mollusc Limax maximus II. Appetitive learning. J. Comp. Physiol, A., in press.

Sahley, C. L., Rudy, J. W. and Gelperin, A. (1981) An analysis of associative learning in a terrestrial mollusc. I. Higher-order conditioning, blocking and a transient US pre-exposure effect. J. Comp. Physiol, 144, 1-8.

Sahley, C. L., Hardison, P., Hsuan, A. and Gelperin, A. (1982) Appetitively reinforced odor conditioning modulates feeding in Limax maximus. Soc. Neurosci. Abstr. 8, 823.

Susswein, A. J. and Schwarz, M. (1983) A learned change of response to inedible food in Aplysia. Behavioral and Neural Biology, 39, 1-6.

Susswein, A. J., Schwarz, M. and Feldman, E. (1986) Learned changes of feeding behavior in Aplysia in response to edible and inedible foods. J. Neuroscience, 6, 1513-1527.

Tempel, B. L., Bonini, N., Dawson, D. R. and Quinn, W. G. (1983) Reward learning in normal and mutant Drosophila. Proc. Natl. Acad. Sci. USA, 81, 3577-3581.

Tennant, W. A. and Bitterman, M. E. (1975) Blocking and overshadowing in two species of fish. J. Exp. Psychol. (Anim. Behav.), 104, 22-29.

vom Saal, W. and Jenkins, H. M. (1970) Blocking the development

of stimulus control. Learning and Motivation, 1, 52-62.
Wagner, A. R. and Rescorla, R. A. (1972) Inhibition in Pavlovian conditioning: Application of a theory. In: R. A. Boakes and M. S. Halliday (Eds.), Inhibition and Learning, London, Academic Press, 301-336.
Walters, E. T. and Byrne, J. H. (1983) Associative conditioning of single sensory neurons suggests a cellular mechanism for learning. Science, 219, 405-408.
Walters, E. T., Carew, T. J. and Kandel, E. R. (1979) Classical conditioning in Aplysia californica. Proc. Natl. Acad. Sci USA, 76, 6675-6679.

CHAPTER 4

Neural Models of Classical Conditioning: A Theoretical Viewpoint

G. Tesauro

Introduction

Computational neural circuit models are now in wide use as theoretical tools by many researchers interested in understanding the cellular substrates of invertebrate learning and memory and, in particular, by researchers studying invertebrate classical conditioning. This has been made possible by two important recent developments: (1), considerable progress has been made in mapping out the circuits and determining the plasticity mechanisms that govern classical conditioning in invertebrates such as *Aplysia* (Carew, Hawkins, and Kandel 1983; Hawkins et al., 1983), *Hermissenda* (Alkon, 1983; Farley and Alkon, 1985), *Limax* (Gelperin, 1986), and *Pleurobranchea* (Mpitsos and Cohan, 1986); (2), the computational resources needed to simulate connectionist models are now readily available in most laboratories. The idea of using such computational models has apparently been independently invented several times, and a wide variety of specific models have been proposed to account for many types of classical conditioning behaviors (Sutton and Barto, 1981; Hawkins and Kandel,1984; Gelperin, Hopfield and Tank, 1985; Moore et al., 1986; Tesauro, 1986; Gluck and Thompson, 1987; Sutton and Barto, 1987; Klopf, 1988; Tesauro, 1988).

In this paper, we shall examine the development and use of computational models from the viewpoint of the theorist who designs and simulates them. The purpose of the paper is twofold: to explain to an audience of biologists and physiologists the important theoretical issues, and the approaches taken by theorists in tackling these issues, and to indicate to biologists how such models can be useful, by providing a novel framework for organizing empirical data, and by suggesting new experiments, at both the physiological level and at the behavioral level.

Considerations of *simplicity* and *generality* are often of prime importance to the theorist. While many biologists would like to gather as much detailed knowledge as possible concerning their specific animal preparation, the theorist often pursues the opposite approach of stripping away from his model as much detail as possible, in order to discover the essential ingredients that account for the phenomena in question. A successful model according to this viewpoint might not explain in exact detail how conditioning would occur in a specific organism, but it should lead to general conclusions about how conditioning must take place in *any* network composed of biological neurons and synapses.

Pursuant to the theoretical virtues of simplicity and generality, we shall focus in detail on the following three key issues: (1) What is the simplest possible circuit diagram which will account for conditioning behaviors? (2) What are the essential ingredients of a model's synaptic plasticity rule, or learning algorithm, in order for the model to behave properly? (3) What are the effects of using *distributed* representations of the sensory stimuli, as opposed to *local* representations? Recall that in a local representation each sensory stiumulus is represented by activity in a single, separate neuron, while in a distributed representation each stimulus is represented by a separate *pattern* of activity in a population of neurons (Hinton, 1984).

Briefly stated, these are the answers to the questions: (1) The simplest circuit that will work uses one neuron for each sensory stimulus and one neuron for the conditioned motor response, with synaptic connections from each sensory neuron to the response neuron. No interneurons or more complicated wiring diagrams are necessary. (2) The plasticity rule must contain the key ingredients of timing sensitivity, and a mechanism of generating sign changes in the synaptic modification under the appropriate circumstances. For biologically plausible local learning rules of a generalized Hebb form, the timing sensitivity comes from using a "stimulus trace" of the presynaptic activity, and the sign change comes from using two postsynaptic terms of opposite sign. (3) With distributed representations, the basic conditioning paradigms still hold, but a much richer class of specific behaviors is possible, as will be seen later.

In the following section, we first review briefly the basic behavioral phenomena of classical conditioning which can be explained by current computational models. In section 3, we discuss the operation of models of conditioning which use local representations, of which there are many specific examples. In section 4, we discuss the use of distributed representations, which have been studied in particular in the LIMAX models of Gelperin, Hopfield and Tank (1985) and Tesauro (1986). In section 5, we examine possible extensions of the model to multi-layer learning systems, along the lines suggested by the "back-propagation" learning formalism. Finally, in the concluding section, we examine the prospects for advances in our understanding of invertebrate classical conditioning through the interaction of computational modeling studies with more traditional cellular and behavioral studies of real animals.

Review of Classical Conditioning

This section presents a brief review of the general phenomenology of classical conditioning. These behaviors are generally seen in vertebrate classical conditioning, and are also seen in at least one invertebrate species, *Limax maximus*. (There are a few vertebrate phenomena such as conditioned inhibition which have not been extensively studied in *Limax*. These phenomena will not be discussed here.)

In the simplest form of classical conditioning, an organism is presented initially with a stimulus, called the conditioned stimulus or CS, which is of no innate significance to the organism, thus the only response it generates is an orienting response. At a later time, the organism receives an unconditioned stimulus, or US, which automatically and invariably causes an unconditioned response (UCR). After repeated pairings of CS with US, the organism comes to respond to the CS with a conditioned response (CR) which is very similar or identical to the UCR (Pavlov, 1927). In the experiments that have been performed on *Limax* (Sahley Gelperin and Rudy, 1981; Sahley, Rudy, and Gelperin,1981; 1984; Sahley, Chapter 3), the conditioned stimuli are various types of foods or food odors (i.e.,

1st-order conditioning :	$CS + US \longrightarrow CS^-$
Extinction of conditioning :	$CS^- \longrightarrow CS^+$
Temporal specificity :	$US + CS \longrightarrow CS^+$
Compound conditioning :	$(CS_1 CS_2) + US \longrightarrow CS_1^-;\ CS_2^-$
2nd-order conditioning :	$CS_1 + US \longrightarrow CS_1^-$;
(type a)	$CS_2 + CS_1^- \longrightarrow CS_2^-$ (a)
(type b)	$(CS_2 CS_1^-) \longrightarrow CS_2^-$ (b)
After $CS_1^- \longrightarrow CS_1^+$, then	
	CS_2^- (a) $\longrightarrow CS_2^-$
	CS_2^- (b) $\longrightarrow CS_2^+$
Blocking :	$CS_1 + US \longrightarrow CS_1^-$;
	$(CS_2 CS_1^-) + US \longrightarrow CS_2^+$

Figure 1: The basic phenomena of classical conditioning which are widely seen across species and stimulus types, and which have been demonstrated in *Limax maximus*.

potato, carrot, mushroom, etc.) and the US is a bitter-tasting substance such as quinidine sulphate. The two types of responses are a feeding response (''eat'') and an avoidance response (''flee'').

A summary of the findings of classical conditioning experiments with panphyletic generality is presented in figure 1. In this figure, the superscripts + and – refer to the type of response generated by the CS: + denotes no response (or an orienting response), and – denotes a conditioned response. Thus in simple first-order conditioning, repeated CS-US pairings result in CS^-, as described above. (This conditioning exhibits a property known as temporal specificity, which means that the CS must occur before or at the same time as the US. If the US occurs first, the CS remains CS^+.) In compound conditioning, two different CSs are paired with the US, with the result that each will generate a CR. In extinction, a previously conditioned CS^- is now presented without pairing to the US, which causes the CS to lose its ability to evoke a CR.

Perhaps the most intriguing finding is in the case of second-order conditioning (Rescorla, 1980). This involves a two-stage learning process: in the first stage CS_1 is paired with the US, and in the second stage CS_2 is paired with CS_1. Two different results are found, depending on whether in the second stage CS_1^- is presented simultaneously with CS_2, or after CS_2. In the latter case, extinction of CS_1^- will not cause CS_2^- to extinguish, while in the former case it will. This means that the latter procedure forms a CS_2-US association, while the former procedure produces a CS_2-CS_1 association.

Finally, the phenomenon of blocking (Kamin, 1969) is again the result of a two-step procedure. In the first step, CS_1 is paired with US, and in the second step CS_2 and CS_1^- are presented simultaneously, followed later by US. Under these circumstances CS_2 does not come to evoke a CR, so one says that the presence of CS_1^- blocks the formation of a CS_2-US association.

One can summarize this discussion by saying that these various training procedures cause the conditioned stimuli to develop a *predictive power* regarding the occurrence of the US. In each case the organism builds up a set of expectations that the US will or will not occur, depending on what particular combination of CS's has been observed. This notion of the predictive power of the CS is very important for successful modeling, as will be seen in subsequent sections.

Conditioning with Local Representations

A natural first attempt at a theoretical model of classical conditioning is illustrated in figure 2. The simplest model circuit uses one neuron for each sensory stimulus, i.e., local representations, and since the phenomena discussed in the previous section involve at most two conditioned stimuli, the model contains one neuron for the first conditioned stimulus (CS_1), one neuron for the second conditioned stimulus (CS_2), one neuron for the unconditioned stimulus (US), and one neuron for the motor response (R). There is a fixed unmodifiable synaptic connection from the US neuron to the response neuron which invariably causes R to fire whenever US fires. There are also modifiable connections from each CS neuron to R that change during the conditioning procedures.

In keeping with our desire for theoretical simplicity, these model neurons are usually assumed to have very simple properties. While the model neurons would be more realistic if they generated action potentials, the existence of action potentials *per se* is not necessary

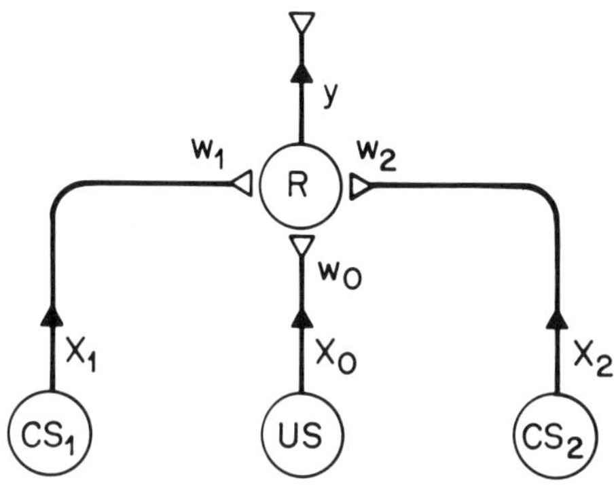

Figure 2: Simplest possible anatomical wiring diagram for a model of classical conditioning, utilizing one sensory neuron for each stimulus (CS_1, CS_2, and US), and one neuron to generate an aversive response (R), with modifiable connections between the conditioned stimulus neurons and the response neuron.

to explain how conditioning works, and thus the states of our models neurons are represented by single real numbers which vary over time between 0 and 1. Specifically, X_1 represents the activity of neuron CS_1, X_2 represents the activity of neuron CS_2, X_0 represents the activity of the US neuron, and Y represents the activity of the R neuron. We think of these variables as representing the short-term average firing rate of more realistic neurons. Similarly, we represent the "strength" of a model synapse by a single real number. The strengths of the synapses from the CS neurons to the R neuron are represented by w_1 and w_2 respectively, which may vary during learning. The synapse from US to R is represented by w_0, which is assumed to be a large, fixed constant. At any point in time, the firing rates X_0, X_1, and X_2 are either 0 or 1 depending on whether the corresponding stimulus is present or absent. The output firing rate of the response neuron is assumed to be deterministically computed from the total input activation from all other neurons which project to it, according to:

$$Y = g(w_0X_0 + w_1X_1 + w_2X_2 - \gamma) \qquad (1)$$

where γ is a threshold variable, and g is a sigmoidal function which saturates at 1 for large positive inputs or 0 for large negative inputs. Time in the model may be either a continuous variable, or it may be broken up into small discrete time steps.

The Hebb rule is a natural first choice for a plasticity rule for modifying the weights during learning. The most literal interpretation of Hebb's hypothesis is that the rate of change of synaptic strength should be proportional to the presynaptic firing rate times the postsynaptic firing rate, i.e.,

$$\dot{w}_i(t) = \varepsilon X_i(t)\, Y(t) \qquad (2)$$

where the dot indicates the time derivative, and ε is a small numerical constant. This rule is initially appealing, because it generates both first-order and second-order conditioning in a rather simple way. It is easy to see that in a first-order conditioning experiment, in which CS_1 is presented and then followed by US, there will be simultaneous activation in both the CS_1 neuron and in the R neuron (because US automatically causes R to fire). Thus the learning rule in equation (1) will strengthen the synapse between CS_1 and R. If this is repeated several times, the synapse will eventually become so strong that firing of the CS_1 alone will cause R to fire, and thus the conditioned response will be generated. Similarly, once CS_1 has been conditioned, it can serve exactly the same role as a US in conditioning a second conditioned stimulus CS_2. Furthermore, biophysical evidence for this type of plasticity rule has been found in mammalian hippocampus (Kelso and Brown, 1986; Kelso, Ganong, and Brown, 1986). However, equation (2) has a number of serious problems associated with it. This rule has no way of distinguishing the time order in which the presynaptic and postsynaptic units activate, and thus would not behave properly in backwards conditioning experiments. Furthermore, the rule has no way to generate negative learning, since the sign of $\dot{w}$

is always positive. Finally, there is a kind of "runaway instability" which occurs once a CS has been conditioned to the point where it can fire, by itself, the response neuron. In this case, presentation of the CS alone will cause both presynaptic and postsynaptic neurons to activate, and the synapse will be strengthened, whereas we know that presentation of CS alone should cause the learned response to extinguish.

To solve the problems of the learning rule expressed in equation (2), it will be necessary to consider more general classes of learning rules. We shall consider here the rules of a "generalized Hebbian" form, which may be written as:

$$\dot{w}_i(t) = \varepsilon F(X_i)G(Y) \tag{3}$$

where the functions F and G are for the moment unspecified functions of the presynaptic and postsynaptic firing rates, respectively. (Strictly speaking, these are functionals, since they may depend on previous as well as current firing rates.)

To solve the problem of timing sensitivity, the function $F(X_i)$ should compute a "stimulus trace" of presynaptic input (i.e., a weighted average over previous times). In discrete-time models, the stimulus trace is defined by:

$$\overline{X}_i(t) = c_1 X_i(t-1) + c_2 X_i(t-2) + c_3 X_i(t-3) + \cdots \tag{4}$$

where the constant c represent the relative weighting of activities at previous epochs. In continuous-time models, the trace is defined by:

$$\overline{X}_i(t) = \int c(\tau) X_i(t-\tau) d\tau \tag{5}$$

Setting $F(X_i)=\overline{X}_i$ in equation 3 yields a model which learns a *predicitve* relationship between previous presynaptic activity and current postsynaptic activity. The use of a presynaptic trace in a classical conditioning model was first proposed in 1972 by Klopf, and is used in one form or another by all successful models of conditioning. It should be emphasized, however, that the trace of other presynaptic variables can also be used in equation (3). In particular, one may also use the trace of the *derivative* of the postsynaptic activity, $\dot{X}_i$. As discussed in Tesauro (1986) and Klopf (1988), in order for the model to work properly in this case, only positive values of the derivative should contribute to the trace, as in the following expression for $F(x_i)$:

$$F(X_i) = \int c(\tau)\sigma(\dot{X}_i(t-\tau))d\tau \tag{6}$$

where $\sigma(z)=z$ for positive z and 0 otherwise. With this presynaptic term, proper conditioning behaviors will also be produced, but the quantitative form of the behavior will be different. In particular, if there is a very long time interval between presentation of the CS and presentation of the US, no learning will take place with equation (6) because the trace of $\dot{X}_i$ will decay to zero by the time the US arrives. On the other hand, the trace of X_i will remain large as long as the conditioned stimulus is present (and for a short time afterwards), so that conditioning can still take place even for very long CS-US intervals.

To enable the model to produce learning of either sign, the postsynaptic dependence G(Y) should consist of the sum of two terms of opposite sign. As discussed in Tesauro (1986), there are many possible ways of doing this. For example, Sutton and Barto (1981) proposed a model utilizing $G(Y)=Y-\bar{Y}$, where the bar notation again denotes a stimulus trace. We shall examine in more detail a model proposed by Tesauro, (1986), which was designed to make explicit contact with the well-known "delta" rule in the psychological literature:

$$\Delta w_i(t) = \varepsilon X_i(t-1)(Y_{actual}(t) - Y_{expected}(t)) \tag{7}$$

where the "expected" postsynaptic firing rate[1] $Y_{expected}(t)$ is a simple

1. We note in passing that this is similar to the notion of learning based on the difference between temporally successive predicted outcomes, which has been proposed for general reinforcement learning by Sutton,(1988).

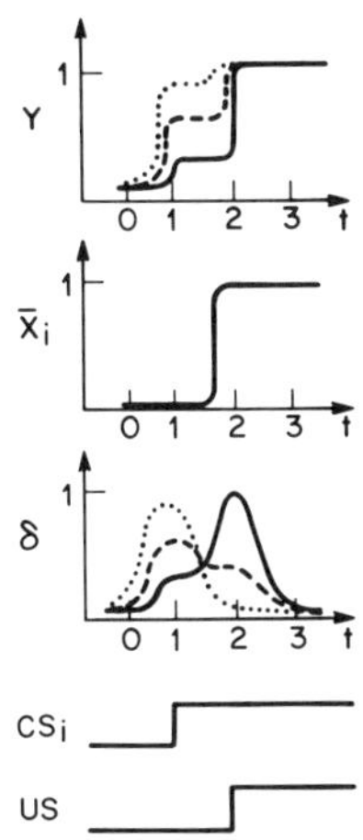

Figure 3: Behavior of the model of (Tesauro, 1986) in first-order conditioning experiment, showing postsynaptic activity Y, trace of presynaptic activity $\overline{X}_i$, and the previously defined quantity δ vs. time. Stimulus CS_i is presented at t=1, followed by US at t=2. Solid curves: trial #1, dashed curves: trial #2, dotted curves: behavior after several trials.

function of the actual firing rate at time (t-1). Specifically, we take $Y_{expected}(t) = f(Y(t-1))$, where $f(x) = \min(x+\eta, 1)$ with η a small numerical constant. This algorithm is closely related to the Sutton-Barto (1981) algorithm, and in fact is equivalent if η=0 and the trace is taken over only one time step; i.e., $\overline{X}(t)=X(t-1)$. The main difference between equation (5) and the Sutton-Barto rule is that stimulus offsets are not required to produce negative learning, whereas they are required in the Sutton-Barto model.

In what follows, graphical representations of the behavior of equation (5) will be presented for each conditioning paradigm. (For specific mathematical details, the reader is referred to Tesauro, 1986).

Figure 3 represents the model's behavior in a first-order conditioning experiment. The conditioned stimulus CS_i is presented at

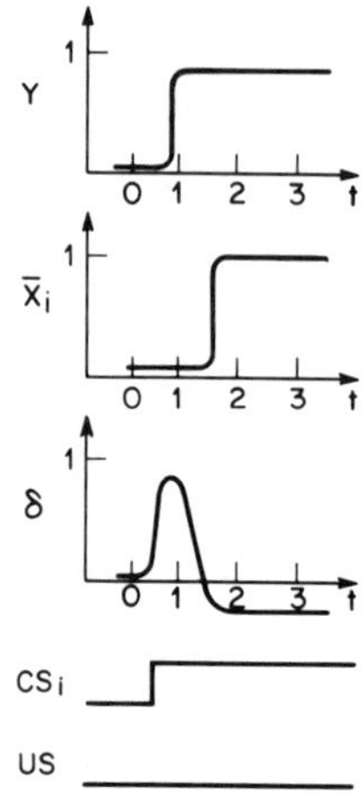

Figure 4: Behavior of the model of (Tesauro, 1986) during extinction of a previously learned association. Stimulus CS_i is presented at t=1 and is not followed by US.

time t=1, followed by the unconditioned stimulus at time t=2. In the first trial, the response neuron firing rate Y is small at t=1, and it jumps to saturation at t=2. From these activity curves, we need to compute two quantities to determine the synapse change: the stimulus trace $\overline{X}_i$, and the quantity $\delta = Y_{actual} - Y_{expected}$. With a simple one-step time delay for the trace, $\overline{X}_i$ is zero at t=1, and is 1 at t=2. The quantity δ, as shown, has a large peak at t=2 on the first trial. Since the weight change is given by the product of $\overline{X}$ times δ, there will be a large increase in synaptic strength on the first trial. In subsequent trials, the initial activation of the response neuron at t=1 will increase as shown. This will have the effect of shifting the peak in δ back to earlier times until finally, after a large number of trials, it will be strongly peaked at t=1, and will be zero at t=2. In that case, no further synapse change will take place because when the trace is large, δ will be zero; conversely, when δ is large, the trace will be zero.

Figure 4 illustrates the behavior of the model during the extinction of a previously learned association. The conditioned stimulus CS_i is presented at t=1, without being subsequently followed by the US. The weight w_i is initially large, so that the response activity at time t=1 is close to saturation. Once again the presynaptic

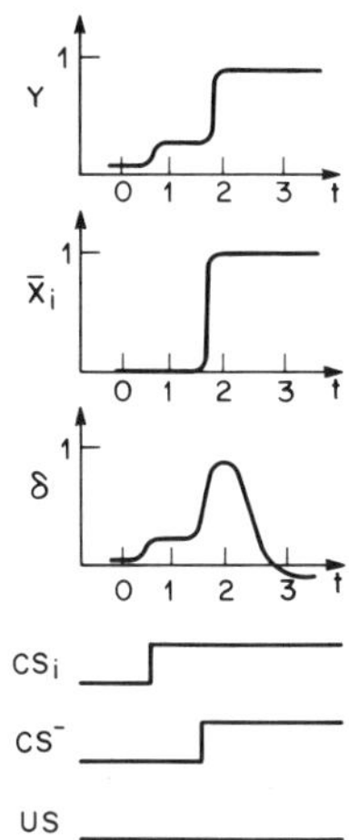

Figure 5: Behavior of the model of (Tesauro, 1986) during the second stage of a second-order conditioning trial. Stimulus CS_i is presented at t=1, followed by the previously trained CS^- at t=2.

trace $\bar{X}_i$ turns on at t=2. However, since Y does not saturate, this time δ has a large postive peak at t=1 and then becomes *negative* at t=2. Thus, the product of $\bar{X}_i$ times δ is negative and the synapse strength will decrease.

Figure 5 illustrates the behavior during a second-order conditioning procedure in which CS_i is presented at t=1, followed by a previously trained CS at t=2. The response neuron's firing rate has a small increase at t=1, followed by a large increase to near saturation at t=2. Delta will then have a large peak at t=2 and will actually go slightly negative at t=3. The net effect will be that second-order conditioning will also work but will be slightly less efficacious than first-order conditioning, depending on how one adjusts the parameters of the model.

Finally, figure 6 illustrates the behavior during a blocking procedure. Both the trained and the untrained CSs are presented at t=1, followed by the US at t=2. The firing rate of R has a large increase at t=1 and then has a further small increase to saturation at t=2. Thus δ will have a large peak at t=1, before the presynaptic trace appears, and will be zero at t=2. Hence no learning takes place.

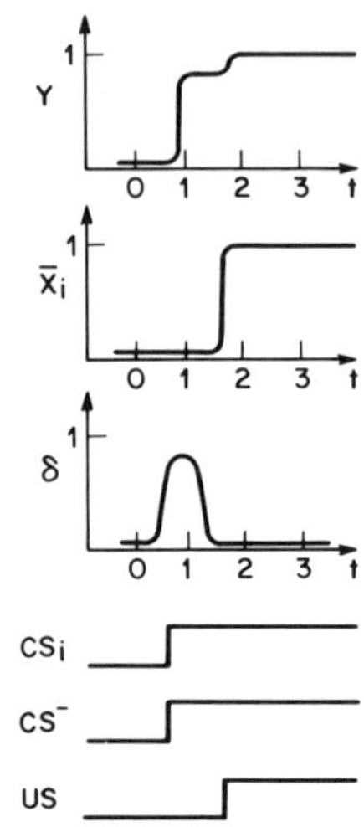

Figure 6: Behavior of the model of (Tesauro, 1986) during the second stage of a blocking trial. Stimulus CS_i is presented simultaneously with the aversively trained CS^- at t=1, followed by US at t=2.

To summarize, we have seen that a computational model consisting of simple model neurons, a simple wiring diagram, and a relatively simple local learning rule is capable of accounting for both first-order and higher-order conditioning paradigms. Although this model does not say in precise detail exactly how conditioning occurs in a particular organism, it has elucidated certain general principles which are likely to apply to most organisms. (The incorporation of timing sensitivity into the presynaptic term and a sign-changing mechanism into the postsynaptic term of the plasticity rule are examples of such general principles.) This class of models is of interest both in terms of its robust behavioral predictions, independent of detail, and in terms of the specific details which can be added to the model to reproduce, for example, the exact shapes of learning curves in a particular animal. Both of these aspects will be discussed in the concluding section.

Conditioning with Distributed Representations

We have seen that simple neural circuit models utilizing local representations of the sensory stimuli are capable of displaying proper conditioning behaviors. In keeping with the theoretical principle of generality, we will now move on to consider the use of distributed representations. While local representations may be valid for modeling of touch receptors and photoreceptors, nature tends to use distributed representations of complex sensory stimuli. Certainly everything that is known about olfaction and taste receptors suggests that odors and tastes are represented in a distributed fashion. Thus if one is interested in modeling, for example, the conditioning experiments performed on *Limax* involving various food odors and food tastes, distributed representations are much more realistic. Furthermore, as one moves higher in the animal kingdom, the use of distributed representations becomes more and more prevalent.

Setting up a model using distributed representations is more difficult than in the local representation case. A number of basic questions with nonobvious answers have to be addressed, such as: (1) Exactly what pattern of activity should one choose to represent a particular conditioned stimulus? (2) To what extent should the patterns of activity which represent two different conditioned stimuli resemble each other? (3) Given representations of two individual stimuli, how should one define a representation of the compound stimulus which is obtained when the two stimuli are presented simultaneously? This task is made more difficult by the lack of concrete empirical data to guide the modeler in his choice of representations.

These questions were first addressed by Gelperin, Hopfield and Tank, in 1985. They suggested that, in the absence of any physiological data indicating how food odors and tastes are represented in *Limax*, the most sensible choice is to represent individual food odors by a *random* pattern of activity in an array of model "taste receptors." In their model array of 100 taste receptors, a particular food odor was represented by activity in a particular set of 20 or so receptors, chosen at random. The choice of the number 20 represented a balance between two opposing considerations. On the one hand, one would like this number to be large in order to maximize both the number of distinct odors that can be represented and the

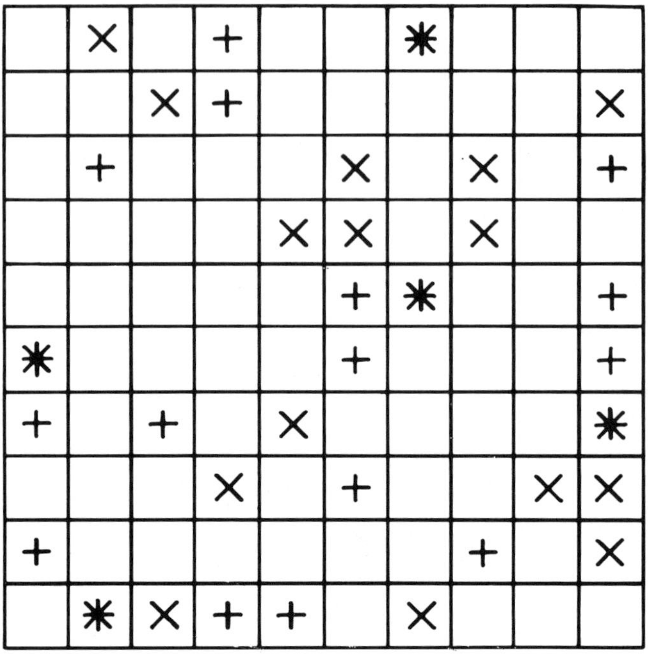

Figure 7: Illustration of distributed representations of two different food categories in an array of 100 model neurons. Neurons which participate in the representation of food category A are marked with a + sign. Neurons which participate in the representation of food category B are marked with a × sign. Neurons which participate in both representations are marked with a star sign. The representations were chosen to be random, uncorrelated representations with a 20% probability for each neuron to be on.

''fault-tolerance'' of the system, i.e., the ability to behave correctly if one of the taste receptor neurons misfires. On the other hand, if it is too large, then the degree of similarity or overlap between different food categories becomes too great. The number 20 turns out to be a satisfactory intermediate value. Figure 7 illustrates two random patterns of activity in a 100-neuron array, representing two different food categories, A and B. This figure shows that, of the 20 or so neurons which represent A, only about 4 or 5 also participate in the representation of B. Therefore the representations are largely non-overlapping.

The representation of the compound stimulus (AB) is a more

difficult matter. Gelperin, Hopfield and Tank (1985) argued that it would be unrealistic to activate all of the neurons that represent A and all of the neurons that represent B, thereby doubling the average level of activity in the array, merely because a food odor mixture is present. Instead, they chose to represent the compound stimulus as a new 20-neuron category which has a substantial overlap with A and B, but which is different from both. Of these 20 neurons, about 6–7 would represent A but not B, 6–7 would represent B but not A, 4–5 would represent both A and B, and 4–5 would represent neither A nor B. With this more sophisticated representation scheme for food odors and food mixtures, a wider variety of behaviors can be obtained, depending on exactly how the various overlaps work out.

Detailed numerical simulations using this representational scheme were reported by Tesauro, in 1986. The most striking differences from local representation models were found for the phenomena involving compound stimuli, i.e., compound conditioning and blocking. Due to the random nature of the categories, significant variability in the degree of overlap between the compound category and the individual food categories is possible, and this variability has important behavioral consequences. For example, in the blocking paradigm, the critical variable which determines the amount of blocking is the degree of initial aversive response to the compound stimulus, which depends on the overlap of the compound stimulus with the aversively trained CS_1. In the simulations reported by Tesauro in 1986, this overlap could vary from as low as 7–8 bits to as high as 14–15 bits. The low overlaps produced an initial aversive response to the compound stimulus of about 0.5, whereas the high overlaps generated an initial aversive response of about 0.9. It was found that if this response level was greater than 0.7 (which covered most food pairs), essentially complete blocking was obtained: the stimulus CS_2 retained nearly 100% of its ability to generate an attractive response. On the other hand, with an initial aversive response of only 0.5, some learning did take place for CS_2, with the result that its attractiveness dropped to only 20–30% of its original value.

Similarly, in compound conditioning the degree of aversive response to the individual components A and B will depend on the overlap between each component and the mixture. For some food pairs the response to both components may be strongly aversive, while

FOOD A	FOOD B	RESPONSE TO A	RESPONSE TO B
5	3	-.331	-.967
4	7	-.995	-.714
4	6	-.916	.000
8	9	-.039	-.996
8	10	-.982	-.927

Table 1: Examples of the behavior of the LIMAX model of (Tesauro, 1986) in compound conditioning experiments, for varying food pairs as conditioned stimuli. The network recognized ten patterns of activity as familiar food categories, labeled 1-10. In each experiment with a particular food pair, both foods were first trained to be attractive, and then were subjected to a single compound conditioning trial. Response afterwards to the individual foods is plotted on a scale from -1.0 (maximal fleeing activity) to 1.0 (maximal eating activity).

for other pairs, one component may be strongly aversive and the other may be only weakly aversive, as illustrated in Table 1. These findings are expected to be representative of the situation in real animals, although the degree of variability has not been systematically studied in *Limax*. Other phenomena involving only single food odors turn out to work in this model similarily as in the local representation models.

The behavior of the model in the compound conditioning paradigm has led to an interesting experimental test in the real animal and recently reported by Gelperin and Hopfield (1988). They studied the response of *Limax* to individual food odors A and B after compound conditioning of the mixture (AB). The results of this experiment indicate that the response to the components depends on the exact nature of the mixture preparation. If the compound stimulus is the odor produced by a mixture in solution of two chemicals characteristic of the individual foods, then the slug will *not* learn to avoid the component odors. However, if the compound stimulus is produced by very closely spaced but physically separate sources of the component odors, then the result is that the components do become aversive. The LIMAX model in its current configuration can only account for the latter finding, and thus needs to be modified in some way in order to reproduce the actual experimental results.

There are two basic ways one might go about trying to modify the model. One is to focus on how compound stimuli are represented in the categorizer front-end of the model, and to develop alternative representations which contain a substantial number of *novel* bits in the representation, corresponding neither to the bits of food A nor to the bits of food B. An alternative and perhaps more intriguing approach is to try to introduce "hidden units" and mulitple layers of weights in the back-end of the model. This would allow the network to learn more complex decision boundaries between stimuli that are aversive and stimuli that are attractive, which might not be obtainable in a single-layer configuration. More will be said about multilayer schemes in the following section.

Multilayer Networks and "Back-propagation" of Errors

The models we have examined so far have been limited to a single layer of connections between the categorizer neurons and the eat-flee decision neurons. Such a system can only learn linear decision boundaries (Minsky and Papert, 1969) and cannot learn an "exclusive-or" task in which single-unit representations of individual stimuli are aversive, but the conjunction of the individual stimuli is attractive. This might be an important limitation to existing models of classical conditioning, and to models of associative memory and more general forms of learning. In addition, single-layer models are less relevant biologically for higher organisms, where multiple layers of processing between input and output are more prevalent.

To go beyond single-layer evaluation modules to systems with multiple layers of weights and internal "hidden" layers, it is necessary to devise a scheme for solving the so-called "credit assignment problem." For example, when there is a difference between the actual and expected postsynaptic activity (so that learning takes place), should one change the weights from the input to the hidden layer, or the weights from the hidden layer to the output units? Recently, a number of solutions to this problem have been proposed, the most notable of which is the so-called "back-propagation" learning rule (Rumelhart, Hinton, and Williams, 1986). Back-propagation changes the weights of a feedforward network according to gradient-descent minimization of a scalar error function defined at the output layer. The learning rule can be expressed as follows: Defining the output error function as the mean-square difference between the actual output state and a "desired" or "teacher" state, summed over a number of input patterns, i.e.,

$$E = \sum_{p} (y_p - d_p)^2 \tag{8}$$

the weights should be changed according to

$$\Delta w_{ij} = -\varepsilon \frac{\partial E}{\partial w_{ij}} \tag{9}$$

where w_{ij} is the connection strength between unit i and unit j. The partial derivative in equation 9 can be computed recursively, layer by layer, using the chain rule, hence the term "back-propagation" of error. The resulting formula for the weight change (we will not reproduce the derivation, but instead refer the reader to Rumelhart and McClelland, 1986) can be concisely expressed as:

$$\Delta w_{ij} = -\varepsilon y_i \rho_j \tag{10}$$

where $\rho_j = \partial E / \partial x_j$, which represents the contribution of cell j to the total error, can be computed recursively as a linear combination of errors of other cells k which receive input from j, i.e.,

$$\rho_j = y_j (1 - y_j) \sum_{k:j \rightarrow k} w_{jk} \rho_k \tag{11}$$

In other words, one first computes the error at the output layer, then the error for all units which project to the output layer, then the error for all units which project to those units, and so on throughout the

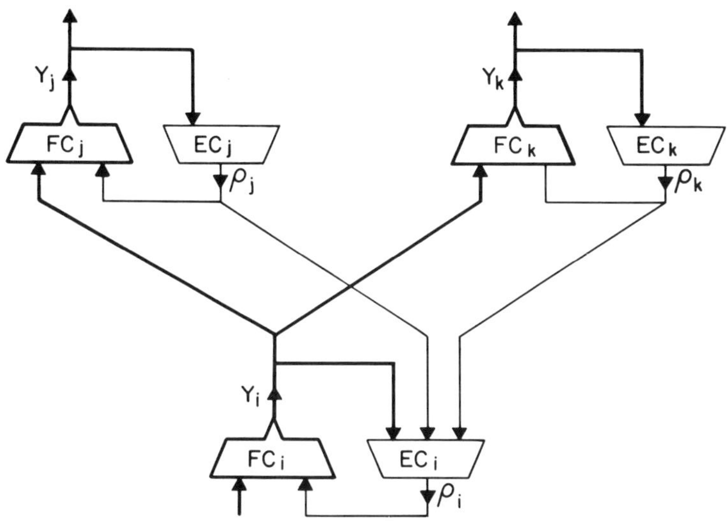

Figure 8: Representation of the scheme proposed in (Parker, 1985) for implementation of back-propagation by function cells and error cells. The usual feed-forward network of function cells is drawn in heavy lines. In addition, for each function cell, there is a corresponding error cell. For each connection in the function-cell network, there is a corresponding connection in the error-cell network.

entire network. The weights are then changed according to the output of the presynaptic unit times the error of the postsynaptic unit.

While the back-propagation algorithm represents an elegant solution to the multi-layer credit assignment problem and has achieved a number of notable successes in learning to solve difficult engineering problems, it appears at first glance to be impossible to implement in a real nervous system, given our understanding of how neurons actually operate. The network must compute in a feed-forward fashion, but in order to modify the synapses, information about the error must flow in the reverse direction. This seems impossible, since information in real neurons always flows from dendrite to soma to axon, and never in the reverse direction.

However, such an impression is misleading. As Parker (1985) first pointed out, back-propagation of the error signal may be done by a second network of "error" cells coupled to the original network of "function" cells, with synaptic connections going in the opposite direction, as illustrated in figure 8. While Parker's scheme does not address the issue of exactly how the error would be computed and used to modify the synapses in the function net, it does show that the error information can flow backwards without the requirement of signal propagation up the axon in the wrong direction.

A "wetware" implementation of Parker's proposal can be formulated utilitizing two biophysical mechanisms which, although somewhat non-standard, are generally accepted. First, a mechanism is needed to compute analog products at least approximately in order to compute the output error signal ρ_i for a given error cell i. This can be seen from the formula for ρ_i in terms of the errors ρ_j in the previous layer, given in equation 11. Analog products may be computed using the mechanism of nonlinear interactions in the dendritic arborization. Evidence from computer simulations (Koch and Poggio, 1987; Rall and Segev, 1987) suggests that voltage-dependent channel conductances may compute analog products at least approximately. In logical terms, excitatory synapses would compute an analog AND operation, while inhibitory synapses would compute an analog AND-NOT operation. A combination of both operations, as illustrated in figure 9, would perform the necessary multiplication of the linear sum of errors from the previous layer by the quantity $y_i(1-y_i)$ (i.e., the derivative of the output of the corresponding function cell).

Second, a mechanism is needed to modify the synapses in the function cell network utilizing this error information. Since the synapses in the error cell network must have the same values as the corresponding synapses in the function cell network, these synapses must be modified during learning as well. Synaptic modification may be accomplished using the mechanism of presynaptic facilitation, which has been extensively studied in the invertebrate *Aplysia* (Hawkins et al., 1983). Note in equation 10 that the synaptic change for weight w_{ij} is proportional to the product of the output of function cell i times the output of error cell j. This suggests the following fortuitous arrangement to modify the synapses in both networks: for the function cell synapses, the error cell axons may serve as the presynaptic facilitators, while for the error cell synapses, the function

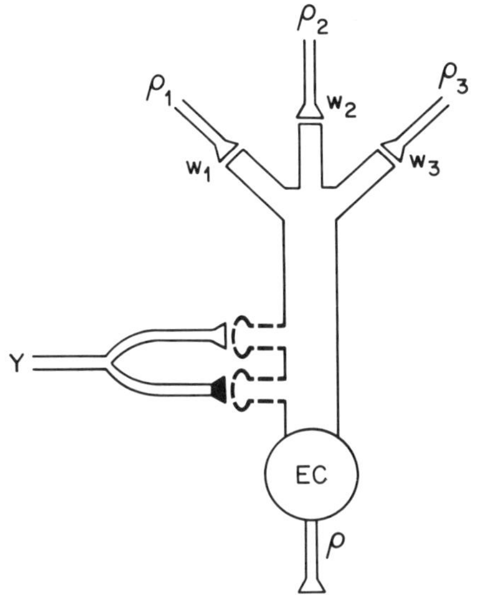

Figure 9: Illustration of error computation in an error cell via known biophysical mechanisms. Inputs from other error cells ρ_1, ρ_2 and ρ_3 are received at the top of the dendritic tree, multiplied by synaptic weights w_1, w_2 and w_3, and summed in the usual way for linear dendrites. Overall multiplication by $Y(1-Y)$ takes place via excitatory and inhibitory synapses (represented by open and solid terminals, respectively) on nonlinear dendrites, drawn in dashed outline. The transfer function of the cell body is linear, i.e., $\rho = Y(1-Y)(w_1\rho_1+w_2\rho_2+w_3\rho_3)$.

cell axons may serve as presynaptic facilitators. If the resulting synaptic change is genuinely symmetric, i.e., depends only on the product of the two signals and not on the identification of which signal is which, then identical weight changes will be computed for both the forward-going synapse w_{ij} and the backward-going synapse Ω_{ij}. Thus if the weights start off at equal values, they will remain equal throughout the course of learning. An illustration of the full system combining both mechanisms is displayed in figure 10.

Admittedly, the proposed implementation scheme suffers from a number of drawbacks in terms of true biological plausibility. An exact correspondence of function cell network with error cell network is

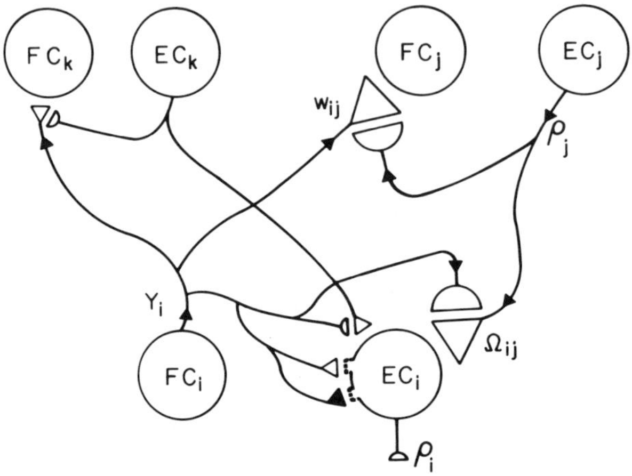

Figure 10: Illustration of the proposed back-propagation scheme combining the error computation mechanism of Figure 9 with a presynaptic facilitator mechanism for changing the weights, as in *Aplysia*. Open triangle: excitatory synapse, solid triangle: inhibitory synapse, semicircle: modulatory synapse. The weight change computed for both the forward-going weight w_{ij} and the backward-going weight Ω_{ij} is the product of function cell output Y_i times error cell output ρ_j.

required at the level of individual cells, and even at the level of individual synapses. Such a degree of structural specificity is not known to exist in mammalian cortex. In the actual biological mechanisms, the nonlinear dendrites do not compute exact analog products, and the presynaptic facilitation is not exactly symmetric with respect to presynaptic input and facilitator input. Three-terminal axo-axonic synapses, while perhaps more prevalent in lower invertebrates, are rather uncommon in mammalian cortex (Shepherd, 1979). The issue of what happens after learning, when the error cell network no longer serves any useful purpose, has not been addressed. Finally, the paradigm of supervised learning requires the basic notion of a "teacher" signal at the output layer of the network and it has not been stated exactly where such a signal would be generated or how it would be conveyed to the network.

Of course, there are a number of obvious modifications to the proposed implementation scheme which could make it more plausible. For example, one could argue that the cells in the model correspond not to individual neurons, but to groups of neurons, and therefore one would need only structural specificity at the group level. One could also have the error cells estimate the error rather than compute it exactly, but a detailed discussion of such changes would distract the reader from the main point of this exercise. The purpose is not to claim that nature actually uses back-propagation, nor that the above implementation scheme is a plausible way for nature to implement back-propagation. Rather, it is to show by explicit demonstration that the algorithm can, at least in principle, be implemented in a biological neural network using known biological mechanisms. Therefore, although back-propagation may be implausible for a number of reasons, and the particular implementation scheme described here may be very implausible, it is not *impossible* to imagine in real biological nervous systems. It is conceivable that a clever genetic engineer of the next century could design an organism using back-propagation–if such organisms do not already exist.

Perhaps the more important general conclusions for biologists to draw from this exercise are these: (a) there are now elegant and powerful theoretical algorithms that solve the multilayer credit assignment problem, and (b) biophysical mechanisms are available that can implement the basic computations required by these algorithms. Back-propagation is only one way of minimizing a global scalar functional (mean-square error) by a particular technique (gradient-descent). There are several other formulations of multilayer credit assignment which involve other minimization techniques, other scalar functionals, and other entirely different approaches. If one seriously believes that nature actually solves the multilayer credit assignment problem, then perhaps it is by one or more of these approaches. It is therefore worthwhile to devote more thought to possible biological implementations of these theoretical algorithms. They should not be summarily dismissed from the domain of biological relevance merely because they do not have immediately obvious neurobiological correlates.

Conclusions

We have seen that computational models of classical conditioning can be constructed which work properly, in the sense that they demonstrate the correct behavior in each of the various conditioning paradigms. Are these models merely interesting intellectual exercises, or can they actually be useful to experimental biologists studying conditioning in real animals? This paper makes the claim that there are a number of ways in which theoretical neural network models of classical conditioning can be useful to experimentalists studying animal learning and memory. The theoretical design concepts and quantitative properties of the "neurons" and "synapses" in the model can provide a new way of organizing one's thinking about the behavioral phenomena of real animals. In the ideal case, a large body of behavioral data would be simply and concisely explained in terms of a few theoretical principles. Furthermore, novel theoretical concepts often lead to the design of experimental procedures which would not be conceived solely by studying animal behavior at the phenomenological level.

The LIMAX models in particular have suggested a large number of behavioral experiments which could provide new insights into the bases of classical conditioning in *Limax*. Indeed, one such experiment pertaining to the representation of compount stimuli (Gelperin and Hopfield, 1988) was described previously. Some of these experiments are suggested by the "robust" properties of the LIMAX models, i.e., predicted behaviors that are always obtained regardless of the precise details of the model design. For example, it appears that the time scale for identification or categorization of the conditioned stimulus must be significantly faster than the time scale of learning a behavioral response. Serious degradation of the performance of the model occurs if this is not the case. Although it is intuitively clear that an animal cannot learn to avoid a particular stimulus on a time scale faster than the time it takes the animal to identify that stimulus, it is nice to see a concrete quantitative explanantion emerge from the behavior of a computational model. Another robust prediction is that the time scale for conditioning an aversive response is much faster than the time scale for extinguishing such a response. This agrees in particular with the *Limax* experiments of Sahley, Gelperin, and Rudy, (1981), which found rapid learning of the aversive response in a single trial, whereas it took several trials to extinguish such a learned response. Finally, the model predicts that blocking would eventually fail after a large number of trials. This is because the categorizer would eventually learn the new category (AB), which would be aversive because of its large overlap with A^-. This has not been studied but could be easily tested by direct behavioral studies in *Limax*.

Another class of experiments is suggested by certain nonintuitive behavioral "quirks" of the model. One such quirk is what might be called an "auto-conditioning" effect: if a food A has been partially conditioned, subsequent presentation of A, without being followed by the unconditioned stimulus Q, could actually lead to a slight *increase* in the aversive response to A. This would occur in the model if the categorization time scale is not rapid, i.e., if not all of the neurons of category A turn on at the same time. Under these conditions, the units which represent A and which turn on early are aversively trained by the large increase in fleeing activity caused by turning on of the later units. Another quirk of the model is that it is possible in some cases that a second-order conditioning procedure $A + B^- \rightarrow A^-$ would actually cause an *increase* in the aversiveness of B^-. This could happen if the extinction rate is low enough, and if the degree of overlap of A and B is higher than normal. Both of these phenomena are usually not seen in animal studies, and it would be most interesting if there were some circumstances under which they could be obtained.

Finally, there are a number of behavioral questions for which the model does not provide an unambiguous answer. Different answers may be obtained by changing, for example, the particular choice of learning algorithm or parameter settings. The most important of these effects has to do with the choice of the presynaptic term $F(X_i)$ in the learning algorithm. If the trace of the presynaptic activity is used, then it will be possible to obtain conditioning by leaving the CS on for an arbitrarily long time, and then following it with a US. Similarly, when a CS has been aversively trained and is presented without US reinforcement, extinction will occur for as long as the CS is left on. On the other hand, using the trace of the derivative of the presynaptic activity, there is only a limited window of time in which conditioning can occur after presentation of the CS, and only a fixed amount of extinction per trial can be obtained. This ambiguity could be resolved directly and quickly by carrying out the appropriate experiments in

Limax. Another ambiguous issue is what happens in multiple-US experiments. For example, if food A is paired with Q_1, and then A and B are presented simultaneously, followed by Q_2, is an aversive respone to B generated? Different anatomical connections in the LIMAX model could provide either a yes or no answer and, once again, experimental data from real animals is needed to guide the modeler. The final ambiguity which we mention here is the issue of learning based on the *offsets* of stimuli, or "trace" conditioning. This has been recently studied in *Limax* (Martin and Sahley, 1986; Sahley, Martin and Gelperin 1988), although the present models do not address this issue.

In addition to new experiments on real animals, interactions between theoretical and empirical studies can also lead to refinements and improvements in the design of the theoretical models, based on experimental data that cannot be explained by existing models. The finding of Gelperin and Hopfield, (1988) that *Limax* can learn to avoid an odor mixture without learning to avoid the components of the mixture is a prime example of a finding which cannot be reproduced by the current versions of the LIMAX models and thus indicates that improvements in the model must be sought. Such improvements naturally fall into two basic categories: improvements in the "front-end" of the model, dealing with the *representation* and *categorization* of sensory stimuli, and improvements in the "back-end" of the model, dealing with *evaluation* of the stimuli and generation of the appropriate motor responses.

Improvements in the front-end of the model certainly seem possible at this point. The categorizer, which was described by Gelperin, Hopfield and Tank, (1985), was based purely on fundamental theoretical considerations of distributed representations, due to the complete absence of anatomical and physiological data concerning the representation of smells and tastes in the *Limax* central nervous system. As more information becomes available concerning sensory representations in the real animal, it is likely that significant alterations in the design of the categorizer will be indicated. For example, it will probably be necessary to incorporate correlations between different food categories, and to devote more attention to the novel components of a compound stimulus (i.e., the neurons which fire in the presence of the compound, but which do not fire when the components are presented individually). In addition to data from

Limax, data from the olfactory bulb and olfactory cortex of vertebrates might also provide important clues to improved models of distributed sensory representations (Granger, et al. Chapter 5).

Improvements in the back-end of the model should also be possible through a combination of empirical and theoretical approaches. Improved plasticity rules should come about as more becomes known about the functional forms and biophysical mechanisms of plasticity in both vertebrate and invertebrate synapses. Behavioral constraints such as precise response topography curves may also provide important clues to possible refinements of existing theoretical learning algorithms. As for the neuronal architecture of the back-end of the model, there is now great interest in models with mulitple layers and hidden units. Theoretical formalisms such as back-propgation have provided intriguing suggestions for learning in such multilayered networks. While back-propagation itself does not appear to be biologically plausible, it is based on the extremely simple and powerful idea of gradient-descent minimization of an error function. It is not unreasonable to suggest that nature might implement gradient-descent learning in some other fashion and that the combined efforts of neurobiologists and network theorists may lead to the elucidation of more plausible schemes for gradient-descent learning.

In summary, the development of computational network models of classical conditioning to date has been the result of a productive interplay between experimental and theoretical studies, and this productive interplay is expected to continue to lead to interesting results in the future. In particular, comparisons of LIMAX with *Limax* have been quite profitable for both sides, as the modelers have produced an interesting list of experiments to be carried out in the real animal, while the experimentalists have in turn found data which cannot be explained within the current versions of the model. Significant progress and understanding should be achieved in both these directions over the next few years.

Acknowledgments

I thank the Editors for critically reading the manuscript and making many helpful suggestions.

References

Alkon, D. L.(1983) Learning in a marine snail. Sci. Am. **249** (1), 70-84.

Barto, A. G. and Sutton, R. S. (1982) Simulation of anticipatory responses in classical conditioning by a neuron-like adaptive element. Behav. Brain Res. **4**, 221-235.

Carew, T. J., Hawkins, R. D., and Kandel, E. R. (1983) Differential classical conditioning of a defensive withdrawal reflex in *Aplysia californica.* Science **219**, 397-400.

Farley, J., and Alkon, D. L. (1985) Cellular mechanisms of learning, memory, and information storage. Ann. Rev. Psychol. **36**, 419-494.

Gelperin, A. (1986) Complex associative learning in small neural networks. Trends in Neuro Sciences **9**, 323-328.

Gelperin, A and Hopfield, J. J. (1989). Differential conditioning to a compound stimulus and its components in the terrestrial mollusc. *Limax maximus.* Behavioral Neuroscience, Vol. 103, pp 329-333.

Gelperin, A., Hopfield, J. J., and Tank, D. W. (1985) The logic of *Limax* learning. In: Model neural networks and behavior. Selverston, A., ed.. New York: Plenum Press.

Gluck, M. A., Thompson, R. F. (1987) Modeling the neural substrates of associative learning and memory: a computational approach. Psych. Rev. **94**, 176-191.

Hawkins, R. D., Abrams, T. W., Carew and T. J., Kandel, E. R. (1983) A cellular mechanism of classical conditioning in *Aplysia*: activity-dependent amplification of presynaptic facilitation. Science **219**, 400-405.

Hawkins, R. D. and Kandel, E. R. (1984) Is there a cell-biological alphabet for simple forms of learning? Psych. Rev. **91**, 375-391.

Hinton, G. E. (1984) Distributed representations. Technical Report No. CMU-CS-84-157, Carnegie-Mellon Univ., Pittsburgh PA.

Kamin, L. J. (1969) Predictability, surprise, attention, and conditioning. In: Punishment and aversive behavior, pp. 279-296. Church, R., Campbell, B. A., eds.. New York: Appleton-Century-Crofts.

Kelso, S. R. and Brown, T. H. (1986) Differential conditioning of associative synaptic enhancement in hippocampal brain slices.

Science **232**, 85-87.

Kelso, S. R., Ganong and A. H. and Brown, T. H. (1986) Hebbian synapses in hippocampus. Proc. Natl. Acad. Sci. USA **83**, 5326-5330.

Klopf, A. H. (1972) Brain function and adaptive systems-- a heterostatic theory. Technical Report No. AFCRL-72-0164, Air Force Cambridge Research Laboratories, L. G. Hanscom Field, Bedford, MA.

Klopf, A. H.: (1988) A neuronal model of classical conditioning. Psychobiology **16**, 85-125.

Koch, C., Poggio, T. (1987) Biophysics of computation: neurons, synapses, and membranes. In: Synaptic function, pp. 637-697. Edelman, G. E. et al., eds.. New York: Wiley.

Martin, K., Sahley, C. (1986) Analysis of associative learning in *Limax maximus*: Excitatory and inhibitory conditioning. Soc. Neurosci. Abstr. 12:39.

Minsky, M., Papert, S. (1969) Perceptrons. Cambridge, MA: MIT Press.

Moore, J. W. et al. (1986) Simulation of the classically conditioned nictitating membrane response by a neuron-like adaptive element: response topography, neuronal firing, and interstimulus intervals. Behav. Brain Res. **21**, 143-154.

Mpitsos, G. J. and Cohan, C. S. (1986) Discriminative behavior and Pavlovian conditioning in the mollusc *Pleurobranchaea*. J. of Neurobiology **17**, 469-486.

Parker, D. B. Learning-logic. (1985) MIT Center for Computational Research in Economics and Management Science, Tech. Report TR-47.

Pavlov, I. P. (1927) Conditioned reflexes. Oxford: Oxford Univ. Press.

Rall, W. and Segev, I. (1987) Functional possibilities for synapses on dendrites and dendritic spines. In: Synaptic function, pp. 605-636. Edelman, G. E. et al., eds.. New York: Wiley.

Rescorla, R. A. (1980) Pavlovian second-order conditioning: studies in associative learning. Hillsdale, NJ: Lawrence Erlbaum Associates.

Rumelhart, D. E., Hinton, G. E. and Williams, R. J. (1986) Learning representations by back-propagating errors. Nature **323**, 533-536.

Sahley, C. L., Gelperin, A. and Rudy, J. W. (1981) One-trial associative learning modifies food odor preference of a terrestrial mollusc. Proc. Natl. Acad. Sci. USA **78**, 640-642.

Sahley, C. L., Rudy, J. W. and Gelperin, A. (1981) An analysis of associative learning in a terrestrial mollusc: higher-order conditioning, blocking, and a transient US pre-exposure effect. J. Comp. Physiol. **144**, 1-8.

Sahley, C. L., Rudy, J. W. and Gelperin, A. (1984) Associative learning in a mollusc: a comparative analysis. In: Primary neural substrates of learning and behavioral change, pp. 243-258. Alkon, D. L., Farley, J., eds.. New York: Cambridge Univ. Press.

Sahley, C. L., Martin, K. A. and Gelperin, A. (1988) Analysis of associative learning in the terrestrial mollusc *Limax maximus* II: Appetitive learning. Manuscript submitted.

Shepherd, G. M. (1979) The synaptic organization of the brain. (2nd ed.) New York: Oxford Univ. Press.

Sutton, R. S. (1988) Learning to predict by the methods of temporal differences. Machine Learning **3**, 9-44.

Sutton, R. S. and Barto, A. G. (1981) Toward a modern theory of adaptive networks: expectation and prediction. Psych. Rev. **88**, 135-170.

Sutton, R. S. and Barto, A. G. (1987) A temporal-difference model of classical conditioning. Proceedings of the Ninth Conference of the Cognitive Science Society, 355-378.

Tesauro, G. (1986) Simple neural models of classical conditioning. Biol. Cybern. **55**, 187-200.

Tesauro, G. (1988) A plausible neural circuit for classical conditioning without synaptic plasticity. Proc. Natl. Sci. USA **85**, 2830-2833.

CHAPTER 5

Unsupervised Perceptual Learning: A Paleocortical Model

R. Granger, J. Ambros-Ingerson, P. Antón and G. Lynch

Introduction

Sensory cortex in mammals is presumed to give rise to powerful perceptual abilities, yet no extant theories address direct links that may exist between the anatomical structure and physiological function of sensory cortical structures and the perceptual computations they perform. This chapter provides a hypothesis of the operation of a particular sensory cortical structure and its inputs (olfactory (piriform) cortex and olfactory bulb), identifies a set of perceptual abilities that have emerged from simulation and analysis of the operation of these structures, and gives a theoretical account of how it is that these perceptual abilities arise from the biological mechanism.

Our approach has been one of relatively strict "bottom-up" modeling, in which well-studied physiological events implicated in a specific form of perceptual learning are modeled as embedded in the known anatomical architecture of the sensory cortical region involved directly in the learning, and simulations of the resulting neural record is examined for its computational behavior. There is a wealth of data on cortical structure and activity in humans and animals, and many studies attempting to identify specific involvements of cortical processing in particular behaviors. It is tempting to believe that there are specific computational tasks to be performed by brain networks which, once formulated, may help guide the search for relevant findings in the neural substrate. Indeed, it is not implausible to assume that some characteristics of brain are there for reasons more of biology than of computation, e.g., for regulatory or maintenance functions, as opposed to information processing. Thus information-processing theories may benefit from abstraction of certain biological features to get at the essence of network function. Even in such a circumstance, however, it is often unclear which of the biological features are central and which peripheral, in part at least because it typically remains unclear what the formal description of the behaviors arising from a given brain region are. For instance, what might be hypothesized as the computational 'function' of sensory cortex? Anatomically, there are a combination of topographic and nontopographic projections to most sensory cortical areas (visual, auditory, and somatosensory); physiologically, it has been supposed that firing rates of neurons may be a code for trasmitting information and computational considerations suggest that sensory cortical regions may perform some type of unsupervised learning. In addition, these anatomical networks are extremely sparse (there is a probability of contact of roughly 0.1 or less between input axons and target dendrites in, e.g., olfactory cortex), and they are composed of both excitatory and inhibitory cells, typically with many more of the former than the latter, arranged into structured assemblies. The cells in these assemblies obey biophysical laws which govern their activity, and there is a large set of transmitters that are known to be active via a large number of channels. Moreover, there are hypotheses that synaptic connections among cells are changed with experience; current results suggest that synaptic strength in cortex can be altered by extremely brief activity, and that there are fixcd rules for the induction of synaptic change. Finally, the "input patterns" of activity that are actually behaviorally relevant to a cortical region are not known. Studies have shown that electrical stimulation of a brain region can be substituted for behavioral stimuli under some conditions, and those artificial stimuli have particular characteristics that might be included in a model.

A number of convergent considerations pointed to the olfactory modality as a good object of study. Behaviorally, Slotnick and Katz (1974) showed the formation of learning sets in rats performing olfactory discrimination, i.e., the ability to learn hundreds of distinct olfactory stimuli very rapidly (five trials or fewer per odor) with long retention, in contrast to other behavioral paradigms in which few stimuli are learned over dozens to hundreds of trials, and longevity of retention is rarely examined (see Staubli et al., 1987). Moreover, electrical stimulation patterns known to optimally induce synaptic enhancement (long-term potentation, or LTP), when applied directly to olfactory cortex in behaving, freely-moving animals, effectively substitutes for an olfactory stimulus in this task and also changes synaptic strength of activated contacts (Staubli, Roman and Lynch, 1985; Roman, Staubli and Lynch, 1987). In addition, the anatomy of olfactory bulb and cortex is unusually well-studied, as compared with more complex neocortical regions (see Price, 1973; Haberly 1985). We have simulated the known physiology underlying LTP in response to this behaviorally-relevant input stimulation, embedded in the known anatomical architecture of olfactory (piriform) cortex, and found that the result is a novel computational mechanism that performs hierarchical clustering of sensory data.

Structure and Functions of Layer II Paleocortex

Our simulation work has focused on olfactory (piriform) cortex. This paleocortical structure can be characterized as a random combinatorial matrix, receiving inputs from the olfactory bulb, which in turn receives inputs from the primary olfactory receptors in the nose; thus the cortex is only two synaptic connections removed from environmental stimuli. The anatomical organization of this cortical region can be abstracted to be a multilayered feedforward network consisting of thousands of layers, each with dozens to hundreds of cells arranged in a "winner-take-all" fashion. The resulting network bears resemblance to "competitive learning" networks proposed by von der Malsburg (1973), Grossberg (1976) and Rumelhart and Zipser (1986), and we will outline some of the similarities and differences among these models.

Combinations of cells with various dendritic and axonal architectures comprise network subassemblies which recur and are themselves combined to form networks. Layer II olfactory cortex is, for instance, composed of subassemblies, or patches, formed by the relatively fixed radial axonal arborization of feedback inhibitory cells in layer II, which therefore densely contact a local radial field of excitatory layer II cells. These excitatory cells contain both apical and basal dendrites, each with distinct architectures: the apical dendrites distally (in layer Ia) receive sparse synaptic contacts from olfactory bulb mitral cells via the lateral olfactory tract (LOT) and more proximally (layer Ib) receive contacts from collateral axons of other layer II cells; before rising up to layer Ib, these collaterals make local excitatory contacts with the basal dendrites of neighboring layer II cells. Initial modeling and results of electrophysiological recording from behaving animals via chronically-implanted electrodes has suggested that the overall operation of each of these patches or subassemblies is that of a 'winner-take-all' circuit, i.e., one in which only the most strongly activated excitatory cell can overcome the local inhibition long enough to exhibit bursting activity. The coherent architecture of layers I and II piriform cortex consists of series of the patch assemblies, each receiving input from bulb via the LOT and input from the collaterals of other layer II cells, and both in a predominantly feedforward-only (rostral to caudal) direction (Schwob and Price, 1978). Figure 1 schematically depicts the architecture of the cortical model.

The cortical layer receives inputs at the fixed sampling frequency exhibited by rats during olfactory learning and memory behavior: the 5 Hz theta rhythm. Our primary findings have been that repeated sampling of stimuli has two major effects: first, multiple samples greatly increase the *information capacity* of a network compared to that for a single sample, i.e., the number of distinct messages such a network can in principle encode; and second, the breaking of the response into distinct samples imposes an organization on the memories thus read out. In particular, a combination of the physiological rules for three types of inhibition, embedded within the cortical anatomy as described, cause the network to give distinct cell-firing responses to a fixed cue on different cycles or samples of the cue, and that those distinct responses convey different hierarchical levels of information about the cue. The interaction of the winner-take-all circuitry, along with cell-specific long afterhyperpolarization (AHP) of strongly firing cells, causes those cells that respond most strongly to the first sample of a cue to become refractory, such that alternate cells are recruited by the net to fire on subsequent cycles.

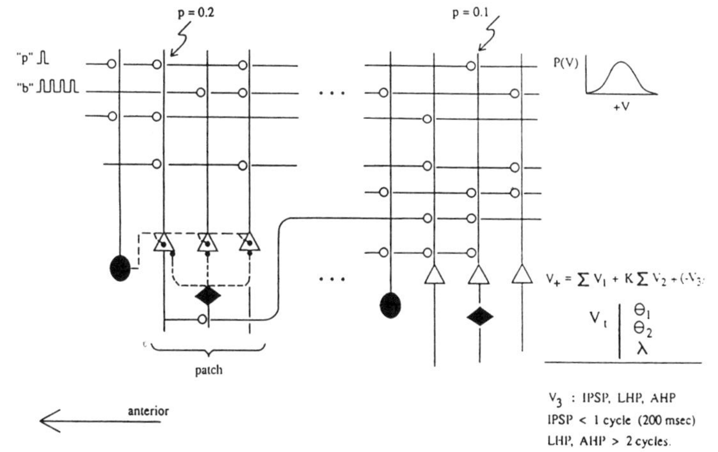

Figure 1: Schematic diagram of simulation of layers I and II piriform cortex. Excitatory layer II cells (white pyramids) contain both apical and basal dendrites, each with distinct architectures and connectivity: the apical dendrites distally (in layer Ia; top portion of diagram) receive sparse synaptic contacts from olfactory bulb mitral cells via the lateral olfactory tract (LOT), which arrive at the anterior portion of cortex (upper left) and become increasingly sparse posteriorly (right-hand-side); layer II cell apical dendrites proximally (layer Ib) receive contacts from collateral axons of other layer II cells; these collateral axons flow in a predominantly feedforward (rostral to caudal) direction (Schwob and Price, 1978). This results in an asymmetric matrix dominated at one end by the LOT input and by associational contacts at the other. Before rising up to layer Ib, the collaterals make local excitatory contacts with the basal dendrites of neighboring layer II cells. Inhibitory cells (black diamonds) organize layer II into patches with dense local connectivity; initial modeling and results of electrophysiological recording from behaving animals via chronically-implanted electrodes has suggested that the overall operation of each of these patches or subassemblies is that of a 'winner-take-all' circuit, i.e., one in which only the most strongly-activated excitatory cell can overcome the local inhibition long enough to exhibit bursting activity. (The essential design of the network has some relation to networks described and studied by Grossberg (1976), von der Malsburg (1973), and Rumelhart and Zipser (1986); see Granger, Ambros-Ingerson and Lynch, 1988, for analytic comparison.) Three types of inhibition are present in the model: IPSPs, long hyperpolarization or LHP, and cell-specific afterhyperpolarization (AHP). Synaptic transmission is probabilistic; there are distinct voltage thresholds for performance (spiking) and learning (NMDA receptor activation).

A Cortical Model

A model of the cortex consists essentially of a sparse random matrix of connections from input lines to simulated cortical cells, with a fixed number of contacts between input lines and target cells. The sparse connectivity creates a hypergeometric distribution of activations across output cells: the probability of **a** active synapses out of **n** synapses on a target cell in response to **A** active input lines out of **N** total input lines to the network is

$$p(\text{a active synapses}) = \frac{\binom{A}{a}\binom{N-A}{n-a}}{\binom{N}{n}} \qquad (1)$$

The formula can be read by noting that the first binomial indicates the number of ways of choosing **a** active synapses on the target dendrite from the **A** active incoming lines; for each of these, the next expression calculates the number of ways in which the remaining **n–a** (inactive) synapses on the dendrite are chosen from the **N–A** inactive incoming lines; the probability of active synapses on a dendrite depends on the sparseness of the matrix (i.e., the probability of connection between any given input line and dendrite); the solution must be normalized by the number of ways in which **n** synapses on a dendrite can be chosen from **N** incoming lines.

The target cells are then arranged into winner-take-all patches, such that, within each patch, only the most strongly-activated cell will respond. The simulation is presented with the same input repeatedly at simulated intervals of 200 msec (the theta rhythm that is both the sniffing rate for rats and the optimal rate for induction of synaptic change). Once a cell has responded, it becomes refractory and cannot respond again for a simulated 1000 msec (corresponding to the afterhyperpolarization (AHP) of cortical neurons). Thus a given cell can respond to only one of a series of five cycles in the model, corresponding to five repeated presentations at the theta rhythm. The learning algorithm corresponds roughly to that proposed by Hebb in 1949: cooccurrence of firing of an input and target result in strengthening of the contact between those two units. The algorithm is based specifically on the physiology of induction of synaptic long-term potentiation (LTP), which diverges from a simple cooccurrence algorithm in a few ways, for instance: synaptic strengths can only increase, not decrease; and each training episode increases them by a small fixed amount, up to a ceiling that is roughly 50% higher than the initial strength (which is constant across all synapses in the network). More important, synaptic change via LTP requires not only cooccurrence of input and target activity, but also has distinct thresholds for target cell firing versus learning: it is possible for a cell to fire without changing its synapses. This distinction leads to the assumption of distinct learning and performance modes for the network and has implications for the interpretation of its behavior (discussed later in this chapter). However, the primary difference is not in the learning algorithm, which can be described as a correlational rule that only increases weights up to a fixed ceiling, but rather in the fact that it requires repetitive stimulation, which, in combination with the winner-take-all local circuits and the cell-firing refractory period, mean that distinct subpopulations of cells and synapses are being strengthened in response to distinct presentations of a stimulus. Our analysis of the resulting sequence of cell-firing responses to repeated inputs has shown that, after learning, the initial responses are coarse-grained category responses and are nearly identical across a range of similar input cues; whereas later responses are increasingly distinct from each other, even for very similar cues, such that the sequence of responses forms a successive approximation of a unique encoding of the cue. The result is that the operation of this cortical network yields the construction of perceptual hierarchies (Lynch and Granger, 1989).

Figure 2 contains a graph of the performance of the cortical network on sets of ten simulated olfactory cues of varying similarity to each other. The x-axis indicates the overlap among the (10) cues presented to the network; the y-axis indicates the average overlap of the output responses to each of those cues, on each of three response cycles in the response sequence. It can be seen that for the first cycle responses, once the input cues are sufficiently similar to each other (about 50% for this example), then the output responses are almost identical (≈ 90%). However, by the third response to a cue, the responses are completely dissimilar to each other, even for inputs that overlap with each other by 90%. This effect is further illustrated by figure 3 in which the network was trained on 10 cues that overlapped with each other by 70% (i.e., the vectors are at approximately 45° to each other). For each of 1000 cells in the cortical simulation, the

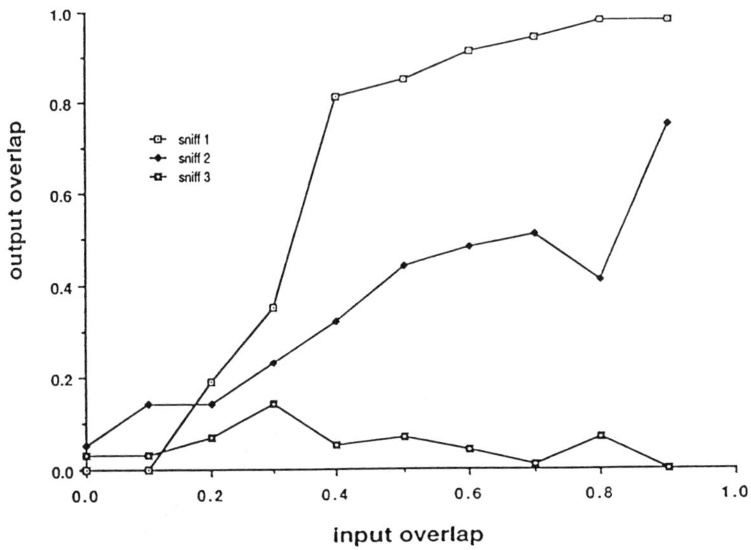

Figure 2: Simulation was trained on 10 groups of input vectors, each of which consisted of within-group overlap ranging from 0% (no overlap) to 90% (i.e., from 90° angles between vectors to less than 26° angles). Average output overlap after training is plotted against input overlap. First-cycle responses (top line) overlap by very little until inputs overlap by roughly 50% (60°), at which point the responses rapidly climb to nearly 100% (nearly 0°). Third-sniff responses (bottom line) overlap by no more than 12% (≈ 83°), even for inputs that overlap with each other by 90% (≈ 25°).

number of cues (out of 10 total) to which that cell responded is indicated by the size of the box at that cell; the largest boxes indicate that the cell fired to all 10 cues. It can be seen that for the first sample, all of the cell firing patterns for the 10 cues were virtually identical (i.e., most cells that fired to one cue fired to all); whereas for the fourth sample, the cell-firing patterns were quite distinct for each of the 10 cues (cells tended to fire to only one of the 10 cues). Finally, in figure 4, first and fourth cycle responses to a set of thirty cues (with 80% overlap) are shown; although the relative distances (nearest-neighbor relations) of 1000-dimensional vectors cannot be preserved in a two-dimensional projection of this kind, the graph depicts the fact that the average distance among the vectors is quite small for the

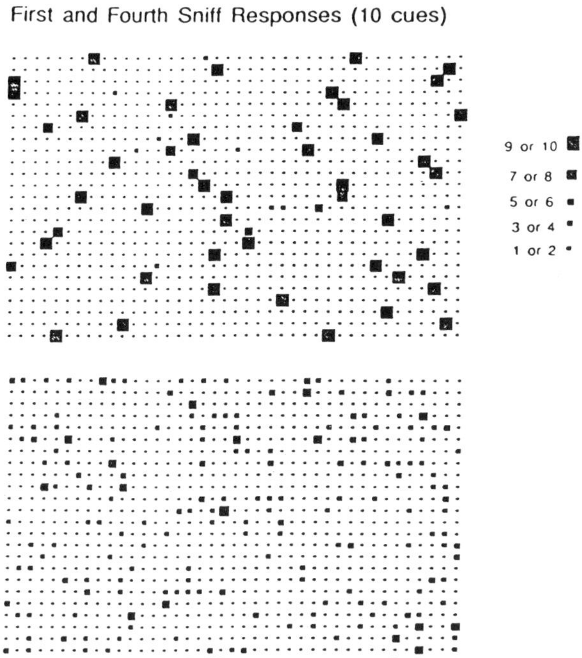

Figure 3: Depiction of first-sniff and fourth-sniff responses of piriform cortical simulation to ten similar learned cues. The ten cues overlap with each other by 70%, i.e., the vectors are roughly at angles of 45° (.79 radians) angles to each other. First and fourth-sniff responses to each of the ten cues in a 1000-cell piriform simulation are overlaid: the size of the box on a given cell indicates the number of cues (out of the ten total) that that cell responded to, on sniff 1 (top) and on sniff 4 (bottom). The first sniff responses overlay almost entirely onto each other; that is, each cell tended to respond either to all ten cues or to none of the cues. This means that these first-sniff responses to each of the ten cues were almost identical; i.e., the response simply denotes the fact that one of the ten members of this category was present, but not which one. In contrast, the fourth sniff responses were all quite distinct; few cells responded to more than one or two of the cues. This means that each cue elicited a unique fourth-sniff response from the network.

first-cycle responses, and quite large for the fourth-cycle responses.

The model can be related to work by von der Malsburg (1973), Grossberg (1976), and more recently by Rumelhart and Zipser (1986) in competitive learning. These approaches use similar architectures containing winner-take-all local circuits, with correlational or "Hebbian" learning algorithms related to the LTP algorithm used here,

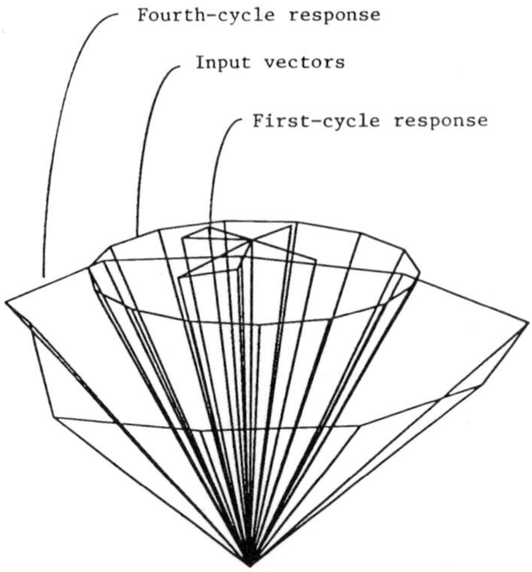

Figure 4: Depiction of first-cycle and fourth-cycle responses of piriform cortical simulation to fifteen similar learned cues. First-cycle and fourth-cycle responses, and the input vectors, are shown as two-dimensional projections of the 400-dimensional input vectors and 1000-dimensional output vectors from the simulation; nearest-neighbor information about the vectors cannot be retained under this dimensionality reduction, but the relative distances of the vectors from a central prototype are maintained and illustrated. The fifteen input cues overlap with each other by 90%, i.e., the vectors are roughly at angles of 26° (.45 radians) angles to each other. The input vectors are illustrated as the roughly circular set of vectors in the figure; first-cycle cortical response produces vectors that are significantly closer together, illustrated by the irregularly-shaped set of vectors clustered near the center, categorizing these cues together, whereas by the fourth cycle, the vector responses are spread out from the initial input vectors, differentiating the cues.

but without the repetitive sampling and refractory periods which result in a series of outputs in our piriform model. Rather, their work can be related to the *first* response of our network to a cue. The LTP learning algorithm can be stated as setting the values of a weight matrix W (P×N) via training on a sequence of binary input vectors $\{L^1, L^2, \ldots\}$.

$$W_{ij}^{t+1} = \begin{cases} \min\{W_{ij}^{t} + \varepsilon, \omega_2\} & \text{if } \overline{W}_{(i)} \cdot L^t \geq \overline{W}_{(k)} \cdot L^t \text{ for } 0<k \leq P \\ & \text{and } \overline{W}(i) \cdot L^t \geq \Lambda \text{ and } \overline{W}_{ij} L_i^j \neq 0 \\ W_{ij}^{t} & \text{otherwise} \end{cases} \quad (2)$$

where P is the number of cells in the network patch, N is the number of input lines, $\overline{W}$ is a binary matrix indicating the location of synapses in the initial (naive) weight matrix W^0, such that $\overline{W}_{ij}=1$ where a synapse exists, i.e., where $W^0 \neq 0$, and $\overline{W}_{ij}=0$ otherwise, $\overline{W}_{(i)}$ is the i-th column of $\overline{W}$, ie., $\overline{W}_{(i)}=(\overline{W}_{i1}, \overline{W}_{i2}, \ldots)$, W^0 is the initial (naive) weight matrix whose entries (synapse weights) are either 0 or ω_1, $\Lambda > 0$ is the minimal number of active synapses needed for a "winner" cell (i.e., the cell that wins the winner-take-all competition in its patch) to potentiate, $\varepsilon > 0$ is the incremental change in synapse weight, ω_2 is the ceiling or saturation weight value, and "·" is the dot or inner product.

Note that the algorithm does not allow synapses with strength 0 to change; a synapse with weight 0 remains zero thereafter. Synaptic strengths cannot be decreased; each weight-change event increases the synaptic strength by a fixed amount ε. As a result, any given synapse can only participate in $\frac{(\omega_2 - \omega_1)}{\varepsilon} = K$ learning episodes, by which time its strength will have been raised irrevocably to its ceiling. Synaptic strength changes for those synapses with initial weight ω_1 are introduced when (1) the input and target cell are both active; (2) the target cell receives sufficient activation (from this input line plus others) so that the cell exceeds a learning threshold distinct from the firing threshold; (3) the synapse is still below its maximal strength 'ceiling' ω_2; (4) the input vector L^i is novel (i.e., learning can only occur once per input cue); and (5) each novel input vector in a category $L^i \in C_\gamma$ is relatively distinct from the others: given $L^i = \gamma + f^i$, then $\forall_{i,j}\, f^i \cdot f^j \ll |f|$.

The learning algorithm given by Grossberg (1976), discretized for comparison, can be stated as follows:

$$W_{t+1} = \begin{cases} W^t + \varepsilon(L - W^t) & \text{if the unit wins} \\ W^t & \text{if the unit loses} \end{cases} \quad (3)$$

The algorithms of von der Malsburg (1973) and Rumelhart and Zipser (1986) are functionally identical to each other, both assuming that weight vectors are of normalized length ($\sum_i W_i = 1$) and the learning rule maintains the normalization over each trial:

$$W^{t+1} = \begin{cases} W^t + \varepsilon\left[\dfrac{L}{\sum_i L_i} - W^t\right] & \text{if unit wins} \\ W^t & \text{if unit loses} \end{cases} \qquad (4)$$

Two key differences between the LTP algorithm and these algorithms are apparent. In the LTP algorithm, the vector is constrained to be in a subspace of the weight space, by the sparse connectivity of the network; this is not true in the Grossberg (1976), von der Malsburg (1973) or Rumelhart and Zipser (1986) models, which assume dense connectivity. Moreover, in the LTP algorithm, the vector is sometimes unable to move towards the training vector (though it will never move away from it), because weights can only increase; the three competitive learning algorithms allow weights to increase or decrease in response to a single input (Granger, Ambros-Ingerson and Lynch, 1989).

Olfactory Bulb

In addition to the above studies of the performance of a cortical simulation, we have studied the interaction of the cortical model with a model of its immediate input structure, the olfactory bulb. Olfactory bulb is a peripheral structure composed of subassemblies entirely distinct from those comprising cortex, in terms of dendritic architectures of the individual excitatory and inhibitory cells involved, their relative placement, and their relative numbers. The main bulb excitatory cells, mitral/tufted cells, contain apical dendrites forming axo-dendritic contacts with inputs from the olfactory receptors, not via a random combinatorial matrix as in cortex, but via a set of topography-preserving dendrite bundles (glomeruli) each containing approximately 92 mitral/tufted cells (Mori, 1987); moreover, these cells also have obliquely-directed dendrites, which form dendro-dendritic synapses with the dendrites of inhibitory granule cells, of which there are roughly ten times more than the number of mitral/tufteds, in contrast with the cortical situation in which it is the excitatory layer II cells that outnumber the inhibitory cells by roughly an order of magnitude. These dendro-dendritic contacts are two-way synapses that transmit direct inhibition of mitrals by granules and direct excitation of granules by mitrals in a graded (continuous) fashion, in contrast to the discretized action (inhibitory and excitatory) afforded by more typical axodendritic synapses, in which the axon must be activated in an all-or-none manner before it can transmit quanta of information to the dendrite it contacts. Figure 5 illustrates the gross anatomy of bulb; not shown are the feedback connections from cortex to bulb, which terminate on the granule cells and on the (also inhibitory) periglomerular cells. Spatial and frequency-coded inputs corresponding respectively to the chemical constituents of an odor and their concentrations, are transmitted to the mitral/tufted cells at the glomeruli. If tufted cells and mitral cells have different thresholds (Mori, 1987) and if, by extension, the mitral neurons vary in this regard among themselves, then the numbers of cells contacting a particular glomerulus that respond to an olfactory nerve input will vary according to the properties of that input. Critically important would be the frequency at which inputs fire, a variable that should reflect stimulus intensity. To the extent that different thresholds are the rule in a glomerular region, the system should work to convert a frequency encoding of intensity into a spatial coding.

This means that for a relatively low concentration of the stimulus, a relatively fixed pool of (mostly tufted) cells will preferentially respond; for higher concentrations, those same cells will be activated, and, in addition, the less readily activated deep tufted and mitral cells will be recruited. We have constructed a simulation of olfactory bulb based on the physiological characteristics of mitral/tufted, granule, and periglomerular cells, as connected in bulbar anatomy; once again this bottom-up approach may be contrasted with other approaches in which, for instance, emergent properties of bulb may become the basis for modeling, as in the case of observed EEG waves and field potentials of assemblies of cells in the model of Li and Hopfield (1989).

Figure 6 illustrates a simulation of the activity of mitral/tufted cells within a single bulb glomerulus; simulated 600-msec traces for two (of the 92 simulated) mitral/tufted cells are shown, illustrating the rhythmic behavior of these cells in response to different frequencies of simulated receptor cell input, in this instance firing at 30 Hz and at 50 Hz. It can be seen that the mitral cells, rather than following the input frequency, fire rhythmically at roughly the theta rhythm, (due in large part to a 200-msec refractory period for inhibitory granule cells in the simulation; see Mori, 1987). The figure gives an example of a cell that fires repeatedly (at the theta rhythm) in response to a 30 Hz input and in response to 50 Hz (cell #11), whereas at least some cells (such as #12) will fire in response to a 50 Hz input but not reliably in response to a 30 Hz input. Figure 7 graphs this effect of increased number of mitral/tufted cells recruited in response to increased input frequency. The graph shows the number of mitral/tufted cells in the simulation that responded with repeated firing at the theta rhythm, in response to various frequencies of input stimulation, ranging from 1 Hz to 60 Hz, which is the appropriate range for receptor cells to fire (Voydanoy, 1989).

Thus, the spatial pattern of activation on the bulb (and hence in the bulb output, which is the input to the piriform cortex) should be relatively deterministic for particular concentrations of given odors, ensuring that the spatial pattern input to the cortex can be thought of as topographically encoding both quality and concentration of the odor, via choice of glomerular patches and numbers of mitral and tufted cells recruited within the patches, respectively.

Cortico-Bulbar System Interaction

A combined cortex-bulb model consists of versions of the two above-described models, with feedback from cortex arriving at bulb granule and periglomerular cells. The model is presented with cues at the simulated olfactory nerve, which in turn gives rise to a bulb response, which becomes normalized by lateral bulb activity and then triggers a sequence of cortical response, followed by feedback to bulb, followed by a new cortical response, and so forth. The following "script" describes the response sequence of the integrated bulb/cortex model to presentation of a mixture of two previously-learned odors, with the task of recognizing the two cues, even though the weak one (W) is

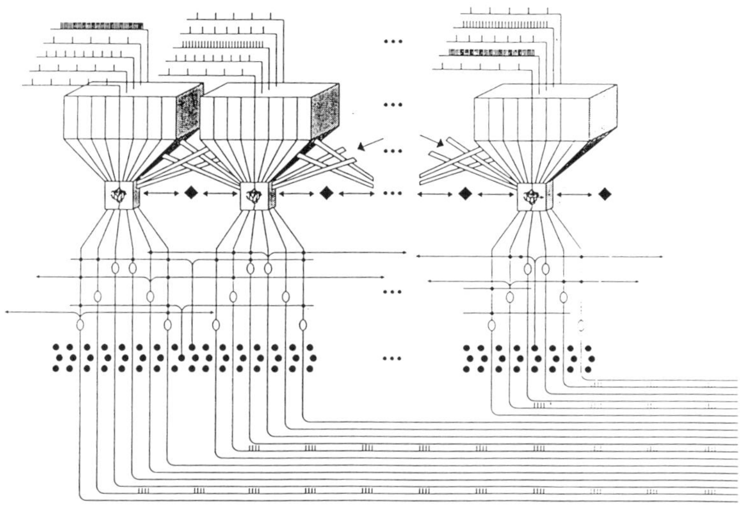

Figure 5: Schematic diagram of simulation of olfactory bulb. The main bulb excitatory cells, mitral/tufted cells (white ovals) contain apical dendrites forming axo-dendritic contacts with inputs from the olfactory receptors via the olfactory nerve (lines at top of diagram), not via a random combinatorial matrix as in cortex, but via a set of topography-preserving dendrite bundles (glomeruli) each containing the dendrites of approximately 92 mitral/tufted cells. The nerve inputs also contact (inhibitory) periglomerular cells (black diamonds). Mitral/tufted cells also have long obliquely-directed dendrites, which form dendro-dendritic synapses with the dendrites of inhibitory granule cells (black circles), of which there are roughly ten times more than the number of mitral/tufteds, in contrast with the cortical situation in which it is the excitatory layer II cells that outnumber the inhibitory cells by roughly an order of magnitude. The dendro-dendritic contacts are two-way synapses that transmit direct inhibition of mitrals by granules and direct excitation of granules by mitrals in a graded (continuous) fashion, in contrast to the discretized action (inhibitory and excitatory) afforded by more typical axodendritic synapses, in which the axon must be activated in an all-or-none manner before it can transmit quanta of information to the dendrite it contacts. Spatial and frequency-coded inputs presumed to correspond respectively to the chemical constituents of an odor and their concentrations, are transmitted to the mitral/tufted cells at the glomeruli. Mitral/tufted cell axons (bottom lines) comprise the lateral olfactory tract (LOT), the primary output from bulb and input to layer I olfactory cortex. In rat, 50 million receptor cell axons make contact with mitral and tufted cell dendrites in the 1,900 glomeruli.

masked by a stronger cue (S). Both odors (W and S) have been previously presented to the simulation, separately, and they have each been learned as individual odors.

1. The simulation is presented with a linear combination of W and S.
2. The mixture generates a mixed spatial pattern in bulb, so that the stronger odor is more represented in the spatial pattern of mitral cell responses than the weak odor is. In spite of the fact that the two components are familiar, the combined spatial pattern is a novel one (if these two odors have never been presented in this combined fashion before) due to the normalization process in the bulb.
3. The novel bulb pattern is nonetheless similar enough to the strong odor to yield a degraded version of the piriform category response for that odor (S).
4. The resulting piriform-bulb inhibitory feedback preferentially masks out the "recognized" odor (S).
5. Renormalization in bulb activates the remaining odor W, together with the few remaining remnants of odor S.
6. Piriform response is a degraded version of the learned category response to the individual odor W.
7. Feedback masks out most of the remaining response; the cycle continues until bulb quiescence.

Figure 8 and figure 9 show the responses of the simulation to a mixture of two simulated odors A and C, both of which previously have been learned individually by the simulation. In figure 6 the spatial pattern of receptor cell activation for A and C are superimposed on the left-hand side of the figure, with the concentration of A (black squares) twice that of C (white squares); the relative sizes of the squares corresponds to the firing frequency of each cell. This simulated receptor response is followed by the simulated piriform cortex response, on two successive cycles or sniffs. For each sniff, the cortical cells that respond are labeled black, white or gray, according to whether they are part of the learned cortical response to odor A presented alone (black), C presented alone (white), or neither (gray). It will be seen that the first sniff cortical response corresponds almost exclusively (90%) to the high-concentration odor (which is A in this example); only four cells out of 50 correspond to the cortical responses of neither odors A nor C, and only one cell corresponds to the response for C. The second sniff cortical response, however, corresponds almost exclusively (96%) to odor C: only two cells out of 50 correspond to neither odor A nor C, and no cells whatsoever correspond to the response for the high-concentration

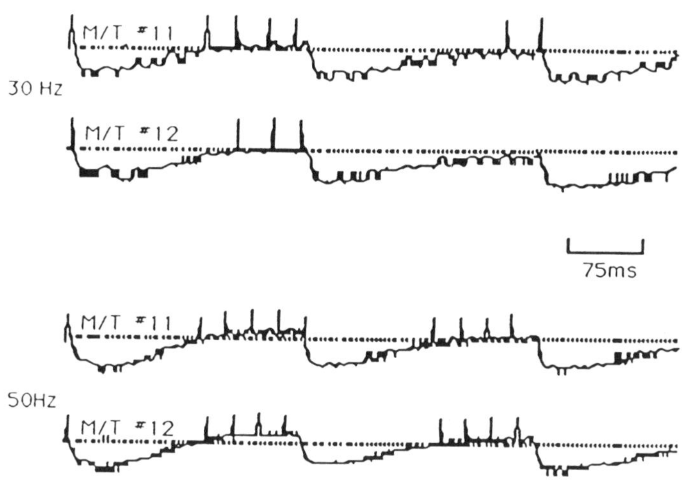

Figure 6: Sample traces of simulated cell potentials plotted over time for two mitral/tufted (M/T) cells over 600 msec. The two sets of traces show the sample M/T cell responses to olfactory nerve (ON) firing frequencies of 30 Hz and 50 Hz, respectively. Under the firing criteria, mitral cell #12s trace would not be considered firing at the theta frequency, since it did not continue firing after the second recovery from inhibition, whereas the other three traces did.

odor, A. Correspondingly, figure 9 displays the simulated receptor response to the same two odors but in the opposite concentration ratio: twice as much C as A. The first-sniff cortical response again corresponds strongly to the high-concentration odor, which this time is odor C; none of the firing cells are part of the representation of odor A and nine cells correspond to neither odor. The second-sniff response contains only three cells out of 50 that correspond to the high-concentration odor, and the majority of the cells correspond to the weak odor (although this response is noisy: 18 cells correspond to

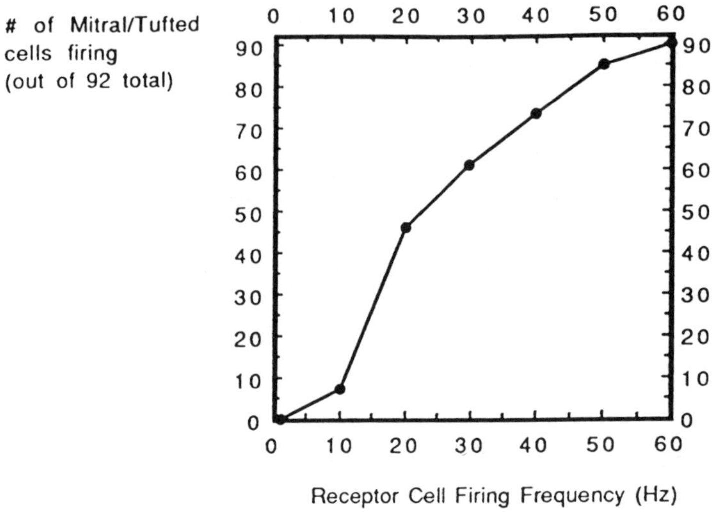

Figure 7: Plot of the number of simulated mitral/tufted (M/T) cells within a single simulated glomerulus firing at the theta rhythm as a function of the input firing frequency of the olfactory nerve (ON) contacting the glomerulus. For a range of ON activity from 1 Hz to 60 Hz, which is the usable range of the ON (Voydonoy, 1989), relatively linear increases in the number of M/T cells recruited are seen for increasing ON frequency. This means that for a relatively low concentration of a stimulus, a relatively fixed pool of (mostly tufted) cells will preferentially respond; for higher concentrations, those same cells will be activated, and, in addition, the less readily activated deep tufted and mitral cells will be recruited.

neither odor). This example illustrates the ability of the simulation to "dissect" or analyze the composition of environmental odors in terms of previously learned constituents.

Clustering by the Bulb-Cortex Model

The operation of hierarchical clustering exhibited by the cortical model alone can be examined in the integrated model, in addition to the integrated model's ability to "dissect" or recognize previously learned constituents of complex mixtures. That is, when trained on sets of similar cues that do not consist of previously-trained subparts, successive cycles of the bulb-cortex model provide successive subpartitioning of the cues into groups as does the isolated cortex model. Figures 10, 11, and 12 show superimposed sequential responses of both bulb and cortex to 10 trained cues on each of three cycles, as well as showing examples of the resposes to one of these cues for comparison purposes. Each of the three figures has four

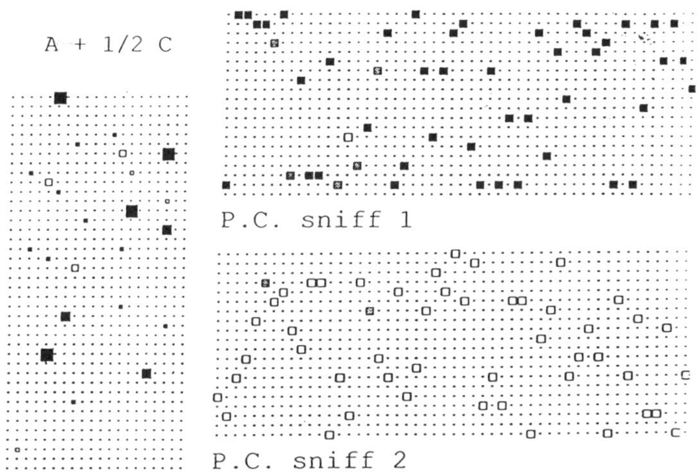

Figure 8: Illustration of operation of simulation when presented with mixture of familiar odors. The input pattern in the olfactory nerve (leftmost side) is coded to show the presence of a strong simulated odor A (black) mixed with a weaker odor C (white); the size of the boxes shows frequency of firing of these axons, indicating simulated concentration of the odor. At right, piriform cortex responses are coded to show the extent to which the firing cells correspond to those that encode odor A (black) vs odor C (white), or neither (grey). Sniff 1 response (top) of the simulation when presented with this mixture corresponds to the learned cortical response for odor A; i.e., this response is what would have occurred if odor A had been presented alone. Sniff 2 (below) corresponds to the learned response for the 'hidden' odor, C.

distinct regions that need description: two vertically oriented blocks on the left side and two horizontally oriented blocks on the right side. The leftmost vertical block is the representation of the current cue in the bulb model: the model consists of 400 simulated mitral/tufted cells, divided into glomerular patches of 10 cells each (each row of the block denotes such a patch), such that more strongly activated glomeruli cause more cells in a patch to fire, as discussed earlier. These cells are modeled as firing at a fixed rhythm, such that during the appropriate time window, any given mitral cell is either active or quiescent, as in the simulated results that were shown in Figure 6.

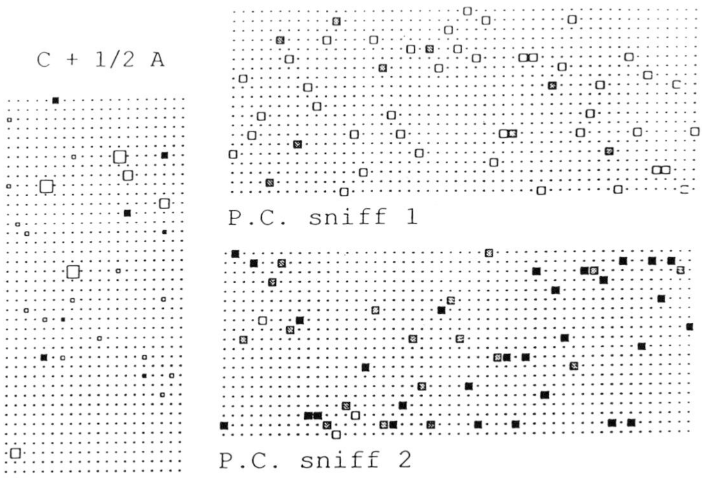

Figure 9: Illustration of operation of simulation when presented with mixture of familiar odors; compare with previous figure. The input pattern in the olfactory nerve (leftmost side) is now reversed relative to the previous figure: the previously-weak simulated odor (C; white) is now strongly present, with the previously-strong odor (A; black) now weakly present. As before, the size of the boxes shows frequency of firing of these axons, indicating simulated concentration of the odor. At right, piriform cortex responses are coded to show the extent to which the firing cells correspond to those that encode odor A (black) vs odor C (white), or neither (grey). Sniff 1 response (top) of the simulation to the mixture now is to odor C (white), since it is the strong odor; whereas sniff 2 corresponds to the now-weak odor, A (black).

Thus these model cells behave as binary units for any given time cycle.

Bulb cells feed nontopographically to the cortical model (top right), consisting of 1000 cells arranged into winner-take-all patches of 20 cells each, as described above. The resulting random cortical activation gives rise to feedback which has been trained via a correlational rule to selectively activate bulb inhibitory cells in signal (i.e., the overall number of mitral/tufted cells that will fire on a given

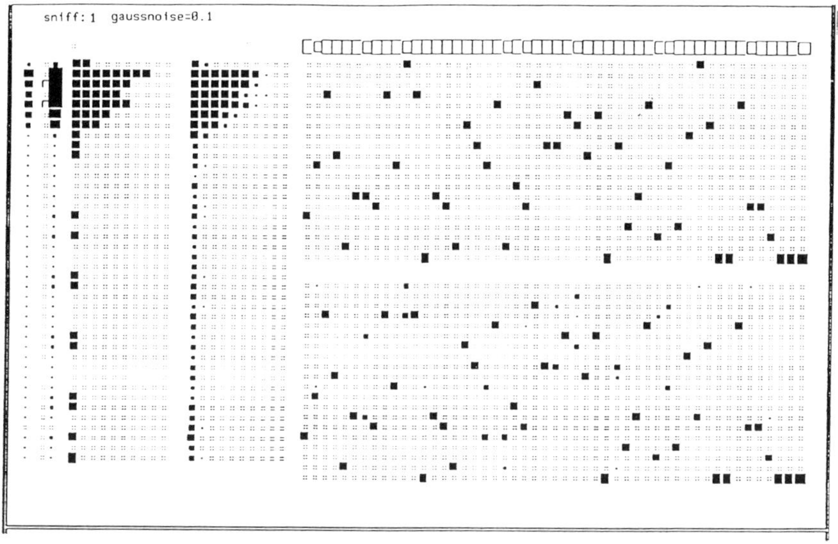

Figure 10: Illustration of operation of integrated bulb-cortex simulation on first cycle response after being trained on populations of similar odors. The input pattern for a single sample cue is shown in the leftmost single vertical column of units, denoting the olfactory nerve signal that is input to the bulb. The bulb response is in the lefthand 10~x~40 block of units which shows active simulated mitral/tufted cells (black boxes) out of 400 such cells arranged into glomerular patches (rows) of 10 cells each. The next block shows the superimposition of ten such first-cycle bulb responses over ten cues which overlap with each other by 90% of their bulb response; larger boxes indicate cells which responded to more of the ten cues. The upper-right box is the (nontopographic) cortical response to the current cue; its (topographic) feedback to inhibitory bulb cells is shown as unfilled boxes in the second to leftmost single vertical column. The lower right box shows the ten superimposed cortical responses, where again larger boxes indicate cells that responded to more of the ten cues. The second and third cycle responses following this one are in the following two figures; see text for further discussion.

cycle), and the resulting spatial pattern of bulb activity is then the input to the next performance cycle. The vertical block to the right of the bulb block shows the superimposed activity of bulb cells over 10 cues, with larger blocks denoting units that responded to more of the 10 cues. Analogously, the bottom right horizontal block corresponds to the superimposed activity of cortical responses over 10 cues.

The progression of the leftmost block (the bulb model) through the three figures 10 through 12 illustrates the inhibitory feedback from cortex to bulb: the initial bulb pattern of response to the cue is shown in figure 10, and the feedback from the resulting cortical response selectively inhibits some of the strongest portions of that bulb pattern, yielding, after renormalization, the leftmost pattern in figure 11; similarly the feedback from that results in the leftmost pattern in figure 12. The patterns to the immediate right of each of these shows

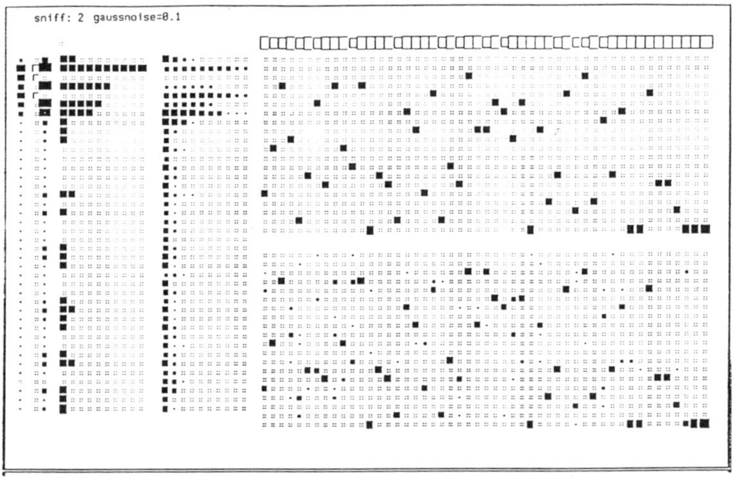

Figure 11: Illustration of operation of integrated bulb-cortex simulation on second cycle response after being trained on populations of similar odors. As in the previous figure, the input pattern is in the leftmost single vertical column, the second-cycle bulb response (after subtraction of cortical feedback inhibition) is to the right, the superimposition of ten second-cycle bulb responses is to the right of that, and the second-cycle cortical response (top) and ten superimposed second-cycle cortical responses (bottom) are to the right of the bulb responses. Compare the previous and following figure, and see text for further discussion.

the 10 superimposed patterns, illustrating their similarities in figure 10 and increasing disparity in figures 11 and 12. The cortical responses at the top right of each of the three figures shows a single cortical response, and the bottom-right responses show the 10 superimposed responses. It can be seen that the 10 responses superimpose almost perfectly on the first cycle (figure 10), but hardly at all by the third cycle (figure 12). The variance (σ^2) among the 10 vectors has been calculated for each of the three cycle responses, via the appropriate formula for vectors of equal length, which is: $\sigma^2 = (N-K\,\|\sum_i X_i\|)/(N-1)$,

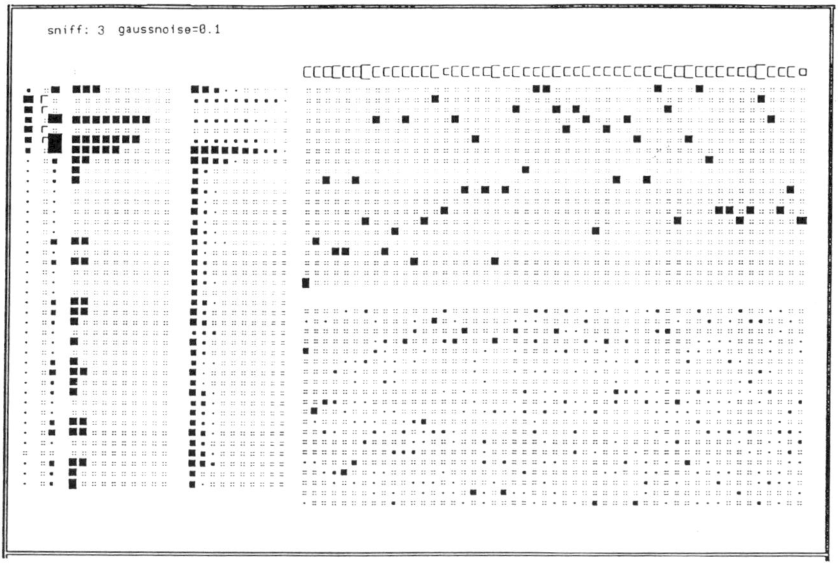

Figure 12: Illustration of operation of integrated bulb-cortex simulation on third cycle response after being trained on populations of similar odors. Arranged as in the prior two figures; see those figures and text discussion.

where N is the number of vectors, $K=1/\|X_i\|$ (for normalization), $\|.\|$ indicates vector norm, and all X_i are assumed to be the same length (Watson, 1983); the variances are 0.063 for the 10 cycle 1 responses, 0.141 for cycle 2, and 0.534 for cycle 3, whereas the variance among the bulb patterns for the 10 input vectors was 0.110. Thus it can be seen that the first cycle response increases the similarity of the vectors whereas the second and third cycles successively orthogonalize them (i.e., $0.063<0.110<0.141<0.534$; σ^2(1st cycle) < σ^2(inputs) < σ^2(2nd cycle) ≪ σ^2(3rd cycle)). For comparison purposes it is worth noting that these latter data are analogous to those produced by the isolated cortical model and shown in figures 3 and 4.

Summary

We have undertaken coordinated empirical and analytical investigations of the computational and behavioral properties of interacting sets of different subassemblies, networks and systems of networks. This is a problem that has not been focused on in most neural network research, which has instead tended to investigate properties of individual nets; in fact, wherever possible, researchers have, for mathematical simplicity, attempted to identify single, complex nets to subsume the functionality of multiple nets. Although it is in general possible to find a (complex) single-net substitute for interacting nets, we have successfully identified (1) individual computational functionality of distinct nets, and (2) novel computational functionality that emerges from network interactions. Moreover, our findings have tended to identify computational (behavioral) functions from interacting nets (such as the unmasking of weak stimuli, as described earlier), that likely would have been overlooked if the focus had been on optimizing a net for just a single purpose. Rather, we have attempted to construct nets derived directly from biological structures, and to identify the multiple functions that can be driven by these nets under different circumstances. One of the goals of the research is the identification of novel network behaviors and novel computational mechanisms to achieve them, rather than the attempt to design a network optimally to perform a predetermined function. As described above, we have identified:

- A novel form of unsupervised learning, in which not a single output is produced by a net, but rather sequential responses are used to encode hierarchical information about the cue environment.
- A network algorithm based on repetitive sampling to support the above behavior, derived from the physiological function and anatomical structure of a cortical layer, and its interaction with a peripheral sensory structure (bulb).
- A form of active sensory analysis of a cue that allows arbitrary mixtures of multiple pre-learned cues to be successfully decomposed and individually recognized.

- A network algorithm to support the above behavior, based on cyclic inhibitory feedback from a cortical-type network to a normalizing network based on the physiology and anatomy of olfactory cortex and olfactory bulb, respectively, and their interconnections.

We intend to continue to study the properties of this cortical-peripheral system and to combine it with a further level of peripheral processing, that of the olfactory receptors, in order to further investigate the potential of these interacting structures to actively sample and analyze the perceptual environment.

The model makes a number of suggestions about cortical functioning that can be tested, such as the notion that cells should respond via spatial patterns of time-locked firing to olfactory cues, and individual units should be sparsely coded for categories of similar odors initially, and that those units initially responding to an odor should become refractory and should be supplanted by other units, which in turn code sparsely for more fine-grained subcategories of odors. Chronic unit recordings are now being collected from layer II piriform cortex in rats engaged in olfactory discriminations. Preliminary results indicate that coding is indeed extremely sparse and is markedly affected by learning in ways compatible with hypotheses based on the simulation and analysis. (McCollum, J., Larson, J., Otto, T., and Lynch, G., 1988). Work is also in progress to test the hypothesis that rats partition their olfactory world and use multiple sniffs to read out information in the manner predicted by the simulation. Finally, the purely computational aspects of the simulation are being actively pursued in more abstract versions of the model to investigate the extent to which the hierarchical partitioning behavior can be applied to difficult clustering situations.

Acknowledgements

This research was supported in part by the Office of Naval Research under grants N00014-87-K-0838 and N00014-89-J-1255, the National Science Foundation under grant IST-85-12419, and by a grant from the J. Howard Pew Trust.

References

Granger, R., Ambros-Ingerson, J., Henry, H., and Lynch, G. (1987). Partitioning of sensory data by a cortical network. In: Neural Information Processing Systems I, American Institute of Physics Publications.

Granger, R., Ambros-Ingerson, J., and Lynch, G. (1989). Derivation of encoding characteristics of layer II cerebral cortex. Journal of Cognitive Neuroscience 1:61-87.

Grossberg, S. (1976). Adaptive pattern classification and universal recoding: Part I. Biological Cybernetics 23:121-134.

Haberly, (1985). Neuronal circuitry in olfactory cortex: Anatomy and functional implications. Chemical Senses 10:219-238.

Hebb, (1949). The organization of behavior. New York: Wiley.

McCollum, J., Larson, J., Otto, T., and Lynch, G. (1988). Single unit activity in piriform cortex of rats during olfactory discrimination learning. Society for Neuroscience Abstracts, 160.11.

Li, and Hopfield (1989). Modeling the olfactory bulb and its neural oscillatory processing. (To appear).

Lynch, G. and Granger, R. (1989). Simulation and analysis of a cortical network. The Psychology of Learning and Motivation 23:205-241.

Mori, K. (1987). Membrane and synaptic properties of identified neurons in the olfactory bulb. Progress in Neurobiology 29: 275-320.

Price, (1973). An autoradiographic study of complementary laminar patterns of termination of afferent fibers to the olfactory cortex. J.Comparative Neurology 150:87-108.

Roman, F., Staubli, U., and Lynch, G. (1987). Evidence for synaptic potentiation in a cortical network during learning. Brain Research 418:221-226.

Rumelhart, D., and Zipser, D. (1986). Competitive Learning. In Rumelhart, D., and McClelland, J., Parallel Distributed Processing, Cambridge: MIT Press.

Schwob, J.E., and Price, J.L. (1978). The cortical projection of the olfactory bulb: development in fetal and neonatal rats correlated with quantitative variation in adult rats. Brain Research 151:369-374.

Slotnick, B.M., and Katz, H.M. (1974). Olfactory learning-set formation in rats. Science 185:796-798.

Staubli, U., Fraser, D., Faraday, R., and Lynch, G. (1987). Olfaction and the "data" memory system in rats. Behavioral Neuroscience 101:757-765.

Staubli, U., Roman, F., and Lynch, G. (1985). Selective changes in synaptic responses elicited in a cortical network by behaviorally relevant stimulation. Sociey for Neuroscience Abstracts 11: 837.

von der Malsburg, C. (1973). Self-organizing of orientation sensitive cells in the striate cortex. Kybernetik 14:85-100.

Vodyanoy, V. (1989). Molecular sensory based on olfactory transduction. In Hong, F.T. (Ed.). Molecular Electronics: Biosensors and biocomputers. NY: Plenum Press,

Watson, G. (1983). Statistics on spheres. NY: John Wiley & Sons,

CHAPTER 6

Biological Constraints on a Dynamic Network: The Somatosensory Nervous System

T. Allard

Introduction

Recent approaches to computer modeling of neural networks have been inspired and guided to some extent by experimental evidence from real neural systems. In particular, biological studies have shown the mammalian brain to be a powerful processor of information that is distributed into different parallel neural pathways mediating different aspects of sensory, motor, or cognitive functions (e.g., Sidtis et al., 1981; Van Essen and Maunsell, 1983; Wise, 1985; Alexander, DeLong, and Strick, 1986; Olson and Jeffers, 1987). Within anatomically discrete parallel systems, physiological attributes emerge from dynamic interactions among populations of neurons and are not held static by the underlying anatomy. Factors shown to influence physiological function of central neurons include selective attention, associative learning, and patterned afferent activity (e.g., Moran and Desimone, 1985; Weinberger and Diamond, 1987; Haenny, Maunsell, and Schiller, 1988; Spitzer, Desimone, and Moran, 1988; Miller, et al., 1989; Miller and Stryker, Chapter 9).

These exciting results in systems neuroscience demand theoretical approaches to underlying neural mechanisms that consider the rules regulating neural interactions. Conversely, observation of the behavior of real neural networks under specific experimental manipulations provides biological constraints for neural modelers interested in self-organizing, fault-tolerant systems.

Recent work on the somatosensory cortex of adult primates has shown that cortical topographies emerge from dynamic interactions among neighboring inputs and are limited, but not specified in detail, by a particular pattern of anatomical projections. In this context, dynamic refers to a process whereby specific physiological attributes of a given neuron depend on the activity of a population of neurons and do not depend solely on the static anatomical connections of that neuron. Cortical topographies representing the body surface of adult primates respond in predictable ways to ongoing changes in the activity of peripheral inputs and patterns of horizontal interactions within and among different cortical fields (Merzenich et al., 1988; Pons, Garraghty, and Mishkin, 1988). The somatosensory nervous system is a dynamically tuned network or set of networks maintaining consistent topographical features on a substrate of diffuse anatomical connections.

In this chapter, I review lessons to be learned from the detailed study of the cortical topography or map of the body surface in adult somatosensory cortex. Many sources of information exist regarding the neural mechanisms regulating dynamic behavior in the somatosensory system:

- Studies of cortical somatotopy in normal animals provide many organizational details that are surprisingly consistent across a wide variety of mammalian species.
- Much has been gained by fine-grained analysis of the cutaneous representations of the body surface following surgical rearrangements of the sensory periphery.
- Finally, behavioral manipulations in neurologically intact animals indicate that physiological changes in map organization reflect normal adaptive processes occurring throughout the lifespan.

General Features of Neocortex

Mammalian neocortex has certain anatomical features in depth or vertical organization that make it amenable to modeling as a 2-dimensional sheet. Neuronal cell density is quite uniform throughout neocortex despite differences in cell size and cortical thickness.

Comparative studies have shown that, with few exceptions, neocortex has approximately 80,000 neurons per sq mm (Rockel, Hiorns, and Powell, 1980). It is a laminated structure with six cell layers whose input and output laminae are consistent in general plan across different cortical fields. The middle cortical layers, e.g., the granule cell layers of sensory cortices, receive thalamic inputs whose terminal arborizations can extend several millimeters across the cortex for a given thalamic cell. This horizontal spread of information across the cortex allows a great divergence of spatial information from restricted portions of the sensory periphery.

The major group of output cells, the pyramidal cells, are excitatory and can send axon collaterals many millimeters across the cortex. Pyramidal cells in the superficial or "supra-granular" layers (above the granule cell layer IV) primarily project to other cortical targets both within the hemisphere and across the corpus callosum to the opposite hemisphere. Pyramidal cells of the "infra-granular" layers project to subcortical targets including the basal ganglia, thalamus, and spinal cord. Pyramidal cells often have recurrent or local collaterals splitting off from the main trunk of the axon that project back to the layer of the original cell body.

In contrast to the long-distance range of the excitatory pyramidal cells, most inhibitory neurons have a more limited sphere of influence, exerting their effects locally (see Jones, 1984, and Rakic, 1976, for reviews). However, pyramidal cells can conceivably exert inhibitory influences over some distance by excitation of local inhibitory neurons. This type of inhibition may be the source of "peri-columnar inhibition" as described by Mountcastle (1978) and modeled by Edelman and his colleagues (1978; Pearson, Finkel, and Edelman, 1987).

There is a general plan of input and output organization throughout all neocortical zones (e.g., Rockland and Pandya, 1979; Shanks, Pearson, and Powell, 1985). Inputs to the middle cortical layers from thalamic nuclei are often termed "feedforward" indicating the direction of information flow from the sensory periphery into the central nervous system. Cortical fields that do not depend on the thalamus for their primary driving inputs may receive "feedforward" inputs from other cortical areas. Superficial and deep layers primarily receive "feedback" inputs from other cortical fields. For example, area 2 of primary somatosensory cortex receives a "feedforward" input in layers III and IV from both areas 3b and 1 and makes a "feedback" projection to layer II. This type of laminar organization is often cited as demonstrating directional hierarchies of information flow among related cortical fields (see reviews by Allard and Merzenich, 1988, Van Essen & Maunsell, 1983).

There is a common vertical organization across neocortical fields both anatomically and physiologically. Morphologically, cell types of signal shape and function common to specific layers are found across the cortex (Jones, 1984). Physiologically, receptive fields (RFs) of individual cortical neurons encountered during single microelectrode penetrations normal to the cortical surface share common "static" properties, such as location on the sensorium and adaptation characteristics (e.g., Powell and Mountcastle, 1959). A receptive field is taken to mean that portion of the sensory periphery whose stimulation affects the activity of a central neuron. (We will see later that some so-called "static" properties of neurons such as RF size and location are malleable and can be changed with sensory experience.)

The notion of common vertical organization has led to the suggestion of functional "mini-columns" in sensory cortex (e.g., Mountcastle, 1978, Hubel and Wiesel, 1977). Mountcastle has speculated that the phylogenetic trend toward encephalization or increased neocortical surface area has been achieved by the proliferation of mini-columns with similar intrinsic processing mechanisms. According to this view, cortical fields are distinguished from one another only by their patterns of connections; obvious histological differences between cytoarchitectonic areas are local responses to different connectional weights of inputs and outputs.

This similarity in vertical organization of cerebral neocortex across different cortical fields has given investigators hope that some general rules of organization will emerge from studying different cortical areas. The generality of neocortical mechanisms implies that certain theoretical problems apply across different neural systems. For example, neural modelers studying the development of ocular dominance columns in visual cortex may also demonstrate probable neural mechanisms underlying cortical topographies in the somatosensory and auditory systems. If general rules of intrinsic cortical organization emerge from these studies of sensory areas whose topographies are well characterized, it may be possible to address questions that are more difficult to formulate in "association" areas whose functional topographies are difficult to imagine.

General Features of the Tactile Lemniscal System

The somatosensory nervous system is organized into functionally distinct anatomical pathways conveying different aspects of mechanoreceptive, kinesthetic, thermal, and nociceptive stimulation of the body to multiple representations in different cortical fields. One major somatosensory pathway, the tactile "lemniscal system" subcomponent, conveys inputs from the skin to the primary somatosensory cortical fields. This subsystem is the principal cutaneous pathway conveying information about discriminative touch. In outline, information from peripheral receptors in the skin is conveyed along primary afferents of the dorsal root system to the ipsilateral nuclei of the dorsal columns in the brain stem. Second-order neurons from these dorsal column nuclei (DCN) cross through the medial lemniscus to the contralateral ventroposterior nuclei of the thalamus. In the primate brain, these thalamic neurons then project to cortical areas 3b and 1 of primary somatosensory cortex.

Neurons in the tactile lemniscal system share certain features: (1) Axons are large and heavily myelinated causing rapid and precise transmission of temporal and spatial information about peripheral stimuli. (2) Functionally similar cell types cluster anatomically forming parallel perceptual channels from the periphery to the cortex. (3) Receptive fields (RFs) are small and, above the level of the medial lemniscus, located in the contralateral sensory field on the opposite side of the body from their respective central neurons. (4) These RFs are topographically organized such that contiguous skin surfaces tend to be represented contiguously in the central nervous system. (5) There exists a tremendous amount of anatomical convergence of peripheral inputs onto central neurons all across the projection axis to the cortex; cortical neurons are normally driven by only a small portion of the total population of potential inputs. In this sense, the anatomical mapping of inputs onto a sheet of central neurons is degenerate rather than point-to-point. The question addressed in this chapter concerns the manner in which detailed somatotopic representation occurs within this diffuse anatomical substrate.

Features of Somatosensory Organization in Normal Animals

Despite the complexity of intrinsic connections and the large degree of anatomical convergence across different levels of the somatosensory system, the representation of the body surface in primary somatosensory cortex shares a number of topographic features across most mammalian species. Like the visual and auditory systems, there is a topographic mapping of sensory epithelium to a cortical target or targets (e.g., Woolsey, Marshall, and Bard, 1942). In the somatosensory system, the right side of the body is represented in the left hemisphere so that the lower extremity, trunk, upper extremity, and face are represented in a mediolateral progression across the primary somatosensory areas of the anterior parietal lobe. This general topographic plan probably reflects developmental mechanisms affecting the projection gradient of peripheral inputs to central targets and can be influenced by manipulations of afferent input early in life (McKinley, et al., 1987).

One particular primate species, the owl monkey (Aotus trivirgatus), has proven an excellent model for studying topographical organization. Owl monkeys have a refined representation of the hand that is similar, in many respects, to the cortical representation of the hand in humans. In addition, owl monkeys have no central sulcus so that the primary somatosensory cortices are easily accessible on a relatively flat surface of the cerebral hemisphere. Because of this configuration of cortical surface, the cutaneous representation of the body surface can be easily sampled and reconstruction of the body map is greatly facilitated. At each penetration site, excitatory RFs are defined as that restricted skin surface that causes an increase in neural activity with near-threshold punctate stimulation.

Recent high resolution studies of somatosensory cortical areas have exploited the features of the owl monkey somatosensory system and revealed important details of functional organization. In these experiments, many closely-spaced parallel microelectrode penetrations are made normal to the cortical surface. At each cortical site, excitatory RFs are defined as that skin surface that causes an increase in neural firing with near-threshold punctate stimulation (e.g., fine glass or monofilament-nylon probes). These RFs are used to reconstruct a highly-detailed map of the contralateral skin surface. Such studies within a given species clearly demonstrate large

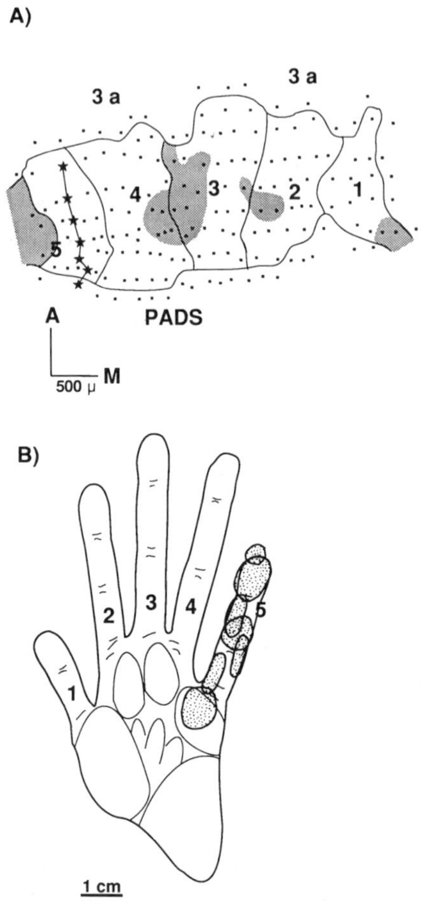

Figure 1: A) Cutaneous representation of the middle digits in primary somatosensory cortex (area 3b) of the adult owl monkey (Dots = microelectrode penetration sites; shading = zone of representation of the dorsal skin; A = anterior, M = medial; bar = 500 microns). B) Glabrous receptive fields showing overlapping fields within digit representations and non-overlapping fields at digit boundaries.

individual variations in local topographies that are probably related to how skin surfaces are used during development and after anatomical connections are fully mature (Merzenich, et al., 1987). Nevertheless, consistent topographic features emerge across individuals. The cutaneous representation of the hand in primary somatosensory cortex, area 3b, of the adult owl monkey exhibits many of these features:

- Within this species, the overall topographic plan is always the same: Individual digits are represented in a segregated fashion in a mediolateral sequence from digit 5 near the wrist medially to digit 1 next to the face laterally.
- Like inputs are segregated topographically: Receptors from the dorsal hairy surface of the hand drive neurons in a different zone from that responding to stimulation of the smooth or glabrous volar skin. RFs with inputs from both hairy and glabrous skin are rare. Similarly, some workers have claimed that there is a banding of functional attributes in area 3b based on adaptation characteristics, rapidly-adapting or slowly-adapting (Sur, Wall, and Kaas, 1984).
- As the cortex is sampled in an anterior-to-posterior direction, a progression of overlapping RFs is observed marching from digit tips across the distal, middle, and proximal phalanges to the representation of the palmar pads. In the medial-to-lateral direction, there is an ulnar-to-radial progression of overlapping RFs across each digit representation and across the palm.
- Inputs from the glabrous skin of individual digits drive cortical neurons in discrete, non-overlapping zones that have abrupt boundaries; individual RFs rarely extend onto two or more fingers. The normal mediolateral progression of overlapping RFs is interrupted at the digit boundary where a small microelectrode movement causes a RF jump from one finger to the next.
- The degree of RF overlap is linearly related to the horizontal distance between two cortical recording sites. This "percent overlap rule" was articulated by Sur, Merzenich, and Kaas (1980) (Fig. 2a). According to this stipulation, RFs do not overlap when separated by more than approximately 600 microns, a physiological "distance limit." This linear relationship holds across the representation of different body parts with wildly different cortical magnification factors (cortical magnification is the ratio of area of cortical representation of a given skin

surface to the area of that skin surface; the cortical representation of lip has a greater magnification factor than the back, for example).

- A corollary to the "percent overlap" rule is that RF size is an inverse function of cortical magnification factor (Sur, et al., 1980; Fig. 2b). That is, the greater the cortical surface area responsive to stimulation of a given skin surface, the smaller the RFs within that cortical zone. This relationship is necessary if the percent overlap rule is conserved. If the percent overlap at a given cortical distance is held constant and the amount of cortex responsive to stimulation of a restricted skin surface is large, then RFs must be correspondingly smaller.

Functional Reorganization after Experimental Manipulations

The tremendous amount of individual variation in local details and proportions of cortical maps alluded to in the beginning of the previous section could result from changes in anatomical connections laid down during development. However, a growing body of evidence across several species suggests that the consistent somatotopic features just listed for cortical representations are determined dynamically by the pattern of activity in peripheral afferents and by the manner in which skin surfaces are used throughout the lifespan of an individual.

Experiments in which driving inputs are eliminated through denervation provide the most convincing evidence that central representations in adult animals are capable of functional reorganization and are not determined solely by anatomical connections. If peripheral receptive fields are in some sense "plugged in" or "hard-wired" to a cortical zone, "unplugging" the zone through denervation should cause the area to become silent. If, on the other hand, the deprived central neurons become sensitive to stimulation of a new skin surface, they must be exploiting previously silent or weakly competitive inputs. Such a result in the anatomically mature

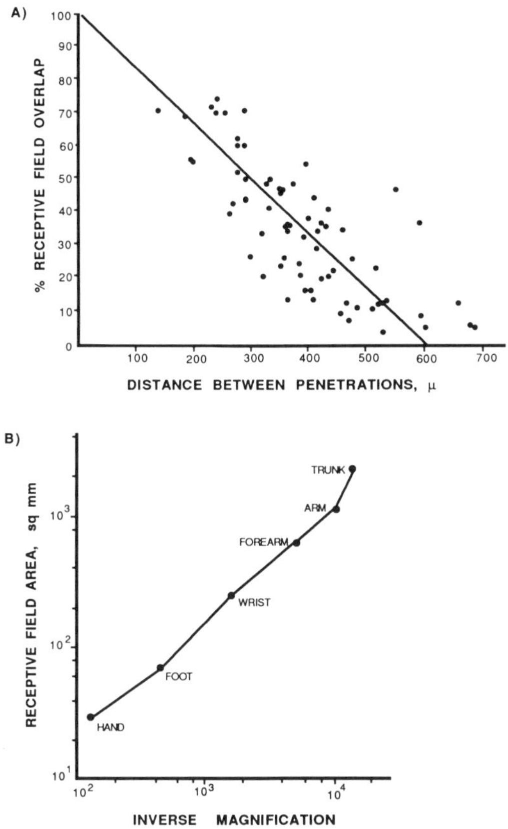

Figure 2: A) Overlap of pairs of cortical receptive fields plotted against the horizontal cortical distance between parallel microelectrode penetrations in primary somatosensory cortex of adult owl monkeys. Receptive fields were located on the hand, forearm, trunk, and foot. B) Mean receptive field area plotted against the inverse of cortical magnification. (A and B adapted from Sur, et al., 1980).

adult brain would demonstrate that a given cortical neuron can be driven by a variety of inputs and combinations of inputs through existing anatomical connections.

The sensory portion of the primate median nerve innervates the glabrous or smooth skin of the thumb to middle finger and the radial half of the palm. The cutaneous representation of the median nerve in area 3b of primary somatosensory cortex takes up almost half the total cortical map of the hand. Section and ligation of the median nerve of adult owl monkeys deprives this relatively large expanse of cortex of its primary driving input. The cortical zone does not become silent after this denervation, however. Instead an immediate unmasking of inputs from the hairy surface of the dorsal hand conveyed through the radial nerve was found in the cortical zone previously responsive to stimulation of the smooth glabrous skin. This unmasking phenomenon demonstrates an active competition between dorsal and volar inputs in the normal cortical representation of the monkey hand (Merzenich et al., 1983).

Immediately after median nerve section, cortical sites separated by as much as 2mm can have overlapping RFs and the relationship between cortical distance and percent overlap of corresponding RFs is disrupted. In contrast, most cortical locations in the normal animal that are separated by more than 600 microns have non-overlapping RFs. This immediate unmasking and increased cortical "distance limit" after median nerve section indicates that there is a large amount of anatomical convergence upon neurons in area 3b that escapes notice with standard neurophysiological mapping techniques. Once again, the specific details of spatial tuning of inputs is physiologically determined rather than specifically hard-wired.

Following these immediate effects of median nerve section, topographic details are progressively modified. Over the next several weeks, the normal relationship between cortical distance and RF overlap is reinstated.

The unmasking result clearly implies an active competition among different inputs and a degree of anatomical convergence from the periphery to its central representation not normally evident by standard mapping techniques. However, the progressive refinement of topography shows that organizational details observed after denervation are not the simple result of the remaining intact anatomical projections. The details of topography must reflect, instead, limiting physiological processes that continuously update the efficacy of particular inputs based on some combination of afferent activity and local neuronal interactions.

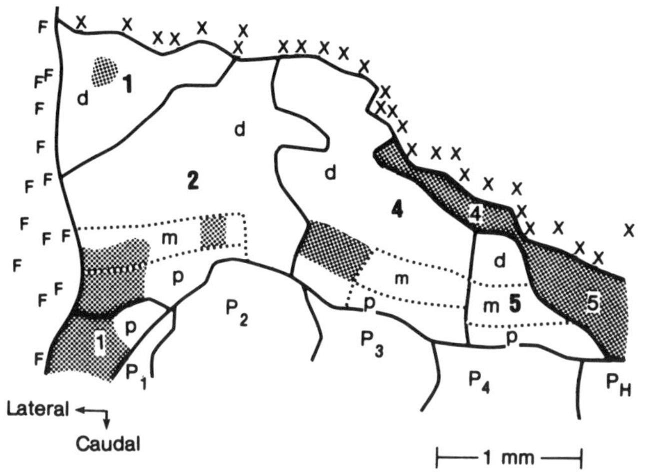

Figure 3: Cutaneous representation of the hand in primary somatosensory cortex of an adult owl monkey 62 days after digit 3 was removed (from Merzenich, et al., 1984).

Similarly, if the anatomy is "hard-wired," complete removal of all inputs from one or two fingers through digit removal should produce a zone of silent neurons. Instead, however, one sees a progressive expansion of the adjacent cortical zones representing the remaining skin surfaces (figure 3; Merzenich et al., 1984). This increase in cortical territory representing adjacent skin surfaces (i.e., an increase in cortical magnification factor) is accompanied by a reduction in RF size.

Features Under Dynamic Control

Which of the general features of topographic organization listed previously is under dynamic control of the somatosensory nervous system?

The overall topographic plan appears to be determined by the pattern of connections laid down during development. McKinley and associates (McKinley et al., 1987; McKinley and Smith, 1989) have shown that spinal transection in kittens affects the representational topography of the remaining innervated body surfaces only if the injury occurs before some critical developmental period. After this period, removing inputs from the lower limbs creates a large "silent" area within the deprived cortical zone. Thus there is a limit to cortical reorganization that is determined by anatomical divergence in the developing nervous system.

The remaining common features of cortical topography mentioned previously are all under dynamic control. When inputs from one finger are eliminated by digit removal, the cortical representation of the adjacent fingers expands into the unoccupied zone. Along with this consequent increase in cortical magnification factor, RFs become smaller in area, preserving the "percent overlap" rule of Sur et al. (1980) and its corollary, the inverse magnification rule.

Despite the radical change in the pattern of inputs following digit removal, there is still a smooth progression of overlapping RFs across the cortical representation of the remaining intact digits. Furthermore, individual digits are still represented in a segregated fashion with abrupt boundaries found even at the borders of digit representations that would never be found next to one another in a normal map.

The immediate unmasking of inputs from the dorsal hairy surface in a cortical zone formerly responding primarily to stimulation of glabrous skin suggests that some inputs (e.g., from the hairy surface of the hand) to a given cortical site might have functional synapses that are actively suppressed by a dominant competing set of inputs, perhaps by some neural mechanism like lateral inhibition. The long interval of topographical refinement that follows this immediate unmasking is a period in which normal rules of topographic organization are reinstated.

Cortical maps of the hand obtained following denervation show that neurons at specific cortical sites can support a large number of possible physiological configurations depending on the pattern of peripheral input activity. Only a small subset of inputs is active at any given time. The cortical RF is determined by the activity of populations of neurons rather than hard-wired by the connectional anatomy. The size and location of cortical RFs, i.e., the spatial tuning characteristics, are under dynamic control.

Consistent with this view, the size and location of cortical RFs can also be altered through behavioral manipulations in neurologically normal individuals. Adult monkeys trained to use a restricted portion of the hand to obtain food show a two- to three-fold increase in the cortical representation of that skin surface in primary somatosensory cortex (Jenkins et al., 1989). These recent data show that the body surface representation in the somatosensory nervous system is continually updated as a normal adaptation to environmental demands on an individual: increased use of a skin surface in a behaviorally adaptive fashion actually increases the amount of cortex directly driven by stimulation of that skin surface.

Cortical RFs form a constantly updated sequence of cortical topographies where individual RFs are created, maintained, and modified in a dynamic fashion. The rules by which a specific subset of driving inputs dominate other inputs are discussed in the following section.

A Temporal Rule Underlying Receptive Field Formation

Merzenich (1987) has hypothesized that a process of input selection underlies the constant updating of cortical RFs and that this selection process is based on the temporal synchronization of inputs converging on a given site in the central nervous system. Combinations of inputs from the large array of anatomically possible inputs can be grouped together to form RFs only if their activity patterns are relatively correlated. According to this view, individual digits are represented at cortex in a segregated fashion not because of anatomical restrictions but because the temporal pattern of peripheral input activity from the skin surfaces of adjacent fingers is relatively independent. That is to say, inputs from a single finger that are mechanically coupled through the elastic medium of the skin will tend to fire more synchronously than inputs from adjacent fingers whose receptors are weakly coupled. Because of the elastic properties of the skin, it is likely that inputs innervating neighboring locations in contiguous skin surfaces will be activated synchronously and that the correlation of activity between inputs will decrease as a function of the distance between them. Similarly, inputs from mechanically isolated skin surfaces will be even less synchronous than their mechanically coupled competitors.

If inputs are combined into candidate RFs on the basis of temporal correlations, the highest temporal correlations would be between inputs from the same skin surface. Therefore the most competitive input combinations would be from contiguous skin surfaces rather than discontinuous patches. The relatively poor correlation of activity across fingers would not produce competitive input combinations from adjacent fingers. Accordingly, RFs that extend onto two digits would be very rare.

As already discussed, the normal rules of topography include the following: (1) a general representational orientation and ordinal topography that is probably due to some developmentally-induced features of the substrate; (2) RFs that move across the skin surface in an overlapping fashion as one moves across the cortex with a recording microelectrode; and (3) this progressive shifting of overlapping RFs obeys a "percent overlap" rule limiting RF size and position according to relative magnification factor. In the normal hand representation in area 3b, the percent overlap rule is not obeyed at the discontinuities between digits. Here a small movement across cortex produces a RF jump from one finger to another. This abrupt representational discontinuity between digits is extremely robust and occurs even when digits that are normally non-contiguous in the cortical map are found next to each other after amputation of the intervening digit. Without a pool of double- digit RFs from which to draw, normal rules of cortical topography must produce digital representations with abrupt discontinuities between adjacent digits.

This hypothesis of RF formation was directly tested by increasing the temporal correlation of inputs from two adjacent fingers in the hands of adult owl monkeys (Clark et al., 1988; Allard et al., 1989). The skin of digits three and four was surgically connected from the distal tips to the proximal web space recreating a congenital condition called syndactyly or webbed finger. The mechanical coupling of the skin of the two digits, accomplished with little change in the anatomy of the projection system, increased the temporal synchronization of stimulation across the newly contiguous skin surfaces much like that observed normally in a single digit.

This alteration of input correlations created a novel class of cortical RFs overlapping both fingers of the syndactyly (figure 4). These unusual RFs were assimilated into a normal cortical topography by replacing

A)

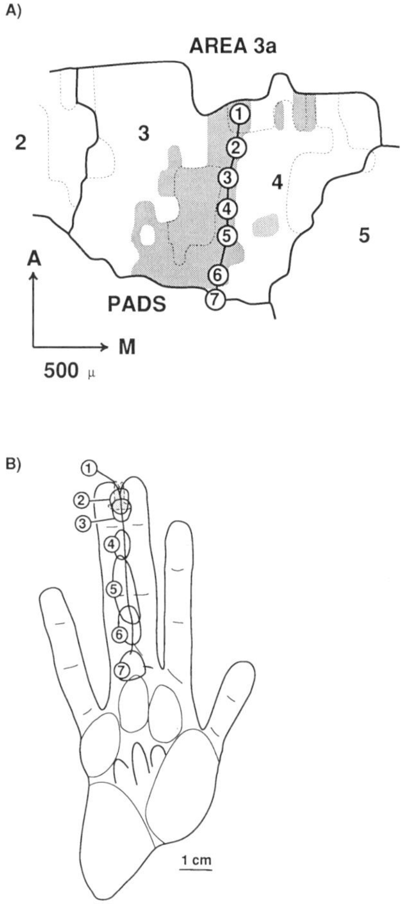

Figure 4: A) Cutaneous representation of the middle digits in primary somatosensory cortex several months after the skin of digits 3 and 4 was connected surgically. B) Representative receptive fields progressing distally to caudally across the line of syndactyly. (A and B adapted from Clark, et al., 1988).

the usual representational discontinuity between fingers with a topographically-appropriate zone of double-digit RFs. Overall, the cortical representation of the two syndactyly digits resembled that of a single normal digit with some interesting exceptions (Allard et al., 1989).

A similar result has been reported after peripheral nerve section and regeneration in monkeys (Wall et al., 1986; Allard et al., 1988). In this type of preparation, single peripheral fibers are disconnected from their original skin locations and randomly assigned to new skin locations. After such peripheral reassignment, the central nervous system is able to reorganize these resorted inputs into novel combinations as evidenced by either single- or multi-component cortical RFs following regeneration. After digital nerve section and regeneration, small RFs align themselves in a normal topographic order, a further demonstration that the pattern of spatial and temporal activity is key in determining which input combinations are likely candidates for selection as central RFs.

Temporal synchronization of afferent input is an important factor in the segregation of ocular dominance stripes in the developing geniculo-striate pathway of the kitten visual system (Chapman et al., 1986; Shatz and Stryker, 1988; Miller and Stryker, Chapter 9). Spontaneous activity of retinal ganglion cells is highly correlated within the eye when compared to the poor correlation across the retinae in a given animal. In the developing visual system, the initial projection of retinal ganglion cells from each eye overlaps in the lateral geniculate nucleus (LGN) of the thalamus and the terminal arborizations of thalamo-cortical axons overlap extensively in layer IV of primary visual cortex. At this early stage, there is no evidence for ocular dominance segregation either in the thalamus or cortex. Later in development, inputs to both the thalamus and cortex differentiate anatomically into segregated monocular regions.

The modeling results of Miller and Stryker (Miller, et al., 1989; Miller and Stryker, Chapter 9) indicate that temporal patterns of afferent activity can drive the formation of ocular dominance bands through relatively simple horizontal computations among neighboring neurons and their inputs. Experimental results indicate that a Hebb-like rule is at work whereby the correlation of activity at presynaptic and postsynaptic sites drives the process of self-organization rather than the absolute level of presynaptic activity. Fortunately, their model can be applied either to physiological or anatomical processes of ocular dominance segregation as the link between correlated afferent activity to the visual cortex; the subsequent pruning of terminal arbors is still unclear. The generality of these results holds out the hope that common network mechanisms underlie topographic organization in both somatosensory and ocular dominance systems.

The role of temporal synchronization in RF formation and cortical topography leads to some interesting speculations. In the somatosensory system, many excitatory RFs are ovoid in shape where all inputs are found within a well-defined boundary. The detailed anatomy of the projection system does not in itself determine which of many possible inputs are grouped together to form a RF. Thus the characteristic shape of the excitatory RF is related directly to the spatiotemporal pattern of inputs which is in turn related to the mechanical properties of the receptors and of the skin in which they are imbedded. One could speculate that multi-component RFs with noncontiguous subcomponents would be possible if the pattern of coactivation was artificially manipulated.

Furthermore, if temporal synchronization of inputs is an important factor in topographical organization, we may begin to understand how like inputs are segregated in cortical maps. For example, most RFs in the area 3b hand representation of the adult owl monkey are comprised of inputs from either hair follicles from the dorsal hand or inputs from the smooth skin of the palmar surface. RFs including inputs from both hairy and smooth skin surfaces are rare. Inputs from hairy skin and smooth skin surfaces are also segregated into non-overlapping zones at cortex (even though the median nerve section experiments show an active competition from both sets of inputs for common cortical territory). In these cases it is possible that the nervous system sorts inputs on their relative coactivation, both in onset and firing profile. The activity profile of afferents innervating glabrous and hairy skin are very different and could contribute to their segregation. In a similar vein, quickly adapting and slowly adapting inputs are reported to appear in segregated zones in both monkey and cat somatosensory cortex (Sur et al., 1984; Sretavan and Dykes, 1983).

Site of Reorganization

One important question regarding the mechanisms underlying functional plasticity in the adult brain concerns the level at which the reorganization takes place. Merzenich and associates (1987) have argued persuasively that cortical systems have the potential for dynamic topographical reorganization, although such changes could occur at other levels of the ascending pathway as well.

Consistent with the view that the cortex is a site for functional reorganization, anatomical studies have confirmed a large degree of spatial divergence in the projection of axons from the ventrobasal complex in the thalamus to primary somatosensory cortex. Landry and associates (1981, 1982, 1987) have shown individual thalamocortical axons to have broad terminal arborizations of 800 to 1500 microns in cat somatosensory cortex. Furthermore, Snow and associates (1988) have demonstrated that such thalamic neurons project to somatotopically appropriate and inappropriate zones mapped physiologically in cat somatosensory cortex. In the latter study, thalamic neurons were identified physiologically and the extent of terminal arborization in the cortex was estimated with physiological techniques with respect to specific details of cortical topography. The results indicated that thalamic neurons with RFs restricted to a single digit projected to a large cortical zone responsive to stimulation of that digit and to stimulation of an adjacent digit. Thus thalamic inputs to topographically "inappropriate" areas are not detected with standard cortical mapping techniques.

Pharmacological studies indicate that local cortical circuitry is critical for the dynamic tuning of the many anatomically feasible inputs to cortical neurons. Local iontophoretic injection of bicuculline, an antagonist of the inhibitory neurotransmitter GABA, into cat somatosensory cortex causes a large increase in RF size of some rapidly-adapting cortical units (Dykes et al., 1984), demonstrating an important role for inhibitory mechanisms in the formation of cortical RFs. Similar effects of bicuculline have been seen on orientation tuning in the visual cortex (e.g., Sillito, Salt, and Kemp, 1985). Modulatory effects of local injection of acetylcholine, norepinephrine, NMDA, and glutamate have been observed in conjunction with RF stimulation in cat somatosensory and visual cortex (e.g., Metherate, Tremblay, and Dykes, 1987; Greuel, Luhmann, and Singer, 1988). All of these pharmacological manipulations indicate that mechanisms exist at the cortical level which could account for dynamic tuning.

Cortical lesion studies also show that cortical processes are involved in some aspects of central reorganization. Restricted cortical lesions of area 3b in owl monkeys cause a redistribution of inputs within the remaining primary sensory cortex (Jenkins and Merzenich, 1987) without changing the anatomy of those inputs. In these experiments, cortical zones adjacent to the lesion progressively take on the function of the ablated cortical territory. Similar changes have been reported after restricted lesions in primary motor cortex as well (Glees and Cole, 1950). Large lesions of the hand and arm representation in area 3b abolish the driving inputs to the second somatosensory area (S II) and produce a subsequent large-scale increase in the representation of the lower extremity (Pons et al., 1988). This expansion of the glabrous foot representation in S II is many times larger than reorganizations reported to date in area 3b, suggesting a greater degree of anatomical divergence of inputs at this level of the projection axis.

In fact, there is evidence that reorganization of inputs is possible at all levels of the somatosensory nervous system. For example, changes in single-unit RFs in the dorsal column nuclei have been reported after cooling of the dorsal root system of peripheral afferents (Dostrovsky, Millar, and Wall, 1976). The topography of dorsal horn neurons in cat spinal cord changes after chronic peripheral nerve denervation (Wilson and Snow, 1987). Although the latter study focused on spino-cervical tract neurons that are outside the mainline lemniscal pathway, it would not be surprising to find similar changes in the topography of second-order neurons in the dorsal column nuclei.

An assumption implicit to recent work in this field is that the area of potential reorganization is limited by the divergence or spread of direct anatomical inputs from the next peripheral level, for example, the spread of thalamocortical terminal arborizations in area 3b of primates. This view suggests a class of models in which potential input combinations are ultimately constrained by the anatomical limits of the ascending projection system. This assumption may be misleading. Evidence from an unusually specialized tactile animal, the raccoon, indicate a different possible route for information flow to primary somatosensory cortex.

The raccoon could be called the "bat" of the somatosensory system. Like the bat with its enormous cortical auditory system, the raccoon has a cortical representation of its glabrous digits that is an order of magnitude larger than that found in primates. Each individual digit is represented by a separate sub-gyrus in primary somatosensory cortex (Welker and Scidenstein, 1959). Presumably this large increase in cortical surface area representing the digits provides the raccoon with its fine manual dexterity when foraging for food in stream beds at night. Removal of the fifth digit in this specialized animal causes an emergence of inputs from the fourth digit in "digit-5" cortex (Rasmusson, 1982). However, double-labeling studies have shown that the ascending thalamocortical pathways conveying information from the digits are largely segregated and cannot account for the reorganization (Rasmusson and Nance, 1986). Furthermore, no obvious changes in the anatomical projection to the cuneate nucleus of the dorsal column nuclei can account for the reorganization (Rasmusson, 1988). Recent anatomical studies have implicated an indirect pathway through a cortical area rostral to the primary cutaneous digit representation as the source of afferent input driving the reorganization in these cases (Doetsch, et al., 1988a,b).

While the raccoon is an extremely specialized animal, it is possible that such indirect routes may be active but overlooked in somatosensory cortex of other mammals. Indeed, relatively long distance intrinsic axon collaterals of greater than 8 mm have been reported in striate cortex of cat (Gilbert, 1983; Gilbert and Wiesel, 1987). Furthermore, in the primary somatosensory cortex of the primate, areas 3, 1, and 2 are all reciprocally interconnected. These long-distance horizontal connections could play an important role in the reorganizational capacities of sensory neocortex.

Formalizations and Limitations

Anatomical Restrictions An assumption underlying most experiments on the dynamics of cortical topographies in the adult somatosensory system is that no long-distance anatomical sprouting occurs following experimental manipulations and that reorganizations exploit existing connections when they do occur. This assumption does not preclude the possibility of local changes in neuronal or synaptic morphology (e.g., Greenough, 1984).

Such anatomical restrictions in the spread of peripheral afferents to central targets inevitably lead to some limiting horizontal distance over which reorganization can occur. We have already discussed the absence of large-scale reorganization in the deprived cortex following spinal transection in adult cats (McKinley and Smith, 1989). On a smaller scale, however, digit amputation results from the Merzenich group (Merzenich et al., 1984) have indicated a distance limit in area 3b of adult owl monkeys on the order of 500-700 microns. Evidence in support of this position is taken from a double-mapping study that shows a small residual silent zone within the cortical representation of the hand 8 1/2 months following the removal of two adjacent digits. This approximate limit of 500-700 microns is consistent with the physiological distance limit described by Sur, and associates (Sur et al., 1980).

However, the immediate effects of median nerve section produce RF overlap at distances up to 2 mm, violating the so-called distance limit, until the normal rules of topography gradually reinstate themselves over a period of several weeks. In addition, the cortical map obtained 8 1/2 months after removal of two digits may represent a slow but ongoing process of further representational refinement rather than the stable endpoint of reorganization. Finally, it is possible that subcortical mechanisms or non-lemniscal pathways through other cortical fields could contribute to reorganization of inputs as is apparently the case in the raccoon. For these reasons, the true horizontal limit to reorganization is still open to debate and may not be specifically limited by the range of thalamocortical arborizations.

Time Course The immediate unmasking of inputs from the dorsal hairy surface in a cortical zone formerly responding primarily to stimulation of glabrous skin suggests that some inputs (e.g., from the hairy surface of the hand) to a given cortical site might have functional synapses that are actively suppressed by a dominant competing set of inputs, perhaps by some neural mechanism like lateral inhibition. The long interval of topographical refinement that follows this immediate unmasking is a period in which normal rules of topographic organization are reinstated. The precise neural mechanisms underlying this reinstatement is at the heart of much current research and is poorly understood.

Receptive Field Formation The creation of a novel class of double-digit RFs after the mechanical coupling of adjacent fingers in the syndactyly study is strong evidence that the relative synchronization of peripheral inputs is a factor of primary importance in the grouping of possible input combinations into potential RFs. Receptors in the skin are mechanically coupled so that those close to each other will tend to be stimulated synchronously and receptors farther away from each other will be activated less synchronously. These input combinations form a pool of possible RFs that are candidates for selection as central RFs on the basis of more global topographic mechanisms.

Receptive field size varies as an inverse function of magnification factor and appears to be a dynamically-determined attribute. When magnification factor increases with the expanded representation of digits adjacent to a peripheral denervation, RF size is decreased. Even normal RFs in the tactile lemniscal pathway to area 3b of primary somatosensory cortex have very abrupt limits to the excitatory field.

How is RF size determined? If combinations of inputs are grouped by their relative correlation and input correlation decreases as a function of distance along the skin surface, then the spatial extent or number of inputs comprising a central RF could be determined by setting a correlation criterion or limit beyond which inputs would not be included in the candidate RF. A high criterion would only allow inputs that were close together on the skin surface and therefore highly correlated by mechanical coupling. Such RFs would be relatively small. Lower correlation criteria would allow more and more inputs to be included as potential input combinations and progressively larger RFs would therefore be possible.

According to this view, there is no *a priori* restriction on which inputs can be combined to form central RFs. Particular RF shapes are determined by the statistics of input coactivation. Given an idealized elastic receptor sheet where input activity varies equally as a function of distance, all possible input combinations would be radially symmetric and contiguous. Receptive fields with multiple components could only be induced by a non-uniform pattern of stimulation. This may have been the case in the syndactyly experiments, where a number of RFs were observed with separate components on the tips of both fingers of the syndactyly. The syndactyly fingers were used as if they were a single finger, and it is likely that the fingertips were coactivated during their normal daily use, that is, reaching a target at the same time.

This simplistic view of the relationship of the temporal pattern of inputs to receptive field formation is a useful way of conceptualizing some experimental observations. We can understand why double-digit RFs are rare in the normal cortical map of the monkey hand. The concept of a criterion of correlation also implies a distinct threshold that would be manifested in the nervous system as an abrupt RF border consistent with experimental results. The dynamic relationship between cortical magnification factor and RF size articulated by Sur and associates (1980) indicates that such a correlation criterion would be linked to the total cortical territory representing a given skin surface.

Receptive Field Topography Denervation experiments have shown that a single site in the central representation of a skin surface has anatomical access to a wide array of potential driving inputs. The temporal coactivation hypothesis can tell us which inputs are likely to be grouped together as candidates for selection as central RFs. However, without a model of lateral interactions among neighboring neurons and their inputs, the temporal correlation hypothesis is insufficient to describe how specific combinations of correlated inputs are in fact chosen from a family of possible input combinations with similar correlations. This problem has been specifically addressed by Finkel and associates (Chapter 7) and Miller and Stryker (Chapter 9) in other contributions to this collection.

Constraints on Connectionist Models

In order to understand the workings of the somatosensory system and probably any functional system in the mammalian brain, it is necessary to understand more than the complex anatomical substrate that supports that system. We must begin to form a model of the way specific functional features emerge from neuronal interactions. This level of understanding is not possible without a combination of computational and biological approaches. An adequate model of the somatosensory nervous system must consider the following questions.

- How is the global orientation of a cortical map determined? At a developmental level, it appears that the ordinal topography of the body surface representation is determined to some extent by the anatomy of the projection gradient of peripheral inputs onto central targets. After spinal transection in 2-week-old kittens, there is a second representation of the trunk and forelimb in the cortical zone normally representing the hindlimb (McKinley, et al., 1987). The pathway of the aberrant inputs to the deprived cortex is unknown but could take place at the level of the dorsal column nuclei. The topography of the representation in the deprived cortex is disrupted. However, many of the RFs are small and form consistent local progressions of overlapping RFs. This developmental preparation may be an example of a normal central mechanism of RF formation and selection superimposed on a disorderly projection of peripheral inputs: local topography is preserved but global topography is disrupted.

- How do given skin surfaces increase their cortical territory? It is apparent that peripheral innervation density is the major determinant of cortical magnification factor. The representation of the face and fingers in primate cortex is greatly enlarged relative to that of the trunk and can be related to the anatomical projection density and distribution. However, the denervation experiments suggest that it is the relative quantity of peripheral activity that determines the specific magnification factor. Skin surfaces on the fingers and pads next to amputated digits greatly increase in cortical magnification factor without any large-scale changes in input activity per se.

  However, "top-down" factors such as attention to a stimulus may also contribute to cortical magnification. One example from the auditory system shows that frequency-tuning curves in auditory cortex can be quickly modified in awake cats learning an aversive association (Weinberger and Diamond, 1987). In the somatosensory system, behavioral training of adult owl monkeys to stimulate a restricted part of the hand causes a great expansion of the cortical representation of the struck surface (Jenkins et al., 1989). It is still unclear to what extent the behavioral relevance of the stimulus potentiates its effective representation at the cortical level.

- How does cortical magnification influence RF size and the corollary attributes of RF overlap and physiological distance limits? This question is really part of the previous question. Factors contributing to cortical magnification must also contribute to spacing of RFs across cortex. Models of cortical topography must ultimately explain how RFs are determined while maintaining a progressively-shifting overlap at neighboring cortical sites.

- What is the rule by which like inputs are segregated in the cortical map? Inputs from dorsal hairy skin and palmar glabrous skin appear to actively compete for the same cortical territory. However, the representation of those competing inputs is segregated into discrete zones. One possible explanation could have something to do with the overall response profiles of the different sets of inputs. For example, fibers innervating hair follicles have different firing patterns than fibers innervating Merkel discs in the glabrous skin. Quickly adapting and slowly adapting inputs may be segregated into clusters or bands in area 3b of cat and monkey on this basis.

References

Alexander, G.E., DeLong, M.R., and Strick, P.L. (1986). Parallel organization of functionally segregated circuits linking basal ganglia and cortex. Annual Review of Neuroscience 9: 357-381.

Allard, T., Clark, S.A., Grajski, K.A., and Merzenich, M.M. (1988). Plasticity in primary somatosensory cortex after digital nerve section and regeneration in adult owl monkeys. Soc. Neurosci. Abstracts.

Allard, T., Clark, S.A., Jenkins, W.M., and Merzenich, M.M. (1989). Reorganization of somatosensory area 3b representations in adult owl monkeys following digital syndactyly. Submitted to J. Neurophys.

Allard, T., and Merzenich, M.M. (1988). Some basic organizational features of the somatosensory system. In von Seelen, W., Shaw, G., an Leinhos, U.M., Organization of Neural Networks. New York: VCH Publishers.

Chapman, B., Jacobson, M.D., Reiter, H.O., and Stryker, M.P. (1986). Ocular dominance shift in kitten visual cortex caused by imbalance in retinal electrical activity. Nature 324: 154-156.

Clark, S.A., Allard, T., Jenkins, W.M., and Merzenich, M.M. (1988). Receptive fields in the body-surface map in adult cortex defined by temporally correlated inputs. Nature 332: 444-445.

Doetsch, G.S., Standage, G.P., Johnston, K.W., and Lin, C-S. (1988a). Thalamic connections of two functional subdivisions of the somatosensory forepaw cerebral cortex of the raccoon. J. Neurosci. 8 (6): 1873-1886.

Doetsch, G.S., Standage, G.P., Johnston, K.W., and Lin, C-S. (1988b). Intracortical connections of two functional subdivisions of the somatosensory forepaw cerebral cortex of the raccoon. J. Neurosci.. 8 (6): 1887-1900.

Dostrovsky, J.O., Millar, J., and Wall, P.D. (1976). The immediate shift of afferent drive of dorsal column nucleus cells following deafferentation: a comparison of acute and chronic deafferentation in gracile nucleus and spinal cord. Exp. Neurology 52: 480-495.

Dykes, R.W., Landry, P., Metherate, R.S., and Hicks, T.P. (1984). Functional roles of GABA in cat primary somatosensory cortex: shaping receptive fields of cortical neurons. J Neurophys 52: 1066-1093.

Edelman, G.M. 1978. Group selection and phasic reentrant signalling: a theory of higher brain function. In Edelman, G.M., and Mountcastle, V.B. The Mindful Brain, Cambridge, Ma.: MIT Press, 51-100.

Gilbert, C.D. (1983). Microcircuitry of the visual cortex. Ann. Rev. Neurosci. 6: 217-247.

Gilbert, C.D. and Wiesel, T.N. (1987). Relationships between cortico-cortical projections, intrinsic cortical connections and orientation columns in cat primary visual cortex. Soc. Neurosci. Abstracts.

Glees, P., and Cole, J. (1950). Recovery of skilled motor functions after small repeated lesions of motor cortex in macaque. J. Neurophys. 13: 137-148.

Greenough, W.T. (1984). Possible structural substrates of plastic neural phenomena. In Lynch, G., McGaugh, J.L., Weinberger, N.M., Neurobiology of Learning and Memory. New York: The Guilford Press, 470-478.

Greuel, J.M., Luhmann, H.J., and Singer, W. (1988). Pharmacological induction of use-dependent receptive field modifications in the visual cortex. Science 242: 74-77.

Haenny, P.E., Maunsell, J.H.R., and Schiller, P.H. (1988). State dependent activity in monkey visual cortex. II. Retinal and extraretinal factors in V4. Exp Brain Res 69: 245-259.

Hubel, D.H., and Wiesel, T.N. (1977). Functional architecture of macaque monkey cortex. Proc. R. Soc. Lond. 198: 1-59.

Jenkins, W.M., and Merzenich, M.M. (1987). Reorganization of neocortical representations after brain injury: a neurophysiological model of the bases of recovery from stroke. In Seil, F.J., Herbert, E., and Carlson, B.M., Progress in Brain Research, Vol 71. Elsevier Science Publishers, 249-266.

Jenkins, W.M., Merzenich, M.M., Ochs, M.T., Allard, T., and Guic, E. (1989). Functional reorganization of somatosensory representations within area 3b of adult owl monkey after behaviorally controlled tactile stimulation. Submitted to J. Neurophysiol.

Jones, E.G. (1984). Identification and classification of intrinsic circuit elements in the neocortex. In Edelman, G.M., Gall, E., and Cowan, W.M., Dynamic Aspects of Neocortical Function. New York: John Wiley and Sons, 7-40.

Landry, P., and Deschenes, M. (1981). Intracortical arborizations and receptive fields of identified ventrobasal thalamocortical afferents to the primary somatic sensory cortex in the cat. J. Comp. Neurol. 199: 345-371.

Landry, P., Diadori, P., Leclerc, S., and Dykes, R.W. (1987). Morphological and electrophysiological characteristics of somatosensory thalamocortical axons studied with intra-axonal staining and recording in the cat. Exp. Br. Res. 65: 317-330.

Landry, P., Villemure, J., and Deschenes, M. (1982). Geometry and orientation of thatlamocortical arborizations in the cat somatosensory cortex as revealed by computer reconstruction. Br. Res. 237: 222-226.

McKinley, P.A., Jenkins, W.M., Smith, J.L., and Merzenich, M.M. (1987). Age- dependent capacity for somatosensory cortex reorganization in chronic spinal cats. Developmental Brain Res. 31: 136-139.

McKinley, P.A., and Smith, J.L. (1989). Age-dependent differences in reorganization of primary somatosenosry cortex following low thoracic (T12) spinal-cord transection in cats. Submitted to J. Neurosci.

Merzenich, M.M. (1987). Dynamic neocortical processes and the origins of higher brain functions. In Changeux, J-P., and Konishi, M., The neural and Molecular Bases of Learning. New York: John Wiley & Sons, 337-358.

Merzenich, M.M., Kaas, J.H., Wall, J.T., Sur, M., Nelson, R.J., and Felleman, D.J. (1983). Progression of change following median nerve section in the cortical representation of the hand in areas 3b and 1 in adult owl and squirrel monkeys. Neuroscience 10: 639-665.

Merzenich, M.M., Nelson, R.J., Kaas, J.H., Stryker, M.P., Jenkins, W.M., Zook, J.M., Cynader, M.S., and Schoppmann, A. (1987). Variability in hand surface representations in areas 3b and 1 in adult owl and squirrel monkeys. J Comp Neurol 258: 281-297.

Merzenich, M.M., Nelson, R.J., Stryker, M.P., Cynader, M.S., Schoppmann, A., and Zook, J.M. (1984). Somatosensory cortical map changes following digit amputation in adult monkeys. J Comp Neurol 224: 591-605.

Merzenich, M.M., Recanzone, G., Jenkins, W.M., Allard, T., and Nudo, R.J. (1988). Cortical representational plasticity. In Rakic, P., and Singer, W., Neurobiology of the Neocortex, New York: John Wiley & Sons.

Metherate, R.S., Tremblay, N., and Dykes, R.W. (1987). Acetylcholine permits long term enhancement of neuronal responsiveness in cat primary somatosensory cortex. Neuroscience 22: 75-81.

Miller, K.D., Keller, J.B., and Stryker, M.P. (1989). Ocular dominance column development: analysis and simulation, Science 245: 605-615.

Moran, J., and Desimone, R. (1985). Selective attention gates visual processing in the extrastriate cortex. Science 229: 782-784.

Mountcastle, V.B. (1957). Modality and topographic properties of single neurons of cat's somatic sensory cortex. J. Neurophys. 20: 408-434.

Mountcastle, V.B. (1978). An organizing principle for cerebral function: the unit module and the distributed system. In Edelman, G.M., and Mountcastle, V.B., The Mindful Brain. Cambridge, Mass: MIT Press, 7-50.

Olson, C.R., and Jeffers, I. (1987). Organization of cortical and subcortical projections to area 6m of the cat. J. Comp. Neurol. 266: 73-94.

Pearson, J.C., Finkel, L.H., and Edelman, G.M. (1987). Plasticity in the organization of adult cerebral cortical maps: a computer simulation based on neuronal group selection. J. Neurosci. 7: 4209-4223.

Pons, T.P., Garraghty, P.E., and Mishkin, M. (1988). Lesion-induced plasticity in the second somatosensory cortex of adult macaques. Proc. Natl. Acad. Sci. 85: 5279-5281.

Powell, T.P.S., and Mountcastle, V.B. (1959). Some aspects of the functional organization of the cortex of the postcentral gyrus of the monkey: a correlation of findings obtained in a single unit analysis with cytoarchitecture. Bull. Johns Hopkins Hosp. 105: 133-162.

Rakic, P. (1976). Local Circuit Neurons. Cambridge, Ma: MIT Press.

Rasmusson, D.D. (1982). Reorganization of raccoon somatosensory cortex following removal of the fifth digit. J. Comp. Neurol. 205: 313-326.

Rasmusson, DD. (1988). Projections of digit afferents to the cuneate nucleus in the raccoon before and after partial deafferentation. J. Comp. Neurol. 277: 549-556.

Rasmusson, D.D., and Nance, D.M. (1986). Non-overlapping thalamocortical projections for separate forepaw digits before and after cortical reorganization in the raccoon. Brain Res. Bull. 16: 399-406.

Rockel, A.J., Hiorns, R.W., and Powell, T.P.S. (1980). The basic uniformity in structure of the neocortex. Brain 103: 221-244.

Rockland, K.S., and Pandya, D.N. (1977). Laminar origins and terminations of cortical connections of the occipital lobe in the rhesus monkey. Br. Res. 179: 3-20.

Shanks, M.F., Pearson, R.C.A., and Powell, T.P.S. (1985). The ipsilateral cortico-cortical connexions between the cytoarchitectonic subdivisions of the primary somatic sensory cortex in the monkey. Br. Res. Rev. 9: 67-88.

Shatz, C.J. and Stryker, M.P. (1988). Prenatal terodotoxin infusion blocks segregation of retinogeniculate afferents. Science 242: 87-89.

Sidtis, J.J., Volpe, B.T., Holtsman, J.D., Wilson, D.H., and Gazzaniga,

M.S. (1981). Cognitive interaction after staged callosal section: evidence for transfer of semantic activation. Science 212: 344-346.

Sillito, A.M., Salt, T.E., and Kemp, J.A. (1985). Modulatory and inhibitory processes in the visual cortex. Vision Res. 25: 375-381.

Snow, P.J., Nudo, R.J., Rivers, W., Jenkins, W.M., and Merzenich, M.M. (1988). Somatotopically inappropriate projections from thalamocortical neurons to the SI cortex of the cat demonstrated by the use of intracortical microstimulation. Somatosensory Res. 5: 349-372.

Spitzer, H., Desimone, R., and Moran, J. (1988). Increased attention enhances both behavioral and neuronal performance. Science 240: 338-340.

Sretavan, D., and Dykes, R.W. (1983). The organization of two cutaneous submodalities in the forearm region of area 3b of cat somatosensory cortex. J. Comp. Neurol. 213: 381-398.

Sur, M., Merzenich, M.M., and Kaas, J.H. (1980). Magnification, receptive field area and "hypercolumn" size in Areas 3b and 1 of somatosensory cortex in owl monkeys. J. Neurophys. 44: 295-311.

Sur, M., Wall, J.T., and Kaas, J.H. (1984). Modular distribution of neurons with slowly adapting and rapidly adapting responses in area 3b of somatosensory cortex in monkeys. J. Neurophys. 51: 724-744.

Van Essen, D.C. and Maunsell, J.H.R., (1983). Hierarchical organization and functional streams in the visual cortex. Trends in Neuroscience 6: 370-375.

Wall, J.T., Kaas, J.H., Sur, M., Nelson, R.J., Felleman, D.J., and Merzenich, M.M. (1986). Functional reorganization in somatosensory cortical areas 3b and 1 of adult monkeys after medina nerve repair: possible relationships to sensory recovery in humans. J. Neurosci. 6: 218-233.

Weinberger, N.M., and Diamond, D.M. (1987). Dynamic modulation of the auditory system by associative learning. In Edelman, G., Hassler, S., and Gall, E., Auditory Function. New York: John Wiley & Sons.

Welker, W.I., and Seidenstein, S. (1959). Somatic sensory representation in the cerebral cortex of the raccoon (Procyon

Lotor). J. Comp. Neurol. 111: 469-501.

Wilson, P., and Snow, P.J. (1987). Reorganization of the receptive fields of spinocervical tract neurons following denervation of a single digit in the cat. J. Neurophysiol. 57: 803-818.

Wise, S.P. (1985). The primate premotor cortex: past, present, and preparatory. Ann. Rev. Neurosci. 8: 1-19.

Woolsey, C.N., Marshall, W.H., and Bard, P. (1942). Representation of cutaneous tactile sensibilty in the cerebral cortex of teh monkey as indicated by evoked potentials. Bull. Johns Hopkins Hosp., 70: 399-441.

CHAPTER 7

A Model of Receptive Field Plasticity and Topographic Map Reorganization in the Somatosensory Cortex

L. H. Finkel

Introduction

A combination of anatomical, physiological, and neurological evidence gathered over the last century has established that the primary sensory cortices of the mammalian brain contain ordered topographic maps of the periphery. The concordance of the maps obtained through these different lines of evidence has led to the view that cortical maps are static structures determined by the anatomical connectivity generated during development. This view, that maps are predominantly anatomical and not physiological constructs has arisen despite the fact that early studies documented an inherent lability in the organization of cortical maps (Leyton and Sherrington, 1917). In the last several years, evidence from a number of species (Kaas, Merzenich, and Killackey, 1983; Wall, 1988), but primarily from the somatosensory cortex of monkeys (Merzenich et al, 1988; Allard, Chapter 6), has shown that cortical maps are, in fact, dynamic structures that undergo a process of continual reorganization. This paper addresses the question of what physiological mechanisms might underlie such dynamic organization in cerebral cortical maps.

The ubiquitous presence of maps in the cortex (and subcortex), as well as the multiple connections between maps, suggests that neural maps serve as the fundamental operational structures in the nervous system. An understanding of the organizational properties of maps is therefore critical for a detailed understanding of cortical function. Thus, we have constructed a detailed computer simulation of the processes underlying dynamic map organization in a small region of somatosensory cortex. These simulations can reproduce many of the experimental phenomena discussed in the preceding chapter. The simulations have been deliberately modeled after known cortical anatomy and physiology. However, a significant degree of simplification has been required due to the complexity of the somesthetic system in particular and of the cortex in general.

The complexity of the nervous system arises not only from the profusion of cells and synapses, but also from the fact that operations are carried out concurrently at several levels of organization and over several temporal scales. Thus, one might account for aspects of representational plasticity at the level of molecules, supramolecular assemblies, synapses, compartments, individual cells, local circuits, individual maps, whole pathways within the somatosensory system, or by invoking aspects of the behavioral repertoire of the animal. Furthermore, changes occur over different time scales from milliseconds to months. A complete explanation must be extensible to all these levels, but any individual model, particularly a preliminary one, must confine itself to a judicious choice of mechanisms at a few key levels. This point notwithstanding, we will nonetheless stress the view that fundamental aspects of the process of map organization emerge from the interactions between mechanisms operating at different structural levels. For example, we have previously found that the functional properties of synaptic modification rules depend upon the architecture of the neural network in which they are embedded (Finkel and Edelman, 1985). As we will demonstrate below, the characteristics of map changes observed may depend upon a complex interaction between the anatomy, cell properties, and synaptic plasticity rules.

Before presenting the results of these computer simulations, we briefly review the physiological observations the model attempts to address and present an overview of the rationale motivating the model. Since the previous chapter has extensively reviewed the experimental data, we will chiefly confine our attention to the possible classes of models which may account for the observed phenomena.

Dynamic Organization in Somatosensory Cortex

The physiological plasticity described in somatosensory cortex involves organized changes in the size and position of receptive fields on the skin surface. Map reorganization has been documented at all levels of the somatosensory system: in the spinal cord (Devor and Wall, 1981), dorsal column nuclei (Dostrovsky, Milar, and Wall, 1976), thalamus (Wall and Egger, 1971), and several areas of primary somatosensory cortex (Kaas, et al., 1983), secondary somatosensory cortex (Pons, Garraghty, and Mishkin, 1988), and higher, multimodal areas of parietal cortex (Hyvarinen, 1982). Similar plasticity has recently been reported in the motor cortex (Sanes et al., 1988). Although similar phenomena have been observed at all these multiple levels and in several different species (Wall, 1988), we will confine our attention to the experiments on the primary somatosensory cortex of owl and squirrel monkeys (Kaas et al., 1983; Mountcastle, 1984).

Cortical areas 3b and 1 of the monkey both contain a complete somatotopic representation of the body surface; neurons in these areas respond to light touch to the skin. Multiple, closely-spaced microelectrode recordings reveal that although the general organizational properties of the map are similar from animal to animal, the finer details of map structure vary greatly among individuals and, furthermore, they vary over time within the same individual (Merzenich, 1985; Merzenich et al., 1987). Dramatic changes in map organization are seen after perturbation of the sensory stimulation to the skin. For example, if a cutaneous nerve is transected, if a local region of skin is continuously stimulated, or if fingers are amputated or sewn together to produce a syndactyly, the locations of map borders (i.e. the borders between representations of different body parts) can move hundreds of microns across the cortex or entirely new representations can emerge. Despite these movements (and in many cases during them), local topography is maintained. There are, however, distance limits beyond which the representation cannot move (Jenkins, Merzenich, & Ochs, 1984; Merzenich et al., 1988; Allard, Chapter 6). Concomitant changes occur in the sizes of individual receptive fields so that, in general, an "inverse rule" holds for the magnification factor of the representation (i.e., the larger the representation of a given skin region, the smaller the receptive field of cells representing that region). This inverse rule appears to be a general property of neural maps relating receptive field size and magnification factor. It has been found to hold true in the normal animal for receptive fields with a wide range of sizes (Sur, Merzenich, and Kaas, 1980) and it continues to operate during map reorganization (Merzenich et al., 1983).

The representational borders found in somatosensory cortex fall into several distinct classes, and it is important to note that very different mechanisms of plasticity may control changes in the locations of each type of border. For example, Dykes and Ruest (1986) have classified the receptive field borders found in somatosensory cortex into four categories: cytoarchitectonic borders, somatotopic borders (involving a discontinuity such as the hand-face, or foot-torso borders), topographic borders (not involving a discontinuity such as between digits), and submodality borders (i.e., rapid adapting and slow adapting units).

Cytoarchitectonic borders are established during development and recent experiments of Rakic (1988) indicate that their location depends upon the characteristics and quantity of thalamic inputs. Most of the results on adult map plasticity reported to date involve changes in "topographic" borders; however, evidence does exist for changes in receptive field properties across several of the other types of borders. For example, movements of representations across the hand/face (somatotopic) border, as well as across the 3a/3b (cytoarchitectonic) border have been reported (Merzenich et al., 1988). In the latter case the cytoarchitectonics have not changed, but units in area 3a normally responsive to deep afferents become responsive to superficial stimulation.

Static versus Dynamic Processes: Since the receptive field of a cell must ultimately reflect the distribution of anatomical inputs received by that cell (through multiple synaptic levels), it seems safe to assume that anatomy determines the constraints on how far a receptive field border can ultimately shift. Conversely, the rapid time course of some components of the observed plasticity (minutes to hours) argues that the acute changes in receptive field maps must be due to a dynamic physiological process.

Two general physiological processes have been proposed to account for receptive field plasticity: (1) "unmasking" of any of several alternate maps which are built into the anatomy and which are chosen between by some process such as the removal of tonic inhibition (Killackey, 1988), or (2) continual reorganization or "generation" of the map within the anatomical substrate by a dynamic process such as the modification of synaptic efficacies (Wall, 1977; Edelman and Finkel, 1984; Mountcastle, 1984; Merzenich, 1988). It is important to note that the distinction between these two proposed processes goes beyond mechanism (inhibition versus synaptic modifications); indeed, it is likely that several different mechanisms (including both inhibition and synaptic modification) are involved in map reorganization. For example, the experiments of Sillito (1975) in visual cortex, and of Hicks and Dykes (1983) and Dykes and associates (1984) in somatosensory cortex support the idea that ongoing inhibitory processes determine certain aspects of receptive field size and structure. In addition, there is substantial evidence from both racoon (Rasmusson and Turnbull, 1983) and rat (Dykes and Lamour, 1988) that deafferentation leads to a release from a form of tonic intracortical inhibition. This inhibition may act to suppress responses to inputs from beyond the recorded receptive field. When the inhibition is released, new receptive field locations may immediately emerge, possibly accounting for some of the rapid reorganizational changes reported (Dykes and Metherate, 1988). It is possible that the level of inhibition may change depending upon the state of "awareness" or other global parameters, thus allowing rapid adaptation of receptive field sizes and tactile discriminability.

It has been proposed that alternate somatotopic maps could be mediated by lemniscal and non-lemniscal (spinothalamic) fibers (Killackey, 1988). However, the "alternate map" hypothesis presents several serious weaknesses. First, the receptive fields of cells driven by the spinothalamic tract are clearly distinguishable from those in the lemniscal system inasmuch as they are larger, have higher thresholds, and are activated by noxious stimuli (Mountcastle, 1984). In contrast, after map reorganization, the characteristics of recorded receptive fields are unchanged (and sizes are occasionally smaller than previously recorded). Second, since map shifts can occur at all cortical locations (and in any direction at each cortical location), there would have to be many alternate maps, and an additional process would be needed to specify which particular map emerges. Finally, in terms of mechanism, it would seem a waste of metabolic energy to maintain one or many alternate maps under tonic inhibition when the same effect can be had by reducing synaptic efficacy.

The major argument in support of an ongoing generative mapping process is the richness of the observed plasticity phenomena. Without such an ongoing process, one would be at a loss to explain the continued topography at all stages of map reorganization. The governing notion in such an explanation is that anatomical divergence and convergence within the lemniscal system (possibly together with corticocortical connnections) provide a large set of anatomical inputs to each cell or cortical column. During normal adult experience only a small fraction of these inputs are "functional", i.e., are of sufficient synaptic efficacy to be detectable in a receptive field experiment. The complement of synapses on the cell are not necessarily silent; they may play a role in more complex aspects of the receptive field (as seen in the visual cortex (Allman, 1988).

Receptive field plasticity is presumed to occur through synaptic modifications that alter the set of functional inputs (and hence the peripheral locus of the receptive field), rather than through anatomical changes. Synaptic modifications are driven by the spatial and temporal correlations of the stimuli received (Clark et al., 1988) - thus accounting for the representational plasticity seen over time in normal (unperturbed) animals, as well as for part of the variance in the fine structure of map organization seen between conspecifics.

A major problem in proposing synaptic modifications as the basis of map plasticity is the difficulty which arises in regulating such changes, particularly since the divergent anatomy will allow a vast number of possible reorganizations. Furthermore, plastic changes in nearby cells must be coordinated if the map is to retain any semblance of order. It has been proposed that map plasticity is regulated by neuronal groups - local assemblies of strongly connected cells, each approximately 50-100 microns in diameter and extending vertically through the cortex (Edelman and Finkel, 1984). These groups constitute the basic units of the map controlling receptive field plasticity, and cooperative and competitive interactions within and between these neuronal groups determine the collective receptive field properties of the individual cells of each group. Selection and competition between neuronal groups in the adult nervous system gives rise to a kind of "ongoing development" in which receptive field properties, topography, and other global organizational parameters are continually adapted to changing environments (Edelman, 1978; 1987).

Any model of cortical map reorganization must depend upon a degree of local corticocortical interconnection in order to explain the continuity of somatotopic representations. However, in principle, the efficacy of these corticocortical connections could be uniformly or randomly distributed and still generate an ordered local topography. The postulation of neuronal groups represents a discretized model in which cooperativity is restricted to local ensembles rather than existing homogeneously and isotropically between cells as a function only of their separation in cortex. A discrete local group of interconnected cells is ideally suited to generate cooperative interactions and as we will show, a large degree of local cooperativity is necessary to produce the dramatic receptive field changes observed.

A major motivation for postulating the existence of discrete neuronal groups is the presence of sharp but moveable borders in the cortical receptive field map. Sharp borders between groups generate sharp receptive field borders and the ability of groups to change their receptive fields allows the location of receptive field borders to shift across the cortical surface. Furthermore, as we will show in the simulations below, group-like structures naturally arise in cortex-like networks. The results of the simulations demonstrate a number of other advantages to organizing the network into groups. Before considering these results, however, we consider the experimental evidence in support of the existence of neuronal groups.

Experimental Evidence for Groups

At present, there is no direct experimental evidence for the existence of neuronal groups, although both rodent whisker barrels (Woolsey and Van der Loos, 1970) and orientation columns in visual cortex (Hubel and Wiesel, 1977) could be considered special cases of neuronal groups. However, there are a number of suggestive indications of a widespread existence of groups. Kaas and associates (1981) made tangential penetrations through the hand region of area 3b in monkey cortex after the median nerve had been cut, resutured, and allowed to regenerate. Several weeks after resuturing they found small islands (40-60 microns) in which all cells shared similar receptive fields (i.e., were either glabrous or dorsal and had highly overlapped positions). A given skin site was often represented in multiple such islands scattered over a wide cortical area. With time, topography returned and the islands were no longer detectable. However, their initial presence suggests the existence of an underlying cortical unit: either structural (Mountcastle, 1978) or functional (Edelman, 1978). In fact, if one Examines the distribution of the "islands" of dorsal representation in area 3b of normal monkeys it is easy to suppose that they are constituted of assemblies of several adjacent groups.

A second source of evidence for groups has recently emerged from recordings in visual cortex. Grey and Singer (1987) have discovered coherent oscillations in the orientation-specific field potentials of cells. Neurons within the same orientation column appear to oscillate in phase at 20-70 Hz and the oscillations of adjacent orientation columns differ in frequency or phase. Similar findings have recently been reported by Eckhorn and his associates (1988). Strongly connected local cell ensembles are well-suited to produce resonant-type oscillations and these studies may provide an important new technique for determining other properties of neuronal groups. In particular, the temporal properties of the oscillation may indicate which groups (in the same or different cortical areas) are linked in a common cortical operation (Grey and Singer, 1987; Eckhorn et al., 1988; Finkel and Edelman, in press).

In this regard, Whitsel and his colleagues (Favorov, Diamond, and Whitsel, 1987) have developed an alternative (but not necessarily incompatible) model of somatocortical operations which bears on the issue of groups. They propose that cortical regions are parcellated into a mosaic of units called "segregates" with each occupying an area of roughly 350-400 microns. Segregates have the property that there is at least one point on the skin surface which is shared by the receptive field of every cell belonging to the segregate. Segregates thus differ from groups in that they are much larger and in that the receptive fields of their cells may only minimally overlap (i.e., perhaps at only one point). Consistent with the existence of segregates, Whitsel and associates (1988) provide evidence from recordings in awake monkeys that receptive fields encountered in quasi-radial penetrations display a large degree of scatter.

Whitsel and his associates have developed a paradigm using 2-Deoxyglucose studies of the activity patterns evoked in monkey somatosensory cortex by stereotyped repetitive stimulation of the hand (Whitsel et al., 1988). They find evidence supporting a very fast, dynamic organization of activity into reproducible patterns of periodic structures. Such activity-generated patterns may affect the receptive field properties of the map, and in fact they argue that traditional receptive field maps are not capturing the most striking dynamic aspects of functional cortical organization. The emergence of activity-generated patterns is not incompatible with map reorganization based on synaptic modifications. It seems reasonable to expect that a series of mechanisms will be found that operate on various time scales - from milliseconds to months - all of which play a role in the ongoing organization of map structure.

Given these general arguments on the rationale for a model of map reorganization based on a divergent anatomy, synaptic plasticity, and neuronal groups, we now turn to a series of computer simulations which show that such a scheme can indeed generate map reorganizations similar to those observed in vivo.

Computer Simulations Figure 1 shows a schematic of the network used in the simulations, the details of which are given elsewhere (Pearson, Finkel, and Edelman, 1987). The network is comprised of 1,536 cells of two cell types: 1024 excitatory cells (triangles) and 512 inhibitory cells (circles). Each cell-type has a characteristic pattern of connectivity (corresponding to its dendritic morphology), but the exact morphology of each cell differs in detail. The connections of each cell are generated randomly according to a probability distribution that decreases with distance between cells. There are excitatory-excitatory, excitatory- inhibitory, and inhibitory-excitatory connections, but no inhibitory-inhibitory connections. Excitatory-excitatory and inhibitory-excitatory connections are relatively short-range (spanning roughly 5% of the dimension of the network); excitatory-inhibitory connections spread, on the average, roughly twice as far. Altogether, there are approximately 70,000 intrinsic ("corticocortical") connections in the network. Excitatory cells receive inputs from an external receptor sheet, which is intended to correspond to the glabrous and dorsal (front and back) aspects of the hand of the monkey.

This receptor sheet contains 1024 receptors and is divided into regions corresponding to four fingers and a subphalangeal region (e.g. there are 96 receptors on the front and 96 receptors on the back of

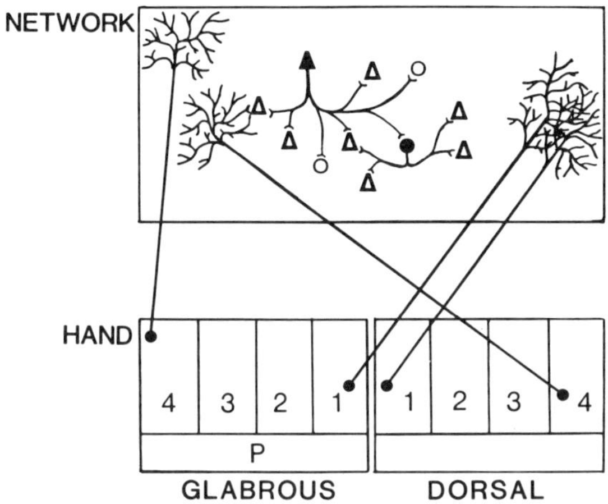

Figure 1: Schematic diagram of network connectivity. The network contains 2 types of cells: excitatory (triangles) and inhibitory (circles). Each cell contacts neighboring cells with a probability dependent upon the distance between them (a few connections of the two darkened cells are shown). Also shown is a receptor sheet representing the hand, which contains glabrous and dorsal surfaces of 4 fingers and a subphalangeal region. 1024 pressure sensitive receptors project to the network in a topographic fashion, each arborizing over roughly 10% of the network. Receptors from the corresponding glabrous and dorsal positions project to the same network location, so that each cell receives equal numbers of connections from the front and back of the hand. The network contains a total of approximately 70,000 intrinsic and 100,000 extrinsic connections.

each finger). As shown in figure 1, fibers from the receptor sheet project topographically onto the network (i.e., adjacent regions of the receptor sheet map onto adjacent regions of the network). Altogether, there are approximately 100,000 extrinsic connections. Note that connections from the front and back of the hand (e.g., the left and right halves of the receptor sheet) are mapped topographically onto the same network. The mapping is arranged such that the front and back of a given digit both map to approximately the same region of the network. It is important to note that each excitatory cell receives an equal number of connections from the glabrous and dorsal aspects of the receptor sheet; yet, as we will show, due to the action of the synaptic modification rule, each cell develops a receptive field which is exclusively glabrous or exclusively dorsal in nature (as is physiologically recorded in the monkey).

The physiological properties of each simulated cell are specified by several parameters (see Pearson et al., 1987, for details) which regulate the rate of voltage decay, the degree of synaptic saturation (modeled as a decrease in driving potential with increasing depolarization), the characteristics of a sigmoidal output firing function, and the properties of a type of shunting inhibition. Shunting inhibition was found to confer a remarkable degree of dynamic stability upon the network over a wide range of stimulation intensities. Unlike hyperpolarizing inhibition, the "multiplicative" properties of the shunt guarantee that the network will remain stable despite substantial strengthening or weakening of various synaptic connections during the course of the simulation.

All anatomical connections are generated at the start of the simulation and, once generated, are fixed (i.e., there is no sprouting or re-wiring). However, the strengths of the connections, both intrinsic and extrinsic, are modifiable according to a synaptic modification rule, that we now briefly describe.

Synaptic Plasticity Most neural models over the last several decades have used some variant of the Hebb rule (1949) in which correlated activity between presynaptic and postsynaptic cells gives rise to an increase in synaptic strength (the original Hebb rule) and uncorrelated activity decreases connection strengths (a modification required for stability). The Hebb rule has the virtue of formal simplicity, it displays a number of useful associative properties, and it is designed to guarantee that repetition of the same stimulus will generate (after synaptic strengthening) a similar but stronger response. However, the Hebb rule has a number of drawbacks that have been less widely recognized. First, there is no experimental evidence for the Hebb rule at the synaptic level. In fact, several experiments directly demonstrate that correlated firing in and of itself does not lead to synaptic strengthening (Carew et al., 1984; Wigstrom and Gustafsson, 1983). Rather than postsynaptic firing, the critical variable appears to be a significant degree of postsynaptic depolarization (Ito, Sakurai, and Tongroach, 1982; Kelso, Ganong, and Brown, 1986; Malinow and Miller, 1986). Presynaptic firing appears less relevant than the biochemical sequelae of transmitter binding to the postsynaptic cell. Thus, instead of a mathematical relationship involving the correlation of activity in connected cells, synaptic plasticity can now be directly tied to a biochemical relationship between events at the postsynaptic terminal.

The most important consequence of the difference between voltage- dependent and firing-dependent rules is the ability to account for context-dependent changes, i.e., changes dependent upon the particular pattern of synaptic inputs. In the Hebb rule all synapses on the postsynaptic cell are equally affected by the firing of that cell and there is therefore no allowance for the individual effect of one synapse upon another. These "heterosynaptic" effects are well known (Rall, 1964; Fleshman, Segev, and Burke, 1988), and play an important role in the function of the neuron, particularly because the electrotonic voltages from one synapse can affect voltage-dependent processes at other synapses on the same cell. More recently proposed synaptic rules such as the "delta rule" (Widrow and Hoff, 1960; Stone, 1986) confer a number of powerful advantages in network learning schemes, but they still do not explicitly allow for heterosynaptic interactions.

For these reasons we have developed several new formal synaptic rules that are based on recent biochemical and biophysical principles of synaptic function and which explicitly incorporate heterosynaptic effects (Finkel and Edelman, 1985; 1987). We distinguish changes in presynaptic efficacy (the relationship between presynaptic depolarization and the release of transmitter) and postsynaptic efficacy (the transduction of transmitter binding into a voltage change). The postsynaptic rule accounts for heterosynaptic interactions and allows one input to a neuron to modulate the efficacy of other inputs to the same neuron.

Figure 2 shows the basic assumptions involved in the postsynaptic rule. This particular scheme is only one member of a large family of such rules that couple electrical and biochemical events within the neuron. Four synapses are shown on a branching dendritic tree; each synapse contains transmitter receptors (rectangles) and voltage-dependent ionophores (triangles). Initially a transmitter,

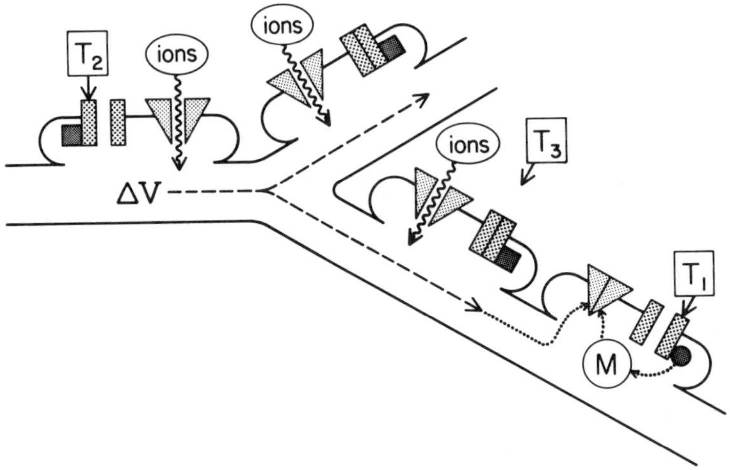

Figure 2: The postsynaptic modification rule. Schematic diagram of four synapses on a branching dendritic tree. Each synapse contains both transmitter receptors (rectangles) and voltage- dependent ionophores (triangles). Binding of transmitter (T1) (lower right synapse) produces a substance, M, which modifies voltage-dependent ionophores (providing that the ionophores are in a modifiable or "depolarized" state). Inputs to other synapses (e.g. T2) cause a depolarization that electrotonically propagates through the cell. If the wave of depolarization reaches the bottom right synapse and increases the fraction of ionophores in the modifiable state when the local concentration of M is still high, a larger change in postsynaptic efficacy will occur. This is the basis of heterosynaptic interactions in the synaptic rule.

T1, binds to receptors at the lower right synapse, evoking a voltage change but also initiating a local biochemical cascade that gives rise to the production of "M", a modifying substance meant to represent an intracellular second messenger such as Ca^{2+} or a cyclic nucleotide. "M" acts on voltage-dependent ionophores at the synapse and increases their conductance. The main assumption of the postsynaptic rule is that the efficacy with which "M" modifies these ionophores depends upon their functional state. In particular, we assume that transitions between the states of the ionophore are voltage-dependent and that the ionophore is modifiable only when the synapse is depolarized (evidence for this assumption is discussed in Finkel and Edelman, 1987). Modifications could consist of any of a number of changes such as an increased maximal conductance, lengthened ionophore open time, shifted current-voltage relationship, or increased effective number of ionophores. Therefore, if at a later time another transmitter, T2, binds to receptors at the upper left synapse, the following sequence of events occurs: A depolarization due to the new input propagates electronically through the cell, changing the state of voltage-dependent ionophores as it reaches them. If the depolarization reaches the bottom right synapse and increases the fraction of ionophores in the modifiable state at a time when the local concentration of "M" is still high (due to the recent arrival of a local input), then the number of modified ionophores, and thus the local postsynaptic efficacy, is increased to a greater extent. By this mechanism, distant inputs can modulate the efficacy of local synaptic changes - a heterosynaptic effect.

Note that this postsynaptic mechanism is very similar to the recently elucidated mechanism of action of the N-methyl-D-aspartate (NMDA) receptor. The NMDA receptor has been shown to play a role in the induction of long-term potentiation (LTP) in the hippocampus (Collingridge, Kehl, and McLennan, 1983; Harris, Ganong, and Cotman, 1984; Collingridge and Bliss, 1987) and provides what is perhaps the best supported candidate for a molecular mechanism of synaptic plasticity in vertebrates. As in our proposed mechanism, NMDA receptors require the concurrent action of transmitter binding and postsynaptic depolarization; but the NMDA receptor incorporates both mechanisms into a single molecule. The NMDA-receptor-ionophore is blocked by Mg^{2+} ions in a voltage-dependent manner so that significant depolarization is required to open the channel (Nowak et al., 1983). If postsynaptic depolarization occurs together with presynaptic transmitter release, the ionophore opens and allows cationic currents to flow, including a calcium current that most likely initiates a second messenger cascade leading to biochemical modifications. The site of these modifications is currently unknown, but the non-NMDA glutamate receptors are leading candidates (the NMDA receptor itself or other membrane proteins are still possibilities).

It is likely that other such molecular mechanisms will be found that share the heterosynaptic property formulated in our model. For example, Dykes and Metherate (1988) have proposed that acetylcholine acting on muscarinic receptors in the somatosensory cortex may give rise to a similar type of modification. A related cholinergic mechanism has also been proposed for plasticity in the visual cortex (Bear and Singer, 1986). The mechanism of sensitization in Aplysia elucidated by Kandel and his colleagues (Hawkins et al., 1983) acts at the presynaptic terminal, but involves a related molecular mechanism. Indeed, it is possible that many of the cortical transmitters and modulatory substances act through such mechanisms, and we have previously discussed how a kind of "transmitter logic" may arise from the combined action of these multiple mechanisms (Finkel and Edelman, 1987).

To a first approximation, both the synaptic rule proposed here and the NMDA receptor mechanism appear to obey the Hebb rule in that presynaptic firing is directly related to transmitter binding and postsynaptic firing is related to postsynaptic voltage. In fact, the NMDA receptor has been widely cited as implementing a Hebb rule (e.g. Kelso et al., 1986). The differences between the implied mechanisms go beyond the differences between voltage and cell firing, however, and are worth pointing out. The major feature of the Hebb rule is that synaptic change depends upon the correlation of presynaptic and postsynaptic firing. Thus, the individual synapse is the unit of interest. In the rule proposed here, the notion of "correlation" is given a biochemical mechanism. But more importantly, the changes at each synapse depend upon the details of events at its neighbors. Thus, the unit of interest involves large populations of synapses. There is still an effective "correlation," but it occurs between the different inputs to a cell, rather than between one input and one output. The differences, although subtle, are responsible for the ability to carry out context-dependent "heterosynaptic" modifications.

The Formation of Neuronal Groups The heterosynaptic property of the synaptic rule is responsible for generating the receptive field properties of the individual cells in the somatosensory cortex model and for the emergence, in the simulations, of neuronal groups. Figure 3a shows the connection strengths of the intrinsic connections in the network at the start of a simulation. The figure has been constructed by drawing a straight line between all excitatory cells that are anatomically connected; the shade of grey of each line indicates the strength of that connection (lighter=stronger; darker=weaker). As can

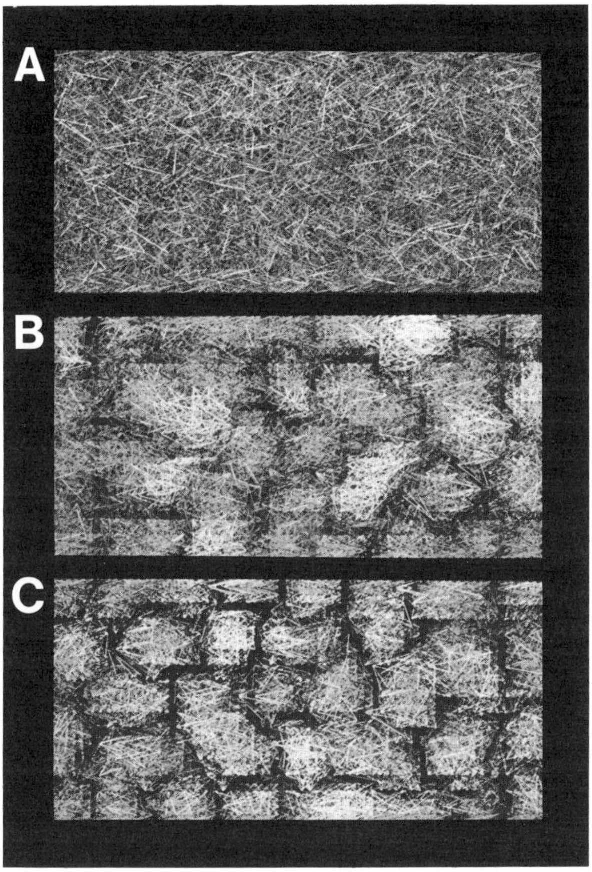

Figure 3: Neuronal Group Formation. (A) Intrinsic connections strengths in the initial network. Straight lines have been drawn between anatomically connected cells (only excitatory cells are depicted in this figure), and the lines have been shaded according to the strength of the connection (lighter=stronger, darker =weaker). Note the absence of any global organization in the connection strengths. (B) Connection strengths after 3 passes of stimulation to the receptor sheet (see text for details). Local regions of have begun to form in which there are increased connection strengths, but there is relatively little synaptic weakening. (C) Intrinsic connection strengths after twelve passes of stimulation. Neuronal groups have formed--local collections of cells with strengthened mutual connections and weakened connections to cells outside the group. Group morphology is relatively stable to further stimulation.

be discerned, there is no global organization to the initial distribution of connection strengths. In fact, the initial efficacies were assigned randomly using a Gaussian distribution. After the generation of connections and the assignment of initial connection strengths, the receptor sheet was stimulated randomly with a small square stimulus (one that simultaneously activated a 3x3 region of receptors) until all locations on the receptor sheet had been stimulated. Separate subroutines allow the receptor sheet to be stimulated with a wide range of stimuli and the activity of every cell to be monitored throughout the simulation (Pearson et al, 1987).

Figure 3b shows the intrinsic connection strengths after three passes of stimulation, i.e., after the stimulus was applied to every location on the receptor sheet three times (each application of the stimulus consisted of 3 repetitions of a sequence of 6 cycles ON followed by 4 cycles of rest). Certain clusters of cells have begun to strengthen their mutual connections (the lighter lines) - these are the incipient neuronal groups.

Figure 3c shows the strengths of the intrinsic connections after 12 passes of stimulation. As can be seen, the neuronal groups have become smaller and more strongly connected. Each group contains approximately 10-20 cells (in vivo, we would predict groups contain hundreds to thousands of cells). Cells within a group have greatly strengthened their mutual connections (compare to figure 3a) and have also weakened connections to cells outside the group. This weakening of connections gives rise to a "moat" around the group that once formed, gives a large degree of stability to the group. Thus, continued stimulation of this network leads to only minor changes in the group structure.

The phenomenon of group formation is robust and occurs over a wide range of stimulation protocols, anatomical architectures, and neuronal parameters. The critical features appear to be the following: (1) local inhibition following a stimulus must affect a larger area than local excitation; (2) stimulation must be relatively even over the receptor surface without excessively favoring local areas; and (3) the synaptic modification rule must depend upon correlated inputs to a cell. It is particularly efficient if higher voltages are required to weaken connections than to strengthen them.

Once neuronal groups have formed, they demonstrate a number of interesting functional properties, preeminent among which is a large degree of nonlinear cooperativity among the cells within a group and a tendency towards competition between cells in different groups.

Groups compete for cells belonging to other groups (some evidence of this can be seen in the intermediate strength connections between some of the groups), and, more importantly, they compete for the location and size of their receptive fields. There is an intrinsic drive for each group to expand its receptive field to the anatomical limits. However, stimulation of one group tends to inhibit neighboring groups (due to the larger effective spread of inhibition) and the ensuing group competition is responsible for many basic aspects of map organization.

Map Organization and Plasticity It is instructive to examine the changes that occur in the receptive fields of individual cells as a result of group formation. The main changes are that, after group formation, receptive fields become smaller and stronger and the receptive fields of all cells in a group become highly overlapped (Pearson et al., 1987). In addition, the receptive fields of all the cells in any one group are exclusively limited to either the glabrous or the dorsal aspect of the receptor sheet. Thus, despite the fact that each cell receives equal numbers of connections from both aspects of the hand, the properties of the synaptic modification rule together with the nonlinear cooperative dynamics of intra-group interactions result in the development of similar receptive field characteristics by all cells in a group. This is an example of how the emergence of groups (a functional organization) changes the relationship between the anatomy and the observed physiological properties.

Another example of the effect of neuronal groups on receptive field properties, and one that directly relates to the recent physiological experiments in somatosensory cortex, involves the phenomenon of group competition. Group competition is best illustrated by constructing receptive field maps of the network. Figure 4a shows the receptive field map corresponding to the network shown in figure 3a at the start of the simulations. The different fingers and subphalangeal area of the hand are distinguished by shading as shown in the small diagram inset on the right. The topographic nature of the underlying anatomy manifests itself in this initial map, in which the four fingers and subphalangeal area are topographic-ally represented. Note that, at this stage, each cell responds to stimulation of regions on both front and back of the hand.

Figure 4b shows the receptive field map after the receptor sheet

Figure 4: Receptive field maps. (A) Before group formation. Receptive fields were determined for each excitatory cell in the network. Receptive fields are indicated by shading the location of the cell in the network according to the corresponding position on the diagram of the hand (right). The initial map shows good topography, but all cells responded to both glabrous and dorsal stimulation. (B) After group formation. Map is now largely segregated into glabrous and dorsal regions. Locations of map borders have changed, but topography is maintained. Black dots indicate glabrous receptive fields, white dots indicate dorsal receptive fields. (Undotted cells retain mixed receptive fields.) (C) Map obtained after repeated, focal stimulation of the glabrous surface of digit 2. Representation of the stimulated surface increased by factor of nearly 14. (D) Map obtained after "median nerve" section. All inputs from glabrous surface of digits 1 and 2 were transected, and the entire receptor sheet stimulated in a normal fashion (see text). Map shows no representation of denervated surfaces, and instead, a new representation of the dorsal surfaces of digits 1 & 2. Large black regions are silent areas, unresponsive to stimulation. With continued stimulation, these silent areas become responsive, yielding an intact topographic representation (see figure on adjoining page).

had been exposed to four passes of stimulation and after groups similar to those in figure 3c had emerged. The map retains its global topographic organization; however, due to competition between groups the exact locations of the borders between fingers and the subphalangeal areas have moved. Also, segregated regions of representation of the front of the hand (black dots) and the back of the hand (white dots) have arisen. The map in figure 4b bears a marked resemblance to the receptive field maps found in monkey somatosensory cortex with segregated regions of dorsal and glabrous representation (Merzenich et al., 1983). The borders between the representations of different hand regions (fingers, palm, etc.) or between glabrous and dorsal representations coincide with borders between neuronal groups, and as previously discussed, the sharpness of these representational borders is, in turn, due to the sharpness of those group borders (which are in most cases 1 cell wide). Some regions of the map in figure 4b contain neither black nor white dots; cells in these regions retain mixed receptive fields and respond to stimulation of both sides of the hand. Most of these cells are located at the edges of the network, and their receptive fields reflect the different density of connections there, a phenomenon not seen in vivo, where the mappings are more homogeneous. Figure 4c shows the results of a perturbation experiment in which continued focal stimulation was applied to the front of digit 2 (indicated by the X in the inset diagram).

A
B
C
D
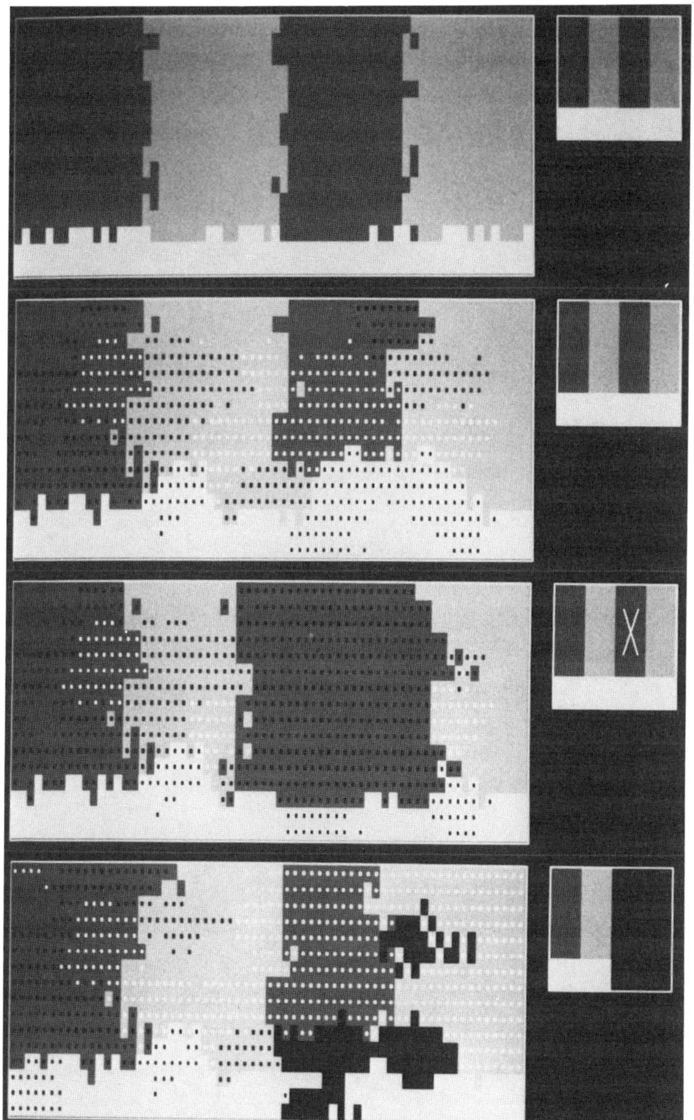

As can be seen, the area of representation of the stimulated surface has greatly increased in the map (by nearly 14 times), occupying regions of the network formerly devoted to the representation of adjacent fingers, the palm, and the back of the hand. If a more uniform distribution of stimulation is once again applied for some time to the receptor sheet, the map again reorganizes back to a configuration resembling that in figure 4b.

The final simulation, shown in figure 4d, emulates a median nerve transection. The connections from the right half of the front of the receptor sheet have been interrupted (denoted by the black area in the inset diagram of the hand), and normal stimulation has been continued to the entire receptor sheet. As can be seen, there is no longer any representation of the denervated surface (i.e., there are no black dots in the right half of the map). In place of the former representation, however, is a new topographic representation of the back of the hand (white dots). In addition, several large silent areas (black regions) have appeared. Cells in these silent regions are unresponsive to all stimulation. However, we have found that with continued stimulation these silent areas gradually disappear, leaving an intact organized map.

These last two simulations illustrate one of the chief advantages of the postsynaptic modification rule. In both the focal stimulation and median nerve section simulations, the changes in receptive fields result from weakening of synapses with high efficacy and strengthening of synapses with very low or no efficacy. Note that, in the Hebb rule, if the synaptic efficacy of a set of connections is so low as to be unable to fire the cell (and hence not be reflected in the receptive field), then no amount of repetitive stimulation of these inputs will lead to strengthening. However, voltage-dependent synaptic rules can "boot-strap" the efficacy to achieve super-threshold values.

Conclusions

These computer simulations, despite their simplifications of the anatomical and physiological properties of somatosensory cortex, capture many of the basic elements of map organization and reorganization. The simulations thus serve as a consistency check on the general idea of selection and competition operating within neural maps. The results in this simplified system would most likely hold as well in more realistic networks. One of the major results concerns the formation of discrete neuronal groups due to a dynamic organization of synaptic efficacies. As a result of their strong internal connection strengths, groups display collective properties such as overlapped receptive fields and coherent firing patterns. Competition between groups (mediated through intracortical inhibition) leads to alterations in the receptive field map, with large changes possible in the location of receptive field borders, while the sharpness of these borders are at all times maintained. The local magnification factor, which initially reflects the density of peripheral receptors (at least in part), can be changed by altering the local distribution of peripheral stimulation. The underlying network anatomy is capable of supporting a large manifold of physiological maps and the particular characteristics of the extant map depend upon the individual variations in the anatomy and connection strengths of the particular network, as well as upon the history of stimulation received by that individual.

It is possible to generate group-like structures in a network using only a Hebb rule; the self-organization into local domains depends most critically upon the use of local excitation and longer distance inhibition. Furthermore, it is possible to reproduce several aspects of topographic map reorganization in networks which are not organized into groups, in particular, receptive field borders can be moved and magnification factors changed by perturbing the stimulation to such networks. The advantage of using neuronal groups and a heterosynaptic rule is that they allow continual reorganization, while at all times preserving a sharp and ordered map. There is a minimization of the time spent in intermediate "fuzzy" states where, for example, receptive fields are half glabrous and half dorsal. Sharp transitions are a result of the high degree of nonlinear cooperativity within each group and the ongoing competition between them.

Neuronal groups, together with a heterosynaptic rule, are also ideally suited for the interactions between multiple maps. As discussed, plastic changes occur in maps at multiple levels of the somatosensory system and any changes in one map must be reflected by all connected maps. In this context, the heterosynaptic properties of the modification rule become particularly important as different classes of inputs must generate separate sets of synaptic changes. It is possible that similar competitive and selective mechanisms operating within the higher-order maps of the temporal, parietal, and frontal cortices may underlie cognitive processes including learning and memory.

These competitive and selective processes provide a number of adaptive advantages for the animal. The continual reorganization of the map ensures that a maximal cortical area is devoted to the more highly stimulated areas of the periphery. This allows maximum use of cortex for those areas that, from an adaptive point of view, are most heavily used. Over the course of development, from birth to adulthood, a large change in the body-form occurs (changes in the relative lengths and masses of different body parts) with relatively minor changes in the structure of the nervous system. It seems very useful to have a degree of map plasticity in order to correlate the changing phenotype with the unaltered central structures. Over evolutionary timescales, changes occur in both body-form and the nervous system. Since the chance of a "double" mutation (in both body-form and nervous system) is highly unlikely, a degree of map plasticity allows gradual evolutionary changes to occur without disastrous consequences for the organism.

Given the complexity of organization in the nervous system, it is likely that additional mechanisms operating at different organizational levels and over different temporal scales could play a role in the mediation of the full range of map plasticity phenomena. The combined actions of a handful of such mechanisms may allow the nervous system to generate its extraordinary adaptive power.

Acknowledgments

The computer simulations discussed here were conducted together with Dr. John C. Pearson at the Neurosciences Institute, New York. Dr. Gerald M. Edelman provided helpful discussions and support.

References

Allman, J. (1988). Variations in visual cortex organization in primates. In P. Rakic and W. Singer Neurobiology of Neocortex. New York, John Wiley & Sons, 29-40.

Bear, M.F. and Singer, W. (1986). Modulation of visual cortical plasticity by acetylcholine and noradrenaline. Nature 320: 172-176.

Carew, T.J.,Hawkins, R.D., Abrams, T.W., and Kandel, E.R. (1984). A test of the Hebb postulate at identified synapses which mediate classical conditioning in Aplysia. J. Neurosci., 4: 1217-1224.

Clark, S.A., Allard, T., Jenkins, W.M., and Merzenich, M.M. (1988). Receptive fields in the body-surface map in adult cortex defined by temporally correlated inputs. Nature 332: 444-445.

Collingridge, G.L. and Bliss, T.V.P. (1987). NMDA receptors - their role in long-term potentiation. Trends Neurosci., 10: 288-293.

Collingridge, G.L., Kehl, S.J., and McLennan, H. (1983). Excitatory amino acid transmission in the Schaffer collateral-commissural pathway of the rat hippocampus. J. Physiol. 334: 33-46.

Devor, M. and Wall, P.D. (1981). Effects of peripheral nerve injury on receptive fields of cells in the cat spinal cord. J. Comp. Neurol. 199: 227-291.

Dostrovsky, J.O., Milar, J., and Wall, P.D. (1976). The immediate shift of afferent drive of dorsal column nucleus cells following deafferentation: a comparison of acute and chronic deafferentation in gracile nucleus and spinal cord. Exp. Neurol. 52: 480-495.

Dykes, R.W. and Lamour, Y. (1988). An electrophysiological laminar analysis of single somatosensory neurons in deafferented rat hindlimb granular cortex subsequent to transection of the sciatic nerve. Brain Res. 449: 1-17.

Dykes, R.W., P. Landry, Metherate, R.S., and Hicks, T.P. (1984). Functional role of GABA in cat primary somatosensory cortex: shaping receptive fields of cortical neurons. J. Neurophysiol. 52: 1066-1093.

Dykes, R.W. and Metherate, R. (1988). Sensory cortical reorganization following peripheral injury. In S. Finger et al.

Brain Injury and Recovery. New York: Plenum Press, 215-234.
Dykes, R.W. and Ruest, A. (1986). What makes a map in somatosensory cortex? In Cerebral Cortex, vol. 5, Sensory-Motor Areas and Aspects of Cortical Connectivity, E.G. Jones and A. Peters, New York: Plenum Press, 1-29.
Eckhorn, R., R. Bauer, Jordan, W., Brosch, M., Kruse, W., Munk, M., and Reitboeck, H.J. (1988). Coherent oscillations: a mechanism of feature linking in the visual cortex? Biol. Cybern. 60: 121-130.
Edelman, G.M. (1978). Group selection and phasic reentrant signalling: A theory of higher brain function. In G.M. Edelman and V.B. Mountcastle, The Mindful Brain. Cambridge, MA: MIT Press, 51-100.
Edelman, G.M. (1981). Group selection as the basis for higher brain function. In F.O. Schmidt, F.G. Worden, G. Adelman, and S.G. Dennis, Organization of the Cerebral Cortex. Cambridge, MA: MIT Press, 535-563.
Edelman, G.M. (1987). Neural Darwinism. Basic Books, New York.
Edelman, G.M. and L.H. Finkel (1984). Neuronal group selection in the cerebral cortex. In G.M. Edelman, W.E. Gall, and W.M. Cowan, Dynamic Aspects of Neocortical Function, New York: John Wiley & Sons, 653-695.
Favorov, O.V., Diamond, M.E., and Whitsel, B.L. (1987). Evidence for a mosaic representation of the body surface in Area 3b of the somatic cortex of cat. Proc. Natl. Acad. Sci. USA 84: 6606-6610.
Finkel, L.H. and Edelman, G.M. (1985). Interaction of synaptic modification rules within populations of neurons. Proc. Natl. Acad. Sci. USA 82: 1291-1295.
Finkel, L.H. and Edelman, G.M. (1987). Population rules for synapses in networks. In Synaptic Function, G.M. Edelman, W.E. Gall, and W.M. Cowan, eds., pp. 711-757, Wiley, New York.
Finkel, L.H. and Edelman, G.M. (In press). The integration of distributed cortical systems by reentry: a computer simulation of interactive, functionally segregated visual areas. J. Neuroscience (In Press)
Fleshman, J.W., Segev, I., and Burke, R.E. (1988). Electrotonic architecture of type-identified alpha-motoneurons in the cat

spinal- cord. J. Neurophysiol. 60: 60-85.

Gray, C.M., and Singer, W. (1987). Stimulus specific neuronal oscillations in the cat visual cortex: a cortical functional unit. Soc. Neurosci. Abstr. 404.3.

Harris, E.W., Ganong, A.H., and Cotman, C.W. (1984). Long term potentiation in hippocampus involves activation of N-methyl-D-aspartate receptors. Brain Res. 323: 132-137.

Hawkins, R.D., Abrams, T.W., Carew, T.J., and Kandel E.R. (1983). A cellular mechanism of classical conditioning in Aplysia - activity-dependent amplification of presynaptic facilitation. Science 219: 400-405.

Hebb, D.O. (1949). The Organization of Behavior. Wiley, New York.

Hicks, T.P. and Dykes, R.W. (1983). Receptive field size for certain neurons in primary somatosensory cortex is determined by GABA- mediated intracortical inhibition. Brain Res. 274: 160-164.

Hubel, D.H. and Wiesel, T.N. (1977). Functional architecture of the macaque monkey visual cortex. Proc. Roy. Soc. Lond. B198: 1-59.

Hyvarinen, J. (1982). The Parietal Cortex of Monkey and Man. New York: Springer-Verlag.

Ito, M., Sakurai, M., and Tongroach, P. (1982). Climbing fiber-induced depression of both mossy fiber responsiveness and glutamate sensitivity of cerebellar Purkinje cells. J. Physiol. (Lond.) 324: 113-134.

Jenkins, W.M., Merzenich, M.M., and Ochs, M.T. (1984). Behaviorally controlled differential use of restricted hand surfaces induce changes in the cortical representation of the hand in area 3b of adult owl monkeys. Soc. Neurosci. Abstr., 10: 665.

Kaas, J.H., Merzenich, M.M., and Killackey, H.P. (1983). The reorganization of somatosensory cortex following peripheral-nerve damage in adult and developing mammals. Annu. Rev. Neurosci. 6: 325-356.

Kaas, J.H., Nelson, R.J., Sur, M., and Merzenich, M.M. (1981). Organization of somatosensory cortex in primates. In F.O. Schmitt et al, The Organization of the Cerebral Cortex. Cambridge, MA: MIT Press, 237-261.

Kelso, S.R., Ganong, A.H., and Brown, T.H. (1986). Hebbian synapses in hippocampus. Proc. Natl. Acad. Sci. USA 83: 5326-5330.

Killackey, H.P. (1988). Static and dynamic aspects of cortical somatotopy: a critical evaluation. J. Cognitive Neurosci. 1: 3-10.

Leyton, A.S.F. and Sherrington, C.S. (1917). Observations on the excitable cortex of the chimpanzee, orangutan, and gorilla. Quart. J. Exp. Physiol., 11: 137-222.

Malinow, R. and Miller, J.P. (1986). Postsynaptic hyperpolarization during conditioning reversibly blocks induction of long-term potentiation. Nature 320: 529-530.

Merzenich, M.M. (1985). Sources of intraspecies and interspecies cortical map variability in mammals: conclusions and hypotheses. In M. Cohen and F. Strumwasser, Comparative Neurobiology: Modes of Communication in the Nervous System. New York: John Wiley & Sons, 138-157.

Merzenich, M.M., Kaas, J.H., Wall, J.T., Nelson, R.J., Sur, M., and Felleman, D.J. (1983). Topographic reorganization of somatosensory cortical areas 3b and 1 in adult monkeys following restricted deafferentation. Neuroscience, 8: 33-55.

Merzenich, M.M, Nelson, R.J., Kaas, J.H., Stryker, M.P., Jenkins, W.M., Zook, J.M., Cynader, M.S., and Schoppmann, A. (1987). Variability in hand surface representation in areas 3b and 1 in adult owl and squirrel monkeys. J. Comp. Neurol. 258: 281-297.

Merzenich, M.M., Recanzone, G., Jenkins, W.M., Allard, T.T., and Nudo, R.J. (1988). Cortical Representational Plasticity. In P. Rakic & W. Singer, Neurobiology of Neocortex. New York: John Wiley & Sons, 41-67.

Mountcastle, V.B. (1978) An organizing principle for cerebral function: The unit module and the distributed system. In G.M. Edelman and V.B. Mountcastle (Eds.), The Mindful Brain. Cambridge, MA: MIT Press, 7-50.

Mountcastle, V.B. (1984). Central nervous mechanisms in mechanoreceptive sensibility. In Handbook of Physiology, Sect. 1: The Nervous System, Vol. 3. Bethesda: American Physiological Society, 789-878.

Nowak, L., Bregestovski, P. , Ascher, P., Herbert, A. and Prochiantz,

A. (1984). Magnesium gates glutamate-activated channels in mouse central neurones. Nature 307: 462-465.

Pearson, J.C., Finkel, L.H., and Edelman, G.M. (1987). Plasticity in the organization of adult cerebral cortical maps: A computer simulation based on neuronal group selection. J. Neuroscience, 7: 4209-4223.

Pons, T.P, Garraghty, P.E., and Mishkin, M. (1988). Lesion-induced plasticity in the second somatosensory cortex of adult macaques. Proc. Natl. Acad. Sci. USA 85: 5279-5281.

Rall, W. (1964). Theoretical significance of dendritic trees for neuronal input-output relations. In R. Reiss, Neural Theory and Modeling. Stanford: Stanford University Press, 73-97.

Rakic, P. (1988). Intrinsic and extrinsic determinants of neocortical parcellations: a radial unit model. In P. Rakic and W. Singer, Neurobiology of Neocortex. New York: John Wiley & Sons, 5-27.

Rasmusson, D. and Turnbull, B.G. (1983). Immediate effects of digit amputation of SI cortex in the raccoon: unmasking of inhibitory fields. Brain Res. 288: 368-370.

Sanes, J.N., Suner, S., Lando, J.F., and Donoghue, J.P. (1988). Rapid reorganization of adult rat motor cortex somatic representation patterns after motor nerve injury. Proc. Natl. Acad. Sci. USA 85: 2003-2007.

Sillito, A.M. (1975). The contribution of inhibitory mechanisms to the receptive field properties of neurons in the striate cortex of the cat. J. Physiol. 250: 305-322.

Stone, G.O. (1986). An analysis of the delta rule and the learning of statistical associations. In D.E. Rumelhart and J.L. McClelland et al, Parallel Distributed Processing: Explorations of the Microstructure of Cognition, Vol. 1. Cambridge, MA: MIT Press, 444-459.

Sur, M., Merzenich, M.M., and Kaas, J.H. (1980). Magnification, receptive field area, and "hypercolumn" size in Areas 3b and 1 of somatosensory cortex in owl monkeys. J. Comp. Neurol. 44: 295-311.

Wall, J.T. (1988). Variable organization in cortical maps of the skin as an indication of the lifelong adaptive capacities of circuits in the mammalian brain. Trends Neurosci., 11: 549-557.

Wall, P.D. (1977). The presence of ineffective synapses and the circumstances which unmask them. Phil. Trans. Roy. Soc. Lond. B278: 361-372.

Wall, P.D. and Egger, M.D. (1971). Formation of new connections in adult rat brains after partial deafferentation. Nature 232: 542-545.

Whitsel, B.L., Favorov, O., Tommerdahl, M., Diamond, M., Juliano, S., and Kelly, D. (1988). Dynamic processes govern the somatosensory cortical response to natural stimulation. In J.S. Lund, Organization of Sensory Processing. New York: Oxford University Press, 84-116.

Widrow, B. and Hoff, M.E. (1960). Adaptive switching circuits. Inst. Radio Engineers, Western Electronic Show and Convention, Convention Record, Part 4, pp. 96-104.

Wigstrom, H. and Gustafsson, B. (1983). Heterosynaptic modulation of homosynaptic long-lasting potentiation in the hippocampal slice. Acta Physiol. Scand., 119: 455-458.

Woolsey, T.A. and Van der Loos, H. (1970). The structural organization of layer IV in the somatosensory region (SI) of mouse cerebral cortex. The description of a cortical field composed of discrete cytoarchitectonic units. Brain Res. 17: 205-242.

CHAPTER 8

Spatial Representation of the Body

C. R. Olson and S. J. Hanson

If I stand holding my pipe in my closed hand, the position of my hand is not determined discursively by the angle which it makes with my forearm, and my forearm with my upper arm, and my upper arm with my trunk, and my trunk with the ground. I know indubitably where my pipe is, and thereby I know indubitably where my hand and my body are, as primitive man in the desert is always able to take his bearings immediately without having to cast his mind back, and add up distances covered and deviations made since setting off.

M. Merleau-Ponty, Phenomenology of Perception.

Introduction

Postural awareness - knowledge of the spatial arrangement of the parts of the body - is so transparent that one hardly ever gives it thought. Nevertheless, the underlying unconscious computations must be enormously complicated. Consider the reduced and simplified postural problem illustrated in figure 1. Given the angles at the wrist, elbow, shoulder and neck joints of the stick figure, one can compute the position of its hand in a head-centered coordinate frame, but to do so precisely involves utilizing a set of long and complex trigonometric formulae. We are able, without thought, to solve multiple problems of this sort virtually instantaneously, computing the spatial position of any part of the body relative to any other part. We do so by using sensory signals from the muscles, joints and skin that bear only a complicated and indirect relation to the angles at the joints. We do so, almost certainly, without applying propositional rules or explicit trigonometric formulae. What is the trick that our brains use? Behavioral and neural studies carried out over several decades have

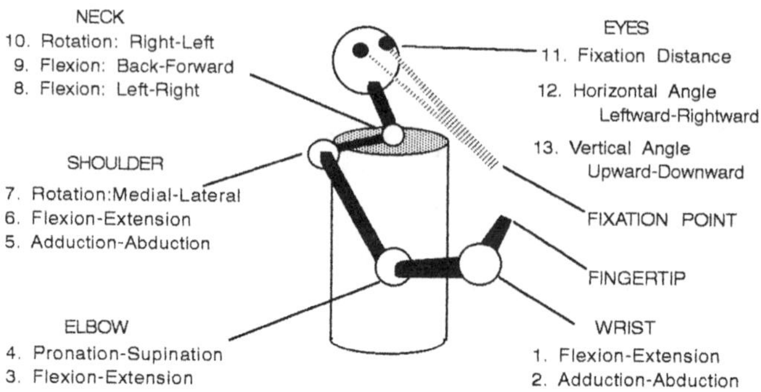

Range of Permitted Angles:

1.	+Pi/2	-Pi/2	8.	+Pi/2	-Pi/4
2.	+Pi/6	-Pi/4	9.	+Pi/4	-Pi/4
3.	+Pi/2	0	10.	+Pi/2	-Pi/2
4.	+Pi/2	-Pi/2	11.	0	8
5.	0	-Pi/2	12.	+Pi/4	-Pi/4
6.	+Pi/2	-Pi/2	13.	+Pi/4	-Pi/4
7.	+Pi/2	0			

Figure 1: A simple articulated system approximating the human body. The arm has seven degrees of rotational freedom and the neck three. There are two degrees of freedom of conjugate gaze and one of vergence, by means of which fixation distance is controlled. The relative lengths of the segments are: Neck: 1; Shoulder: 2; Upper arm: 2; Forearm: 2; Hand: 1.

resulted in a mass of factual detail related to brain mechanisms of postural awareness, but have provided no general answer to this question. We will summarize the major experimental findings because they form a constraining framework for theory. Then we will develop a theory of postural information processing based on modeling studies showing that postural problems are soluble by modifiable networks composed of neuron-like elements.

Human Performance

Human subjects have been tested on many tasks that require postural perception. Some tasks draw on rudimentary postural abilities without necessarily requiring spatial analysis, while others cannot be performed without explicit analysis of bodily spatial coordinates. We summarize the results of these studies with two aims: first, to establish a baseline against which to compare the performance of artificial systems, and, second, to consider whether patterns of human behavior afford any clue to the nature of the brain mechanisms underlying postural analysis.

Kinesthesia A simple and very widely used test of postural sensation is simply to require the subject to identify the direction of a movement imposed on a joint. This test is spatial only in a trivial sense, because the relevant information comes entirely from a single joint, but it establishes levels of sensitivity that must ultimately constrain more complex forms of spatial analysis. At the interphalangeal joint of the middle finger, normal subjects perform above chance for very small and very slow movements (2° at 2°/sec). Performance improves as either magnitude or speed is increased and is virtually perfect for 5° at 4°/sec (Gandevia and McCloskey, 1976). Despite its wide use for testing postural sensation, this procedure actually taps an ability (appreciation of movement) that is distinguishable on several grounds from postural awareness narrowly defined (appreciation of static position). The following observations suggest that static and dynamic postural awareness depend on dissociable mechanisms. (1) Movements so slow as to be undetectable in themselves eventually result in detectable changes of perceived angle (Horch et al., 1975). (2) The ability to detect a change of angle, as opposed to the ability to detect movement, depends critically on muscle receptors (Clark et al., 1985). (3) The illusions of movement and position produced by muscle vibration are independent, so that subjects may simultaneously perceive the arm to be moving and to be in a fixed position (Sittig et al., 1985). (4) The two sorts of illusion are differentially affected by factors such as the load placed on the muscle and the frequency of vibration (McCloskey, 1973). To capture the distinction between the two forms of perception, it has been proposed that the term "kinesthesia", traditionally applied to all postural sensations, be restricted to sensations of movement (McCloskey, 1973; Roland and Ladegaard-Pedersen, 1977).

Matching of angles A simple test of postural awareness proper (appreciation of static position) is to have the subject match with the joint of one limb the angle imposed on the corresponding joint of the other limb. This task tests spatial ability only in a very limited sense because the information derives from a single joint and because perfect performance could be achieved without reliance on spatial analysis simply by matching unprocessed sensory signals from the two limbs. The mean standard deviation of matches produced on successive trials is six degrees or less at the knee (Horch et al., 1975), the interphalangeal joint of the index finger (Rymer and D'Almeida, 1980), the elbow (Worringham et al., 1987), and the shoulder (Paillard and Brouchon, 1968). Roughly comparable results are obtained when subjects free to make shoulder, elbow and wrist movements are required to bring the fingertips into apposition on opposite sides of a narrow barrier placed in the midsagittal plane, a task solved by assuming symmetric arm postures (Slinger and Horsley, 1906; Crowe et al., 1988; Paillard and Brouchon, 1974). In addition to the variable error, there is often a constant error, of uncertain origin, whereby the active matching limb consistently overshoots or undershoots the passive matched limb by several degrees (Slinger and Horsley, 1906; Paillard and Brouchon, 1968, 1974; Horch et al., 1975; Soechting, 1982; Worringham and Stelmach, 1985; Worringham et al., 1987).

Hand-hand matches Some postural tasks cannot be solved simply by matching the angles of corresponding joints. For example, a blindfolded subject may be asked to bring the tips of the index fingers together by moving one hand while the other is held at a fixed location. If the fixed hand is outside the midsagittal plane, then correct performance requires placing the arms in asymmetric postures. Slinger and Horsley (1906) found that the mean error in this task is around 3 cm (roughly one twentieth of the length of the arm). The general pattern of constant errors is consistent individuals: the hand crossing the midline tends to extend too far beyond the midline and to be placed too close to the body. The details of the pattern of constant errors vary among individuals and even between arms of the same individual. The level of accuracy observed in the early study has been confirmed recently but without any systematic analysis of the directions of constant errors (Crowe et al., 1987). It would be of interest to know whether performance on tasks requiring explicit spatial analysis, as these do, is predictable from performance on tasks of simple joint sensibility or is qualitatively different. The requisite comparisons have not been carried out.

Stereognosis Subjects may be required to judge the size of an object by palpating it. This is essentially a test of spatial appreciation of hand posture. The level of sensitivity evinced in such tasks is high.

Performance is above chance when spheres several tens of millimeters in diameter differ by only around a millimeter (Roland, 1976). This is of interest because it suggests that there is greater refinement in perceiving the relative spatial arrangement of the parts of one hand than in perceiving the arrangement of the hands relative to each other. This finding is incompatible with the simple form of postural scheme in which points on the body are assigned coordinates in a single reference frame and errors arise from noise that enters each computation independently.

Matching of inclination A subject may be asked to hold the forearms so that they are parallel (Soechting, 1982) or so that they assume the same inclination relative to gravity (Worringham et al., 1987). If the upper arms are fixed at nonparallel orientations, then to perform the match requires more than simply equating the angles of corresponding joints. The mean standard deviation of successive attempts to match a given orientation is in the range 5-7 degrees. This is commensurate with the variable error obtained in simple joint-angle matching. Subjects may actually be less variable when matching inclinations relative to gravity than when matching joint angles (Soechting, 1982). Subjects are more accurate at interlimb matching of some aspects of orientation (e.g., angular elevation) than of others (e.g., angle relative to a parasagittal plane) and it has been suggested that the angles permitting easy matching correspond to the axes of the coordinate system in which limb angles are expressed by the brain (Soechting and Ross, 1984). The proposal that there is a single set of orthogonal axes in terms of which limb inclinations are expressed in the brain is provocative but arbitrary.

Eye-hand matches Eye-hand coordination has been analyzed far more extensively than any other expression of postural awareness (Howard and Templeton, 1966; Howard, 1982). In the task employed most commonly, subjects viewing a target on a horizontal opaque table are required to reach beneath the table and bring the unseen hand to a point directly under the target. Alternatively, subjects may be required to place a mark on the table directly above the passively positioned hand (Gross et al., 1974). For a given subject, the positions attained on successive reaches for the same target are highly consistent. The standard deviation of the mediolateral angle is only a few degrees (Bock, 1986). Certain constant errors of considerable magnitude do occur. The mediolateral constant error, having a magnitude up to 10% of the length of the arm, is in a direction that varies among individuals and depends as well on position (Loemker, 1930; Foley and Held, 1972; Bock, 1986). The constant error of distance is nearly always such that the fingertip is placed farther from the body than the fixation point (Foley and Held, 1972; Gross et al., 1974). The average magnitude of the constant error of distance has been estimated variously at around 10% of the length of the arm (Gross et al., 1974) and at nearly 25% (Foley and Held, 1972), but the large figure obtained by Foley and Held may reflect the impoverishment of visual distance cues in their procedure. The fact that humans make constant errors is potentially of great interest because the deviations from perfect performance arising from different algorithms are distinctive. However, no systematic effort has been made as yet to relate human constant-error patterns to predictions based on particular computational strategies.

Postural awareness as an automatic process Is posture computed automatically or only when spatial information about the body is behaviorally relevant? This question is of theoretical importance and also is relevant to the choice of behavioral conditions under which to study single neurons that carry out spatial analysis. It has been addressed by studies on the "spatial" stimulus-response compatibility phenomenon. The essence of S-R compatibility is that sensory stimuli elicit with particular speed certain innately matched motor responses. There are anatomical S-R pairings: for example, human subjects respond more quickly to a light in the right visual hemifield when using the right hand than when using the left hand (Nicoletti et al., 1982). There are also spatial S-R pairings: for example, the response of either hand to a visual stimulus in the right hemifield is faster if the responding hand is currently placed to the right of the midline of the body (Nicoletti et al., 1982). Even when the index finger and middle finger of the right hand are used to deliver responses in two-choice paradigm, the index-finger response to a right-hemifield light is more rapid with the hand supine (placing the index finger to the right) than with the hand prone (placing the index finger to the left) (Heister et al., 1986; 1987). Comparable but weaker spatial S-R effects depend on the vertical position of the hand (Nicoletti and Umilta, 1985). Spatial compatibility effects occur even when vision is excluded (Attneave and Benson, 1969; Wallace, 1972). These results demonstrate that when we use a limb we are influenced inexorably by its spatial location. They fall short of showing that spatial sign is computed automatically because performing a motor act is a deliberate process that in all likelihood involves directing attention to the limb in its current position. The question remains: does tactile stimulation of an unattended and unused limb produce spatially specific effects? This question could be answered by measuring the effect on visual reaction time of a distracting cutaneous stimulus delivered to the hand when it is not used for responding. The approach of using a task-irrelevant distracting stimulus has been employed previously in studies of auditory-visual interaction (Simon and Craft, 1970).

Summary of human performance Human levels of accuracy have been established in a variety of tasks that require rudimentary postural awareness or explicit spatial analysis. The error data provide a benchmark against which to assess the performance of artificial systems and may ultimately throw light on the nature of the computational strategies implemented in the brain. However it will probably not be possible to make sense of systematic tendencies in human error data until the error patterns associated with competing models of postural analysis are clearly formulated. Despite the heavy emphasis that has been placed on studies of accuracy, alternative approaches (for example, reaction-time studies of spatial stimulus-response compatibility) may be better suited to bridging the conceptual gap between behavioral measurements and underlying brain mechanisms.

Peripheral Sources of Postural Information

Information from numerous sources might be used by the brain in constructing a spatial image of the body. Potential sources include the motor system, receptors in the muscles and tendons, joint receptors and skin receptors. The evidence for and against each source has been weighed over many times (Matthews, 1977;1982; McCloskey, 1978; Clark and Horch, 1986). We will summarize the properties of each potential input pathway with emphasis on features that pose a particular challenge in the context of spatial analysis and will conclude by considering how pieces of information derived from different systems might be welded into a unitary postural message.

The motor signal Muscles are composed of narrow multinucleated contractile fibers. A typical muscle, the soleus of the cat, contains altogether around 25,000 fibers (Burke, 1981). The muscle fibers are under the control of alpha motor neurons whose cell bodies are located in the ventral horn of the spinal cord and the cranial motor nuclei. The feline soleus is innervated by axons from approximately 140 alpha motor neurons (Burke, 1981). Each of the axons innervating a muscle terminates on a unique set of motor fibers; together with the axon, these form a so-called motor unit (Burke, 1981; Henneman and Mendell, 1981). The number of fibers in a motor unit ranges from around ten to many hundred (Burke, 1981). The amount of force produced by a motor unit is proportional to the number of fibers it contains. Small motor units are much more common than large ones. The global state of contraction of the muscle is determined both by how many motor units are active and by the frequency of firing of the active motor units. During graded increases of force produced under voluntary control, motor units begin to fire in ascending order by size, and each unit, once recruited, contributes to increases of force above its threshold by firing at a higher rate (Milner-Brown et al., 1974). At a given level of activation, the behavior of the muscle can be described in terms of a length-tension curve. There is a length at which it exerts no force. When stretched beyond this length, it exerts a resistive force which is a function of the degree of stretch. The higher the level of activation, the shorter the default length and the stronger the force opposed to small stretches (Rack and Westbury, 1969; Houk and Rymer, 1981; Partridge and Benton, 1981). The details of the length-tension curve vary complexly with activation-level in a way which can be adequately modeled only by numeric solution of a set of differential state equations (Hatze, 1977; 1978). For the brain to use the activity of motor neurons as a basis for postural sensation would be difficult both because the motor command establishes a length-tension curve rather than a unique muscle length and because, with tension precisely known, computing the length of the muscle still requires taking into account the complexly nonlinear dependence of length on firing level.

Muscle Spindles The spindle organs are intramuscular stretch receptors (Matthews, 1981; Prochazka, 1981; Burgess et al., 1982, Hulliger, 1984). The feline soleus possesses approximately 60 spindle organs (Barker, 1974), a number roughly commensurate with the number of its motor units. Each spindle consists of 4-12 "intrafusal" muscle fibers organized into a narrow bundle with a length of approximately a centimeter (Barker, 1974). The spindle is attached to the surrounding intramuscular connective tissue at its poles and undergoes passive stretching as the surrounding muscle lengthens. A spindle may be differentially sensitive to the contractions of different motor units (Binder et al., 1976), presumably favoring those with fibers in its immediate neighborhood. Sensory axons, which contact the intrafusal fibers within a capsule surrounding their equatorial region, are excited by stretching. A typical spindle gives rise to one large (primary) sensory axon and a few small (secondary) sensory axons (Barker, 1974). Primary fibers are especially sensitive to dynamic stretch whereas secondary fibers exhibit marked sensitivity to maintained length (Boyd, 1980). In fibers sensitive to static stretch, the rate of firing is a nearly linear function of length (Jami and Petit, 1978). The assumption of precise linearity is made in a recent model of spindle sensory responsiveness (Hasan, 1983). The static signal slowly declines during maintenace of a fixed posture, undergoing a change equivalent to around 5° of joint angle over 15 seconds (Burgess et al., 1982; Wei et al., 1986) with the result that central compensation would be required for accurate perception of a maintained posture.

The intrafusal fibers of the spindle organ are contractile at their poles and their level of contraction is controlled by activity in 8 to 15 gamma motor axons (Barker, 1974; Murthy, 1978). The importance of fusimotor innervation is reflected by the very high ratio of gamma axons to intrafusal muscle fibers (1:1) as compared to the very low ratio (1:100) of alpha motor axons to extrafusal muscle fibers (Burke, 1981b). When fusimotor fibers are active, the spindle polar region is stretched and the sensory fibers consequently are excited. A more subtle effect results from changes in the differential activity of static and dynamic fusimotor fibers, as a result of which spindle sensory fibers undergo shifts of relative sensitivity to active as opposed to maintained stretch (Boyd, 1980). With the muscle at fixed length, the level of activity in spindle sensory fibers is a monotonic saturating function of activity in fusimotor axons (Lewis and Proske, 1972; Cheney and Preston, 1976; Jami and petit, 1978). With muscle-length and fusimotor firing-level permitted to vary independently, their effects are approximately additive but nonlinear facilitation sometimes is observed (Jami and Petit, 1978). Fusimotor axons follow comparatively closely the rule of coactivation, whereby their firing parallels that of skeletomotor axons innervating the muscle proper, and they appear even to undergo recruitment in an order comparable to that of skeletomotor axons (Burke, et al, 1978; Burke, 1980). As a result, with the muscle held at fixed length, the level of firing of spindle sensory fibers closely tracks the level of force being produced by the muscle as a whole (Vallbo 1970b; 1974; Vallbo et al., 1979; Burke, 1980). With force held constant and length varying under voluntary control, the firing of an individual spindle fiber may either increase or decrease with length (Burke et al., 1978; Vallbo et al., 1979;1981), the sign of the effect being contingent in part on the speed of the movement (Prochazka, 1981) and in part on the fiber's individual balance of sensitivity to passive and fusimotor stretching (Burke, 1981a). Only in the relaxed muscle, with motor fibers nearly silent, does the firing of spindle afferents consistently increase as a function of muscle length.

Golgi tendon organs. The tendon organs are force-sensitive receptors located at the musculotendinous junctions of muscle (Barker, 1974; Proske, 1981). In the soleus of the cat, there are around 90 tendon organs, each interposed in series between the tendon and some 25 muscle fibers. Collectively, the muscle fibers that exert force across the tendon organs constitute around 1% of all fibers in the muscle and derive from around 10% of all motor units (Stephens et al., 1975; Burke, 1981b). Tendon organs are excited when they are stretched by forces acting across them. They are exquisitely sensitive to active contraction of the muscle fibers with which they are in series (Houk and Henneman, 1967; Stephens et al., 1975), and also respond to forces produced by passive stretching of the entire relaxed muscle (Wei et al., 1986). Their sensitivity to force exerted when the relaxed muscle passively resists stretch may not be as great as their sensitivity to the force produced by active muscular contraction because parallel elements in the muscle (connective tissue and noncontracting skeletomotor fibers) share the resistive load (Proske, 1981) but direct tests of this notion in human subjects have proven difficult for technical reasons (Vallbo, 1970a). The firing rate of tendon-organ afferents has been shown to vary as a function of force in experiments where force is controlled by varying the rate of electrical stimulation of the whole motor nerve; the function is monotonic and slowly saturating and can be fitted reasonably well by a power function with an exponent slightly less than one (Proske, 1981). However, when force is produced by voluntary contraction of the muscle, the shape of the curve must depend on the order of recruitment of the motor units in series with the particular tendon organ. The shape of the curve under these circumstances is not known. The tendon organs, although they sense force, may contribute to length analysis, as discussed below.

Joint receptors Deformation-sensitive nerve-endings are insinuated into the fibrous capsule and ligaments surrounding each joint (Skoglund, 1973; Grigg and Hoffman, 1982). These receptors seem to be ideally positioned to monitor the angles at the joints and were once thought to be the major source of postural sensation. Now it is realized that even those joint receptors anatomically best suited for postural analysis (the Ruffini afferents) carry only a very crude representation of joint angle (Proske et al., 1988). Most increase their firing rate only as the joint is moved into an extreme or even unphysiological position; most are sensitive to rotation about more than one axis; roughly half are excited when the joint moves to either of the opposed extreme positions about a given axis (Clark and Burgess, 1975; Grigg, 1975; Grigg and Greenspan, 1977; Burke et al., 1988). The difficulty that would attach to basing postural sensations on the firing of these receptors is heightened by their sensitivity to changes in torque imposed at constant position (Grigg, 1975; Grigg and Greenspan, 1977).

Skin receptors It has been suggested that cutaneous mechanoreceptors play a role in the perception of joint angle (Moberg, 1983). This is reasonable because the skin overlying the joints is distorted as a result of rotations. However, in the human hand, most cutaneous receptors are sensitive to changes in the angles of underlying joints only at extreme positions (Burke et al., 1988). Thus, like joint receptors, cutaneous receptors provide to the central nervous system information about the angles of the joints that is both incomplete and inelegantly coded.

Relative importance of various sources of postural information Experiments aimed at determining which sensory receptors give rise to postural sensations have been carried out in human subjects. The approach has been to silence or stimulate selectively receptors of a particular type. The results indicate that muscle receptors play a central role in postural awareness and that receptors of several other kinds make modest contributions (Matthews, 1977;1982; McCloskey, 1978; Clark and Horch, 1986). Stimulating muscle receptors by applying vibration to the belly of a muscle leads to robust illusions of position and motion at the joint across which the muscle acts (Goodwin et al., 1972; Matthews, 1977). The illusions unquestionably are due to stimulation of receptors in the muscle because they occur even when stimulation is applied to a part of the muscle distant from the joint and the illusory displacement and motion are invariably in a direction associated with lengthening of the muscle. The mediating receptors might be primary spindle afferents, secondary spindle afferents or Golgi tendon organs, as receptors in all three classes respond to vibration of the muscle (Burke et al., 1976a,b), but it is generally assumed that muscle spindles play a dominant role (Matthews, 1977; Clark and Horch, 1986). Illusions of movement and position are affected differently by various experimental conditions in a manner suggesting that they depend respectively on primary and secondary spindle afferents (McCloskey, 1973). The prepotent role of muscle receptors has been demonstrated by experiments involving silencing as well as stimulation. On one hand, if muscle receptors alone are active, following anesthesia of the skin and joints (Gandevia and McCloskey, 1976; Roland and Ladegaard-Pedersen, 1977) or surgery (Cross and McCloskey, 1973), then postural sensations persist although they may be blunted. On the other hand, if muscle receptors alone are blocked, by anesthetizing the muscle or mechanically disconnecting it from the joint, then sensations of posture are grossly reduced (Gandevia and McCloskey, 1976) and the impairment is especially pronounced for the sense of static posture as opposed to detection of movement (Clark et al., 1985).

In contrast to sensory signals, corollary discharges from the motor system appear to play only a minor role in postural awareness (McCloskey, 1981; Matthews, 1982). The lack of importance of motor signals for assessing the state of the muscle is indicated by the following experimental results. Subjects with the muscles of one hand temporarily weakened, so that, to produce a criterion force, they must generate a more intense motor signal, may perceive their effort to be greater but do not perceive the resulting force to have changed (Roland and Ladegaard-Pedersen, 1977) and subjects with all sensory feedback from a finger acutely blocked fail to perceive it as moving even when they successfully execute a movement (Goodwin et al., 1972). Against these findings from normal subjects, one must place, tentatively, the fact that patients with chronic sensory neuropathy are able to produce complex movements with an arm that is permanently anesthetized (Rothwell et al., 1982). It seems unlikely that they could do so without some accompanying postural sensations.

The necessity for cross-source integration It is a curious fact that muscle spindles, although clearly implicated in postural sensation, give rise to a highly ambiguous signal which is roughly the sum of a muscle-length component (useful for spatial analysis) and a component related to motor drive and force (useless for spatial analysis). The motor-command component must somehow be removed if the spindle signal is to be used as a basis for accurate postural sensation. The removal has been thought to occur within the brain as a result of subtractive interaction between the sensory signal from the spindle and a corollary discharge arising from the motor system (McCloskey, 1978;1981; Matthews, 1982; Clark and Horch, 1986). However this explanation is in conflict with the observation that motor signals in themselves do not produce postural sensations. There are several alternative purely sensory mechanisms by which the spindle signal could be transformed into an accurate indicator of muscle length. These are worth consideration although they do not figure in traditional accounts. (1) At fixed length, any increase in the strength of the motor command is accompanied by an increase in force exerted by the muscle. The signal arising from the force-sensitive Golgi tendon organs might thus be used in place of corollary discharge for cancellation of the motor-command component of spindle discharge. This scheme is appealingly simple. Weighing against it are the apparently low sensitivity of tendon organs and the nonlinear relation that pertains between force and fusimotor activation even under isometric conditions. (2) A signal reflecting solely the rate of change of the muscle length might be created by subtracting the firing of secondary (static) spindle fibers from that of primary (dynamic) spindle fibers. The resulting signal might then be integrated to give length. This scheme, suggested by Matthews (1977), is clever but seems better suited to digital than to analogue computation and would not necessarily eliminate the signal component related to fusimotor drive. (3) Because there is variation among spindles in relative sensitivity to muscle stretch and fusimotor outflow, the firing of a population of spindle fibers in fact carries unambiguous information about muscle length. To extract this information would require bringing together signals from the entire population of spindles via synapses of sign and magnitude such that the motor components cancel but the length components do not. The appropriate adjustment of synaptic weights might be difficult to arrive at by hard-wiring but could result from a local learning mechanism if an external source of feedback about muscle length or posture were utilized.

The superadded complexity of musculoskeletal structure Even if a sensory signal linearly related to muscle length were available, the relation of this signal to postural coordinates would be highly nonlinear. This becomes clear on cursory consideration of how the positions of head and arm are controlled. Movements of the head are accomplished by rotation at the joints of the seven cervical vertebrae (Richmond and Vidal, 1988). Lateral flexion and forward and backward flexion depend on movements distributed across all cervical joints. The movement commonly referred to as rotation (rotation of the head about a vertical axis) depends primarily although not exclusively on the specialized joint between the two cervical vertebrae closest to the skull (the atlas and axis). Within the mass of muscular tissue that controls the posture of the head, some twenty individual pairs of muscles commonly are distinguished (Warfel, 1985). The actions of most individual muscles are not confined either to a single joint or to a cardinal axis of rotation. For example, the splenius capitis, which has its origin in the thoracic vertebrae and its insertion in the skull, acts across all the intervening joints to produce backward flexion, lateral flexion and rotation. Some muscles, for example, the longus colli, are unitary in name only, having multiple points of origin and insertion along an extended series of vertebrae. The complexity of the arrangement of the neck muscles is indicated by studies of spindle and tendon-organ afferents (Richmond and Bakker, 1982) indicating that the preferred movement directions of individual receptors vary continuously over a wide range (Chan et al., 1987).

Movements of the arm are accomplished by rotation about a succession of joints running in sequence from the sternum (breastbone) to the clavicle (collarbone) to the scapula (shoulder blade) to the humerus (arm) to the ulna and radius (forearm) and on to the carpals, metacarpals and phalanges of the wrist and hand. The shoulder blade is capable of considerable translation and rotation as a result of movements about the scapula-clavicle and clavicle-sternum joints. Motions of the shoulder blade may occur independently (as in shrugging) but are also an important synergistic component of most movements of the arm. At the shoulder-joint proper (between scapula and humerus) a ball-and-socket arrangement permits rotation with three degrees of freedom (by convention, the cardinal rotations are defined as: abduction vs. adduction; forward flexion vs. backward flexion; medial rotation vs. lateral rotation). At the elbow, the ulna rotates about the humerus on a hinged joint with a single degree of freedom (flexion vs. extension). A second degree of forearm rotational freedom (pronation vs. supination) results from rolling movements of the radius about the ulna. Movements at the wrist (flexion vs. extension and abduction vs. adduction) occur primarily between the radius and the carpals. Control of the shoulder and arm is the function of some forty muscles (Warfel, 1974). These generally are better segregated and have more limited and precise actions than muscles of the neck, but their arrangement is by no means simple. Several muscles exercise significant actions across more than one joint. For example, most muscles acting on the scapula originate directly from the thoracic skeleton and, acting across two joints, produce complex combinations of scapular translation and rotation. Furthermore, during large changes in the angle at a joint, the mechanical actions of a muscle may change significantly. Thus, as the elbow flexes, the biceps brachii contributes relatively more to flexion and relatively less to supination (Gielen and Zuylen, 1986).

As a result of complicating factors such as these, the lengths of individual muscle spindles must bear a very indirect relation to the angles of the joints. It should be possible, given the anatomy of the arm, to derive a function crudely establishing the length of each muscle in relation to posture, but this formidable task has not to our knowledge been undertaken systematically. The actions of individual muscles have been characterizd by electromyography primarily in relation to the question: during what movements or postures is the muscle active (Basmajian, 1967).

Summary of peripheral sources Signals from a variety of peripheral sources are used for constructing a central spatial representation of the body. Classic accounts of how the joint angles are computed from the medley of peripheral signals have placed emphasis on simple operations - subtraction; integration - by which signals from various sources could be transformed and combined. However, the assumption that an approach based on simple formal calculations could be made to work seems optimistic in light of the properties of receptors and the complex structure of the musculoskeletal system in which they are placed. Golgi tendon organs and muscle spindles carry signals that are related almost linearly to the force and length of the muscle but that bear a nonlinear relation to torque and angle at the joint because of complex musculoskeletal anatomy. Joint and skin receptors and the motor system carry signals that are highly nonlinear and even nonmonotonic functions of joint angle. Networks able to learn arbitrary mappings of inputs onto outputs may represent the best approach to processing information presented in this form.

Cortical Mechanisms

Signals originating from deep receptors are relayed to the cerebral cortex via a series of subcortical stations spanning the spinal cord, medulla and thalamus (see Allard, Chapter 6). Relatively little is known about the integrative as opposed to relay functions of these structures. Accordingly, our account of postural neurophysiology will begin at the level of the cerebral cortex, which is known to play a critical role in postural awareness. Questions to which we will devote particular attention include the following: are there neurons that process information from deep receptors specifically; are there neurons that carry pure muscle-length signals; and are there neurons that encode the spatial coordinates of the limb rather than the lengths of muscles?

Cortical somatosensory areas The somatosensory cortex of macaque monkeys occupies a large and continuous zone spanning the postcentral gyrus, the superior and inferior parietal lobules, and opercular and insular cortex within the sylvian fissure. At least eleven areas have been distinguished within this territory: 3a, 3b, 1, 2, SII, 5, 7b, Ri, PA, Ig and the supplementary sensory area (SSA) (Jones and Porter, 1980; Robinson and Burton, 1980; Murray and Coulter, 1981; Kaas, 1983; Mountcastle, 1984). These areas have been demarcated variously by analyzing cytoarchitecture, by identifying independent maps of the body, and by noting distinctive patterns of neuronal sensory responsiveness. The somatosensory areas can be divided into two broad categories: low-order areas (3a, 3b, 1 and 2), which are innervated by the ventrobasal nucleus, contain comparatively precise maps of the body and correspond collectively to Area SI as defined in classic studies; and high-order areas (SII, 5, 7b, Ri, PA, Ig and SSA), which are innervated by thalamic nuclei outside the ventrobasal complex and contain comparatively crude maps of the body. The high-order areas are thought to constitute a second processing stage for information that has passed through the low-order areas (Jones and Burton, 1976; Bowker and Coulter, 1980). At each level of the hierarchy, areas processing postural information may be distinguished from those that process primarily cutaneous information.

Postural neurons in Area SI There is overwhelming evidence that two divisions of Area SI - 3a and 2 - receive most of their sensory input from "deep" receptors, in contrast to divisions 3b and 2, which are dominated by cutaneous input (Dreyer et al., 1975; Hyvarinen and Poranen, 1978; Kaas et al., 1979; Jones and Porter, 1980; Soso and Fetz, 1980; Tanji and Wise, 1981; McKenna et al., 1982; Kaas, 1983; Pons et al, 1985). The deep input to Area 2 apparently derives from joints and muscles in roughly equal numbers of neurons, as determined qualitatively by monitoring neural activity while palpating muscle bodies and joint capsules (Hyvarinen and Poranen, 1978; Soso and Fetz, 1980). Area 3a receives a strong projection from muscle-spindle afferents, as indicated by the occurrence of large surface potentials following electrical stimulation of muscle sensory nerves at a current level so low as to excite only the largest (primary spindle) afferents (Philips et al., 1971; Schwarz et al., 1973). Whether significant inputs to Area 3a derive from the joints is not known. Golgi tendon organs may be an additional source of deep input to areas 3a and 2 but this has not been tested.

Regardless of the identity of the mediating receptors, it is certain that neurons in areas 3a and 2 are sensitive to changes in the angles of the joints, to angular force at the joints and to static joint position. Neuronal kinaesthetic responsiveness is apparent in monkeys moving spontaneously or subjected to passive movement by the experimenter (Hyvarinen and Poranen, 1978; Costanzo and Gardner, 1981; Gardner and Costanzo, 1981) but studies of animals trained to produce repeated standard limb movements have been especially enlightening (Yumiya et al., 1974; Tanji, 1975; Jennings et al., 1983). They have indicated that areas 3a and 2 are generally similar (Jennings et al., 1983) and that the signals carried by neurons in them are not radically different from muscle-spindle signals. Like spindles, the cortical neurons exhibit direction selectivity for stretches and loads applied to the passive muscle (Yumiya et al., 1983; Tanji, 1975; Jennings et al., 1983). Like spindles, they exhibit both phasic (during movement) and tonic (during maintained position) activation, with the relative strength of dynamic and static effects varying from cell to cell (Tanji, 1975). Like spindles, they differ in their susceptibility to shortening of the muscle when it results from the subject's voluntary contraction, with some neurons undergoing an decrease of firing and others (presumably with stronger fusimotor drive) undergoing an increase (Yumiya et al., 1974; Tanji, 1975). Like spindles, nearly all neurons are sensitive to the force produced by the muscle as well as to its length (Tanji, 1975; Jennings et al., 1983).

The most striking feature differentiating cortical neurons from peripheral receptors is the occurrence of sensory convergence, either across body loci or across modalities. Approximately a fourth of neurons sensitive to deep stimulation respond to rotation about more than one joint (Soso and Fetz, 1980; Costanzo and Gardner, 1981; Gardner and Costanzo, 1981). This might reflect input from a single biarticular muscle when the joints are adjacent on the same limb, but not when the joints are on different limbs (Soso and Fetz, 1980). The degree to which deep and cutaneous inputs converge on single neurons is not clear. It is traditional to classify neurons of Area SI into mutually exclusive deep and cutaneous categories (Dreyer et al., 1975; Soso and Fetz, 1980; Tanji and Wise, 1981; McKenna et al., 1982) but this may be no more than a conservative response to the extreme difficulty of establishing whether or not a neuron responsive to skin stimulation is independently responsive to stimulation of underlying structures. Hyvarinen and Poranen (1978) report deep-cutaneous convergence in a substantial minority of Area SI neurons.

Postural neurons in Area 5 The high-order areas (SII, 5, 7b, Ri, PA and Ig) differ from SI in that neurons responsive to deep and to cutaneous stimulation tend to be intermingled in each area. However, in areas SII, lateral 7b, Ri, PA and Ig, as characterized in an exhaustive study by Robinson and Burton (1980), neurons sensitive to cutaneous stimulation outnumber those sensitive to deep stimulation by a large factor (around 4:1). In contrast, in Area 5, as described by Sakata et al. (1973) and Seal et al. (1982), and in the medial part of Area 7b as characterized by Leinonen et al. (1979), neurons responsive to deep input are markedly preponderant. Thus areas 5 and 7b are particularly plausible candidates for sites of high-order postural information processing. Of the two areas, Area 5 has been much more heavily studied. The properties of neurons in Area 5 may be summarized as follows.

(1) Sources of sensory input. The distinctive trait of Area 5 is its domination by sensory input from deep receptors sensitive to the angles of the joints. Joint-angle-sensitive neurons form a majority of cells in Area 5 whereas neurons responsive to cutaneous stimulation are encountered only rarely (Duffy and Burchfiel, 1971; Sakata et al., 1973; Mountcastle et al., 1975; Seal et al., 1982; Kalaska et al., 1983). Some neurons in Area 5 appear to differ from the majority of neurons in area SI in exhibiting deep-cutaneous convergence, in possessing large bilateral cutaneous receptive fields, and in responding to stimulation of joints on several limbs, but estimates of the percentage of such cells have ranged from very high (Duffy and Burchfiel, 1971) to quite low (Mountcastle et al., 1975). Visual or visual-somatosensory neurons constitute a few percent of the population in Area 5 (Seal et al., 1982).

(2) Motor signals. From 10% to 30% of task-related neurons in Area 5 begin to fire well before the time (-80 msec: Fromm and Evarts, 1982; Burbaud et al., 1985) at which the earliest signs of EMG activity preceding the movement are thought to occur (Seal et al, 1982; Kalaska et al., 1983; Chapman et al., 1984; Burbaud et al., 1985). Because these cells begin to fire at a time when no muscle activation and so no reafferent somesthetic stimulation has occurred, they are assumed to receive input from the motor system and perhaps from primary motor cortex, where firing begins earlier still (Fromm and Evarts, 1982; Kalaska et al., 1983). The notion that some Area 5 cells receive corollary discharges from the motor system is supported by the finding that task-related firing persists in 20% to 30% of Area 5 cells after deafferentation of the operant arm (Bioulac and Lamarre, 1979; Seal et al., 1982). The degree to which these properties set Area 5 apart from divisions of SI with deep input is not clear. On one hand, there are neurons in SI which begin firing before the first signs of

EMG activity associated with reaching movements (Soso and Fetz, 1980; Fromm and Evarts, 1982). On the other hand, it is claimed that deafferentation abolishes task-related activity in Area SI (Bioulac and Lamarre, 1979; Seal et al., 1982).

(3) Insensitivity to force. Although neurons of Area 5 carry a motor signal, it is not as strong as that carried by neurons in motor cortex. This was demonstrated by Kalaska and Hyde (1985) in experiments requiring monkeys to maintain a given arm position against loads varying in magnitude and direction. The modulation of firing level with respect to force was moderately lower in Area 5 than in Area 4. This has been taken as evidence for the specific involvement of Area 5 in posture-related as opposed to force-related functions, but to draw such a conclusion firmly would have required independently assessing the position sensitivity of the two populations of neurons.

(4) Selectivity for direction of voluntary movement. During simple voluntary flexion and extension movements at wrist or elbow, most task-related neurons in Area 5 exhibit direction-selectivity, firing a burst in conjunction with one movement and undergoing weaker excitation or inhibition during the opposite movement (MacKay et al, 1978; Seal et al., 1982; Chapman et al., 1984). Direction-selectivity is also apparent in a radial-movement task employed by Georgopoulos and colleagues. The monkey, in performing this task, moves a grasped manipulandum across a nearly horizontal surface, starting from a central point and aiming toward one of eight targets located on a circle with an 8 cm radius (Kalaska et al., 1983; Georgopoulos et al., 1984; Georgopoulos and Massey, 1985, 1988). Most neurons in Area 5 undergo phasic activation during task-performance and most are tuned for direction (Kalaska et al., 1983). The tuning curves are broad, possessing a mean half-width at half-height of around 180 °.

(5) Sensitivity to limb position. A substantial fraction of Area 5 neurons, and perhaps a majority, carry tonic signals related to the static position of the arm (MacKay et al., 1978; Georgopoulos et al., 1984; Georgopoulos and Massey, 1985). In the radial-movement task employed by Georgopoulos and Massey (1985), position-tuning (with the hand held at one of eight targets) was approximately as sharp as direction-tuning (with the hand moved toward one the eight targets). Positional preference is gradiental rather than positional: the firing level of the neuron is a monotonic function of the position of the hand

with respect to a particular spatial axis. The orientation of the critical axis varies from cell to cell and all possible orientations are represented by the population as a whole. In some 40% of neurons, the direction of movement eliciting a maximal burst and the position at which tonic firing is maximal are noncongruent (Georgopoulos and Massey, 1985).

"Spatial" encoding of hand position? Georgopoulos and colleagues have interpreted data from Area 5, as described above, and comparable data from primary motor cortex (Area 4) (Schwartz et al., 1988; Georgopoulos et al., 1988; Kettner et al., 1988) as indicating that cortical neurons encode the position of the hand and its direction of movement in spatial coordinates. The observation that neuronal firing levels depend on the position of the hand is unquestionable but it is important to ask whether this observation justifies concluding that cortical neurons encode the spatial coordinates of the limb as opposed to joint angles or muscle lengths. When hand movements are carried out within a comparatively restricted range, the coordinates of the hand bear a nearly linear relation to the angles of the joints and the lengths of the muscles. As a result, it is not possible, without great care, to disentangle sensitivity to the coordinates of the hand from sensitivity to the states of the joints or muscles. To demonstrate definitively the presence of selectivity for hand-position would require experiments and results such as the following:

- placing the hand in the same position with different postures of the arm and showing that firing is related to position rather than posture;
- moving the hand over a range of positions so large that the linear relation between position and joint-angle clearly breaks down and showing that firing is related more closely to position than to joint-angle;
- demonstrating that the identical analyses applied to electromyographic recordings from the arm muscles indeed reveal joint-angle rather than hand-position sensitivity.

A dramatic finding bearing on the spatial encoding issue has recently been published by Georgopoulos et al. (1989). They report that neurons in primary motor cortex are activated during the period of presumed mental rotation intervening between stimulus and arm movement in a task where monkeys reach toward a locus displaced by a fixed number of degrees from a visual cue. Neuronal firing shifts smoothly from the pattern associated with reaching toward the cue to that associated with reaching toward the displaced location. This seems to imply that Area 4 is involved in a high-level process (imaging of the hand) and thus provides implicit support for the notion that neurons in Area 4 represent hand position in a spatial rather than muscle- or joint-based coordinate system. There is a less interesting potential interpretation, however, namely that neurons of Area 4 participate in widespread subthreshold activation of the lower motor system during the period when the spontaneous tendency to reach for the cue is being replaced by the trained intention to reach for the displaced locus. In reaction-time tasks involving suppression of an invalid response and execution of a valid response, muscles that would be used for the aborted invalid response undergo EMG activation (Eriksen et al., 1985). Should it be the case that sub-movement-threshold EMG activation occurs in the "mental rotation" task, then the results could be explained by recourse to the classic view that each neuron in Area 4 carries a motor signal related to activation of one muscle or a few related muscles. They would no more justify regarding Area 4 as a substrate for mental rotation than they would justify regarding the muscles in this way.

Third-tier areas Moving posteriorly beyond the high-order somatosensory areas, one arrives at a third tier of cortical areas. These include Area 7a (Hyvarinen and Shelepin, 1979; Hyvarinen, 1981), Area Tpt (Leinonen et al., 1980) and the superior temporal polysensory area (STP) in the anterior bank of the superior temporal sulcus (Bruce et al., 1981; Hikosaka et al., 1988). These areas possess the collective distinguishing trait of receiving some somatosensory input but of being dominated by input from other sensory modalities: visual in the case of Area 7a and STP; acoustic in the case of Area Tpt. Systematic parametric studies of somesthetic function have yet to be carried out in these areas.

Summary of cortical mechanisms The somatosensory cortex appears to analyze information from the muscles and joints in at least two stages. Information is processed first by neurons in restricted subdivisions (3a and 2) of Area SI and then in high-order Area 5. The properties of neurons at the first stage are not dramatically different from the properties of muscle spindles, although some neurons do appear to put together information from multiple muscles and to combine deep and cutaneous inputs. Neurons at the second stage show more pronounced signs of integration, carrying out joint-joint and cutaneous-joint convergence over comparatively large parts of the body, and may encode limb position in terms of spatial coordinates, although the evidence for spatial encoding is limited. In areas forming a third tier, somatosensory and visual signals are brought together.

Neuropsychology of Postural Awareness and Related Functions

There are numerous clinical disorders that seem to affect the internal representation of the body. These include gross losses of proprioceptive sensation, astereognosis, visuomotor ataxia, autotopagnosia, and hemiasomatognosia. The question that we will ask with respect to these disorders is whether any involves a selective degradation of spatial perception of the body.

Injuries at stages of the somatosensory system up to and including primary somatosensory cortex certainly are capable of impairing proprioceptive sensation. Head and Holmes (1911) described patients who, following cortical injury, could localize stimuli relative to the skin surface, by pointing out on a model the spot being stimulated on an unseen limb, and yet could not match with the good limb a position imposed passively on the affected limb. However, such patients continue to perform poorly when required to do no more than detect passive movement at a joint. The most plausible explanation is that pathways carrying joint-angle information have been blocked at an early level, producing a deficit of joint-angle sensation rather than of spatial integration.

Astereognosis is defined generally as an inability to recognize familiar objects by palpation (Corkin et al., 1970; Hecaen and Albert, 1978). This might be founded on any of several underlying deficits including an inability to appreciate the posture of the hand. Astereognostic patients indeed almost invariably display deficits on tests of joint and skin sensitivity, and their lesions are very commonly found to encroach on primary somatosensory cortex (Corkin et al., 1970; Roland, 1965). Semmes (1965) has described a population of patients who fail on tests requiring them to compare by palpation complex two- and three-dimensional shapes and whose simple sensory deficits cannot explain their failure. However, the difficulty experienced by these patients is in building up a complex spatial image of the object rather than in sensing the posture of the hand; the simple caliper function of the hand is unaffected as shown by accurate judgement of object size. Thus studies of astereognosis have provided no evidence for a condition in which joint-angle sensitivity is normal but spatial appreciation of the hand is impaired.

A disorder of visually guided reaching (optic or visuomotor ataxia) has been described in connection especially with injury to posterior parietal cortex and its fiber connections (Rondot, 1989). Patients exhibiting this condition experience under natural circumstances problems such as missing their food when attempting to eat with knife and fork. In an experimental setting, they commit systematic errors with respect to distance and direction when pointing or reaching. It is tempting to speculate that visuomotor ataxia is based on disruption of the spatial schema of the arm but this is probably not the case. Patients with visuomotor ataxia are able, by anecdotal report, to use their arms accurately when visual guidance is not required, as when reaching in the dark for a touched part of the body (Rondot et al., 1977; Damasio and Benton, 1979; Perenin and Vighetto, 1988). Even the patient of Denes et al. (1982), although unable to point to body parts on command or imitation, could readily brush away a lock of hair from her eyes. Further, in cases of lateralized injury, signs of visuomotor ataxia are much more likely to be restricted to the contralateral visual field than to the contralateral arm (Perenin and Vighetto, 1988; Rondot, 1989). Finally, visuomotor ataxia often is dependent on presentation of the target in the periphery of the visual field and may virtually disappear if the patient is permitted to direct eyes and head toward the target, although the requisite arm movement remains the same.

Patients exhibiting the rare condition of autotopagnosia are severely impaired on tasks requiring them to point to parts of their own bodies whether the commands are given verbally or by means of a chart on which the body parts to be touched are numbered (Semmes et al., 1963; De Renzi and Scotti, 1970; Poeck and Orgass, 1971; Sauguet et al., 1971; Denes et al., 1982; Ogden, 1985a). The disorder, in some cases, may reflect aphasia (Poeck and Orgass, 1971; Sauguet et al., 1971) or a closely related weakening of the conceptual representations of body parts, as reflected by failure to identify them even in isolated drawings (Semenza and Goodglass, 1985). However, it would be difficult to explain, on the basis of aphasia alone, the performance of autotopagnosic patients who are able to name body parts when the examiner touches them and are able to identify mispointing when the physician tests an accomplice who mimicks autotopagnosic behavior (De Renzi and Scotti, 1970; Ogden, 1985a). Autotopagnosics may have difficulty even in imagining and describing the topology of body parts and their lack of appreciation of the arrangement of parts may extend to drawings of the body or to inanimate objects (De Renzi and Scotti, 1970; Ogden, 1985a). Finger agnosia is a condition similar to autotopagnosia in which patients experience difficulty primarily in comprehending the anatomical arrangement of the fingers (Kinsbourne and Warrington, 1962; Ettlinger, 1963; Poeck and Orgass, 1969; Gainotti et al., 1972). Regardless of how autotopagnosic deficits are explained, the most noteworthy point in the present context is that the impairment affects appreciation of how the parts of the body are connected; not awareness of their momentary spatial arrangement as determined by the angles of the joints. Thus the deficit is not of spatial analysis in the sense implied by the term as we use it.

Hemiasomatognosia is a condition, arising from damage to the right inferior parietal lobule, in which patients ignore the left limbs. Unawareness of the limbs may be profound, as in patients who deny that the left half of the body is theirs even when looking at it, who move the right arm on being requested to move the left, and who experience sensory stimulation of the left half of the body as occurring in a corresponding location on the right (Critchley, 1953; Hecaen and Albert, 1978; Heilman and Valenstein, 1979; De Renzi, 1982). Hemiasomatognosia is best regarded as an expression of a generalized failure to direct awareness to the left half of space, because it virtually always occurs in conjunction with visual hemifield neglect (Bisiach et al., 1986). In visual hemifield neglect, the "left half" of space is not delineated on strictly anatomical lines (at the midline of the retina) but rather falls at the center of the seen or imagined object of regard even when this is confined to one hemifield (Bisiach and Luzzatti, 1978; Ogden, 1985b; Ladavas, 1987). Blindfolded neglect patients, when searching with the right finger for a marble hidden in a small rutted maze, take longer to find the marble when it is on the left side of the maze (De Renzi et al., 1970). The notion that hemiasomatognosia might likewise be organized with respect to a spatial rather than to the anatomical midline has not been tested.

Summary of neuropsychology The somatosensory cortex of the postcentral and parietal lobes can be thought of as forming three tiers: primary somatosensory cortex (Area SI), high-order somatosensory cortex (Area 5) and cortex with somatosensory but also with visual functions (the association cortex of the inferior parietal lobule, corresponding to monkey Area 7). Destruction of the first tier (Area SI) results in a profound loss of deep sensation reflected by failure even on tasks with no spatial component such as detecting movement about a joint. Destruction of the third tier (inferior parietal association cortex) is at the root of most so-called body-schema disorders, but these disorders turn out on close inspection to concern the body only insofar as it is used for visually guided reaching (visuomotor ataxia) or is visualized (autotopagnosia) or is attended to (hemiasomatognosia). It is in the middle tier (Area 5) that one would expect, on grounds of location, connectivity and physiology, to find mechanisms underlying the spatial representation of the body. Yet injury to human Area 5, instead of producing distinctive body-spatial symptomatology, seems to cause only a mild impairment of cutaneous and joint sensation (Corkin et al., 1970). This paradox has yet to be resolved.

Models of Postural Function

Analytic models of ideal systems The position, orientation, velocity and acceleration of the segments of a multiarticulated system can be computed exactly if the dimensions of the system are known and if one is given the angles of the joints and their derivatives with respect to time (Craig, 1989). However, in any reasonably complex system, the computations are slow and storage of precomputed values places an excessive load on memory. Various computational strategies have been proposed to enhance speed and reduce the load on memory (Raibert, 1978; Benati et al., 1980a,1980b; Hinton, 1984). These schemes share the feature that they store information about the jointed system and break down the calculations in a way that permits rapid parallel processing. They may throw light on how parallel analysis of postural information is accomplished in the brain. However, it is open to question whether computational strategies based on the equations of motion of an ideal jointed system can be made to generalize smoothly to organisms, with their nonlinear sensory signals and irregular anatomy, or, for that matter, to robots incorporating unconventional design features. The equations of motion quickly become intractable as structural irregularities are introduced. This consideration weighs in favor of non-analytic computational systems, including neural nets, that treat postural analysis as a generalized function-approximation problem without regard to the complexity or analytic solubility of the underlying equations. In the following sections, we will consider the advantages and disadvantages of three connectionist approaches to postural computation.

Strictly linear connectionist models In linear neural networks, the activation level of each output unit is a weighted sum of the activation levels of the input units. Each weighting coefficient corresponds to the strength of a synapse between an input unit and an output unit. By appropriate training, the synaptic weights of such a network can be adjusted so that it computes any linear function of the outputs on the inputs or achieves a planar least-squares best-fit to any nonlinear function (Kohonen, 1988). The operations of linear networks are subject to metadescription in terms of "metric tensors" (Pellionisz and Llinas, 1982) but no obvious practical benefit attaches to so describing them (Robinson, 1982). Linear analysis has been applied to simple biological processes including the vestibulo-ocular reflex, where the outputs (contractions of the various extraocular muscles) are very nearly linear functions of the inputs (activation levels of hair cells associated with the various labyrinths) (Robinson, 1982; Pellionisz and Graf, 1987; Villis and Tweed, 1988). However the applicability of linear analysis to postural problems is severely limited because the underlying functions are nonlinear. Given the force (Jongen et al., 1989) or velocity (Ritter and Schulten, 1988) to be produced at a given point on the limb, a linear network can compute the required levels of muscle activation, but only under the very severe constraint of local neighborhood linearization: the parameters embodied in a specific network are valid for only one posture of the limb and for each new posture a separate network must be created. Thus the central problem of postural analysis is skirted rather than solved.

Combinatorial connectionist models Simple postural problems have been solved by connectionist networks that associate a different output with every possible combination of input values. In essence, these networks confront postural analysis as a problem in associative memory. The approach is similar in spirit to that embodied in CMAC models (Albus, 1975). It is effective so long as the number of input variables remains small but scales up poorly as task complexity increases.

Kuperstein (1986;1987), using a modified delta rule, has trained two-layer networks composed of linear units on a simple form of the inverse postural problem. The system is described most thoroughly in Kuperstein (1987), on which our commentary is based. The inputs to the network convey information about the horizontal and vertical angles of the eyes in the head, while the outputs govern muscles that determine the position of an arm with three degrees of rotational freedom (two at the shoulder and one at the elbow). The head is fixed on the otherwise rigid body. The network learns, given the angles of the eyes, to bring the tip of the limb to the point in space that the eyes are fixating. At asymptote, the mean distance error is around 4% of the arm's length. The coordinate transformation mastered by this network is relatively simple but still is intrinsically nonlinear. One is forced to ask: how can a network in which the values on the output lines are a strictly linear function of the values on the input lines represent a nonlinear input-output function? The answer is: by a careful choice of how to represent the input. There are two eyes and each rotates vertically and horizontally. Thus eye position could be described exhaustively by values on four input lines. However, in the actual network, there are 2,400 input units. Each input unit is maximally activated when a given eye is pointing in a particular direction or when the difference between the angles of the two eyes is at a particular value. The oculomotor receptive fields of the input units, considered collectively, cover evenly the state space of possible binocular eye positions. Assigning one input unit to every possible eye position permits associating a unique arm position with every eye position subject only to limits imposed by the grain with which the input space is quantized. This is a memory-intensive method with the inherent restriction that networks based on it must expand geometrically with the number of inputs. To compute a nonlinear function of ten joint angles with the same level of angular resolution achieved by the Kuperstein network would require 20^{10} input units: 1000 times the number of neurons in the human brain.

Mel (1988) has described a superficially different but fundamentally similar approach to postural information processing. Mel's network (MURPHY) solves the problem of forming a visual image of an arm given the angles of the joints. The arm has one degree of rotational freedom at the shoulder, one at the elbow and one at the wrist. The rotational axes are parallel so that the segments of the arm remain at all times within a single plane. The state of the arm is encoded by 273 input units (91 units are associated with each joint and each is activated maximally when the joint assumes a particular preferred angle). At the output layer, there are 4096 units, each of which corresponds to a certain locale on a 64x64 grid subdividing the plane in which the arm moves. Training is carried out over a series of trials. On each trial, the arm is placed in a new position, whereupon three input units (whose preferred angles correspond to the angles imposed on the three joints) become maximally active. Training is accomplished by creating a functional linkage between these three units and any output unit corresponding to a grid-point on which a visible part of the arm now lies. The resulting sigma-pi synapse is an and-gate with the property that if all three joint units are maximally active then the output unit will be activated. At the end of training, the network can generate, given any state of the joints, an image of the visible parts of the arm in their resulting positions. Aside from its high degree of biological implausibility (there is no evidence for the existence in the brain of sigma-pi synapses let alone for their ability to appear *ex nihilo* on a single trial) this system suffers from a major weakness closely similar to the one that afflicts the Kuperstein model. The complexity of the network must explode exponentially as the number of degrees of rotational freedom is increased to a realistic level because for every possible combination of joint angles a new set of sigma-pi synapses is required.

Function-approximating connectionist models Multilayered acyclic networks containing hidden units with sigmoidal transfer functions are a powerful alternative tool for carrying out postural coordinate transformations. These networks construct an approximation to any desired input-output function by summing the sigmoidal transfer functions of a set of hidden units. The sum-of-

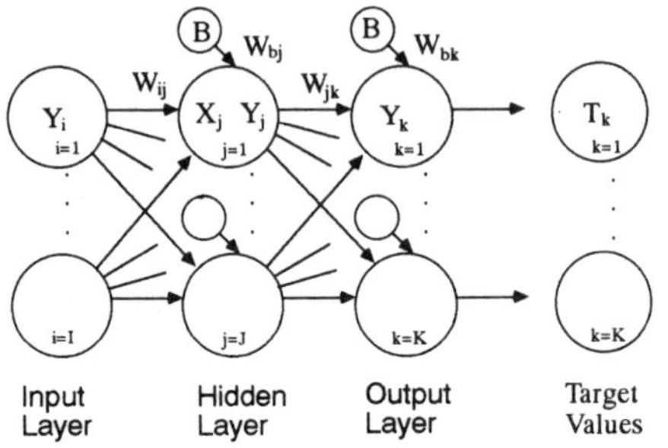

Forward Pass Rules

$$X_j = W_{bj} + \sum_i Y_i * W_{ij}$$

$$Y_j = \frac{1}{1-e^{-X_j}}$$

$$Y_k = \sum_j (Y_j * W_{jk}) + W_{bk}$$

Weight Modification Rules

$$E_k = T_k - Y_k$$

$$\Delta W_{jk} = \eta * E_k * Y_j$$

$$\Delta W_{bk} = \eta * E_k$$

$$E_j = \sum_k E_k * W_{jk}$$

$$W_{ij} = \eta * E_j * (1-Y_j) * Y_j * Y_i$$

$$W_{bj} = \eta * E_j * (1-Y_j) * Y_j$$

Figure 2: Three-layered forward-pass networks were trained by standard back-propagation procedures (Rumelhart et al., 1986a). y_i: input-unit activation level (range: can assume any real value). x_j: hidden-unit input function (range: can assume any real value). y_j: hidden-unit activation function (range: 0 to 1). y_k: output unit activation function (range: can assume any real value). w: synaptic weights (range: can assume any real value). t: the value the output line should produce in response to the current input. e: error functions. η : learning-rate parameter. Δw : change imposed on synaptic weight.

sigmoids approximation is superior to a linear approximation because it can capture the nonlinear inflections of the target function. It is more efficient than a combinatorial solution, when applied to postural coordinate-transformation functions, because it does not dedicate resources to point-by-point input-output associations, which are of little use in a case where the outputs change comparatively gradually over the inputs. The sum-of-sigmoids solution is biologically plausible, as compared to other smooth funtion-approximating algorithms, because sigmoidal transfer functions (reflecting threshold and saturation effects) are typical of neurons. The fact that sum-of-sigmoids networks do not require fine quantization of the input space

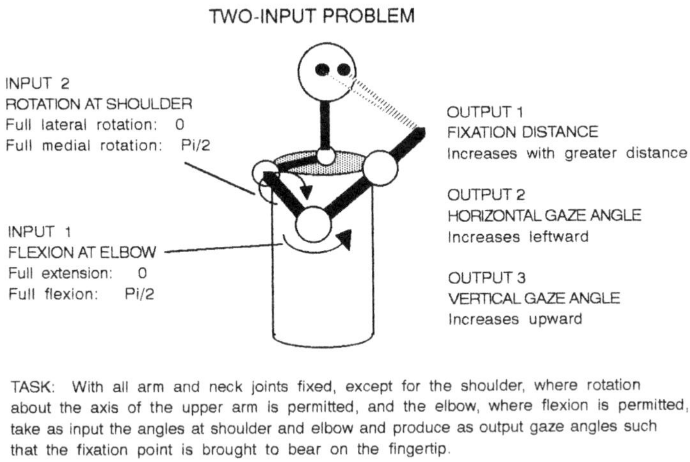

Figure 3: The two-input problem.

further encourages biological comparison because the joint-angle selectivity of cortical neurons generally is coarse.

To study the postural-analysis capabilities of multilayered networks, we have trained them on spatial problems involving an arm-neck-eye system possessing altogether thirteen degrees of rotational freedom (figure 1). The simulated system has two degrees of freedom at the wrist, two at the elbow, three at the shoulder, three at the neck and three at the eye. The shoulder is treated as a fixed ball-and-socket joint incapable of motions such as shrugging that result from scapular translation. This is the most complex articulated system to have been subjected to analysis with connectionist networks.

The networks that we have studied possess a standard three-layered architecture as illustrated in figure 2. Each input unit (i) projects to every hidden unit (j) via a synapse with a specific weight (w_{ij}) and each hidden unit projects to every output unit (k) via a synapse with a weight w_{jk}. The function relating the activation of a hidden unit (y_j) to the activation of the input units (y_i) is sigmoidal with the particular shape dependent on the weights of the synapses w_{ij} and w_{bj}. The function relating output-unit activation (y_k) to hidden-unit activation (y_j) is linear, with the slope and offset dependent on weights w_{jk} and w_{bk}. The biasing synapses w_{bj} and w_{bk} can be

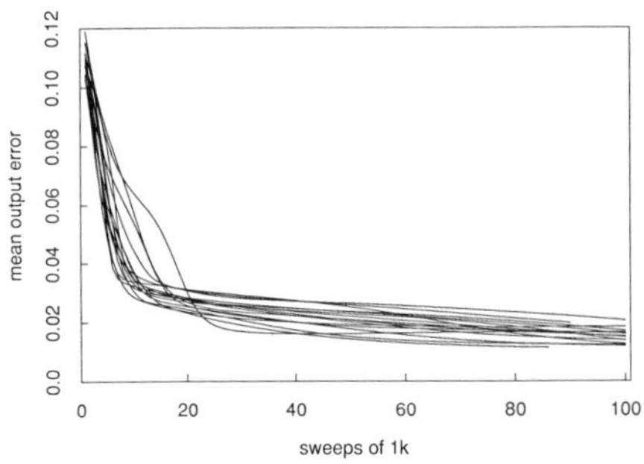

Figure 4: Learning curves of 20 ten-hidden-unit networks trained on the two-input problem. The output error was computed by expressing the distance between the fixation point and the fingertip as a fraction of the length of the arm from shoulder to fingertip. The mean error was computed across all 1000 randomly generated arm postures that made up the training set. By the end of the first sweep through the training set, considerable learning had already occurred, with the result that the initial mean error (around 11%) is much lower than the mean obtained from an untrained network (around 90%). The mean error after 100 presentations of the training set was consistently around 2% of the length of the arm.

thought of as originating from units that are tonically active at a fixed level of 1.

Networks are trained by use of the back-propagation algorithm as described by Rumelhart et al. (1986a) and detailed in figure 2. At the start of training, the synaptic weights are set to random values. The network is then put through a series of supervised learning trials. On each trial, the input lines are set to predetermined values (y_i), the levels of activation assumed by the output lines (y_k) are compared to the desired activation levels (t_k) and the synaptic weights are adjusted by use of the error-reducing learning rule. All synaptic weights, including the bias weights, are subject to modification. The weights converge on a solution over thousands to millions of trials according to the complexity of the problem.

To illustrate principles underlying the operation of these networks, we will first present results obtained on a simple postural transformation, the two-input problem illustrated in figure 3. This problem is pedagogically useful because the small number of variables makes it easy to analyze network properties. The essence of the problem is that the network must learn to direct the eye to the fingertip on being given two inputs: the rotational angle of the

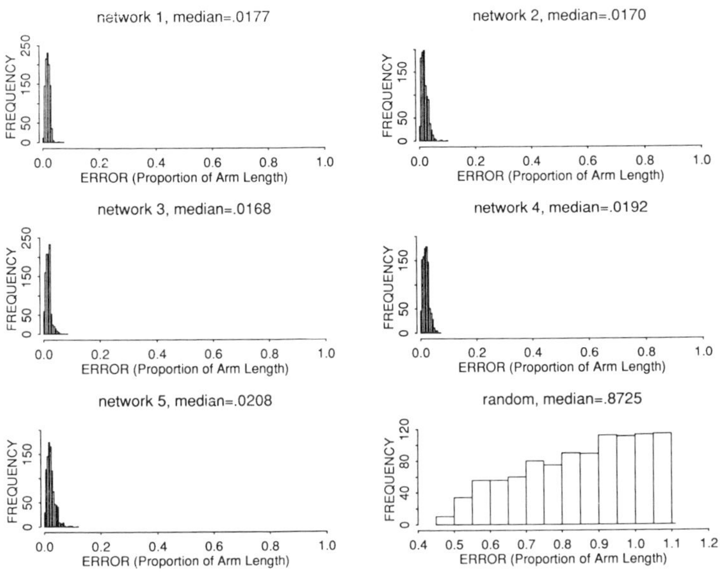

Figure 5: Networks trained on the two-input problem performed as well when given a transfer set of 1000 new randomly generated arm postures as when assessed on the training set. The frequency histograms of error magnitude were obtained from five trained networks and from one untrained network with randomly generated weights (lower right).

shoulder (from thumb up to thumb left over a range of 90°) and the flexional angle of the elbow (from extension to flexion over a range of 90°). The other arm and neck angles are fixed throughout training and testing: the head points straight ahead on the thorax, the upper arm points straight ahead on the shoulder and the hand is colinear with the forearm. The eye is controlled by three output lines encoding fixation distance, horizontal gaze angle and vertical gaze angle respectively.

Training was carried out by repeatedly exposing the network to a set of 1000 input-target pairs generated by randomly stipulating the arm angles and then computing the eye angles required to bring the fixation point to the fingertip. The set was presented 100 times for a total of 100,000 trials. As a general measure of performance, we used the mean distance of the fixation point from the fingertip expressed as a percentage of the total length of the arm (from shoulder to fingertip).

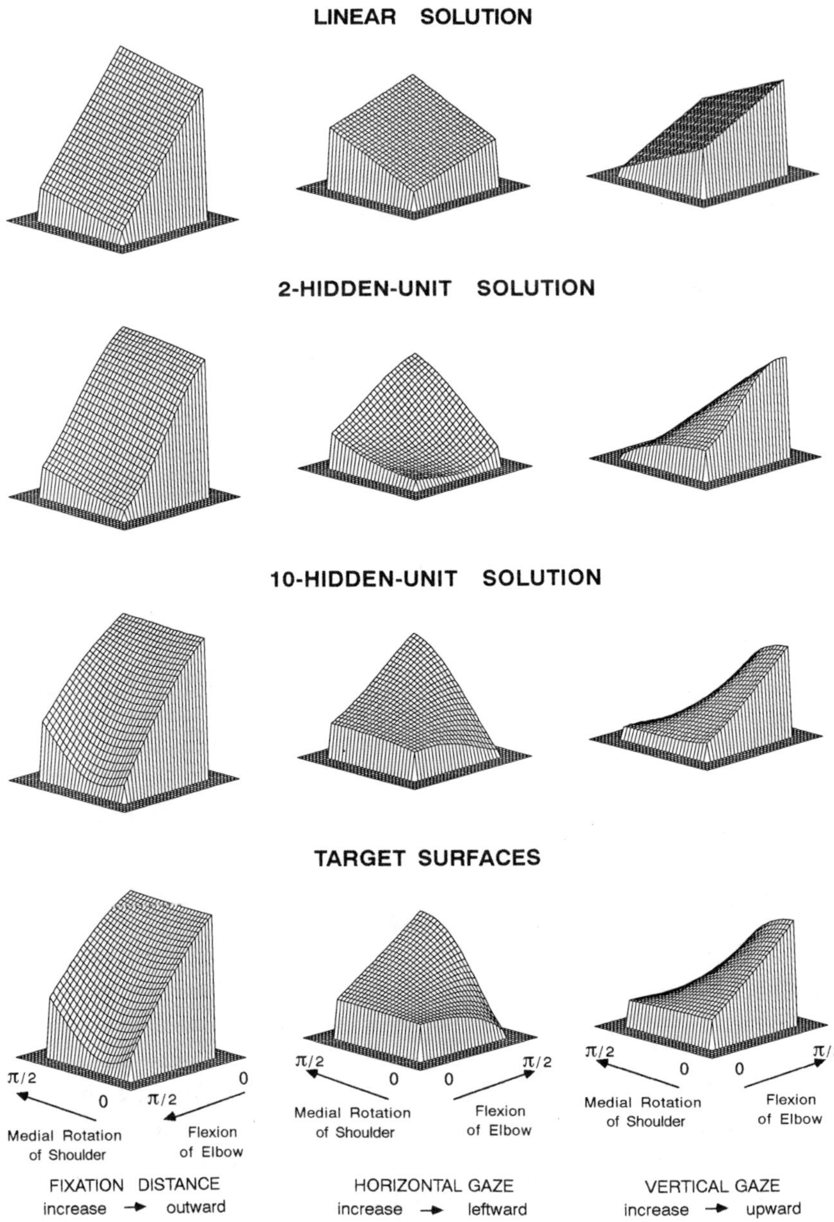

LINEAR SOLUTION
2-HIDDEN-UNIT SOLUTION
10-HIDDEN-UNIT SOLUTION
TARGET SURFACES
π/2
0
π/2
0
Medial Rotation of Shoulder
Flexion of Elbow
FIXATION DISTANCE
increase → outward
HORIZONTAL GAZE
increase → leftward
VERTICAL GAZE
increase → upward

Figure 6: Each surface represents the activation level of one output unit as a function of the activation levels of the two input units. The target surfaces represent the outputs of an ideal network with perfect performance on the two-input problem. The output activation surfaces of linear, 2-hidden-unit and 10-hidden unit networks form successively better approximations to the target surfaces. The linear network (a perceptron) was constructed by omitting the hidden layer, connecting input units directly to output units, and training by use of the delta rule. Note that a special choice of axes has been made for the distance surface to ensure its visibility.

Performance quickly improved over the first few thousand trials and then gradually approached asymptote (figure 4). The asymptotic level of performance was around 2% of the length of the arm for twenty networks with 10 hidden units each. Networks with fewer than ten hidden units performed much more poorly, but networks with more than ten hidden units were only marginally better. The training set covered the state space of the system sufficiently densely so that the network was forced to learn a generalized input-output function, not just to cope with the training items as special instances, as indicated by the fact that performance was excellent when networks were tested with an entirely new set of one thousand items (figure 5).

How do networks solve this problem? This question is best approached by noting that the required input-output functions can be represented as a set of three target surfaces (figure 6, bottom row). Each surface represents the desired output (fixation distance, horizontal gaze angle or vertical gaze angle) as a function of the two input variables (elbow flexional angle and shoulder rotational angle). The input-output functions of trained networks can be represented as surfaces and compared directly to the target surfaces. In a linear network (one where the level of activation of each output unit is simply a weighted sum of the levels of activation of the inputs) each output surface is a plane giving a good least-squares fit to the corresponding target surface (figure 6, top row). In a nonlinear network, each output surface is curved. As the number of hidden units increases (figure 6, second and third rows) so does the subtlety with which the the output surface captures the inflections of the target surface.

What confers on the output surface its particular pattern of curvature? Each output surface is a weighted sum of the activation

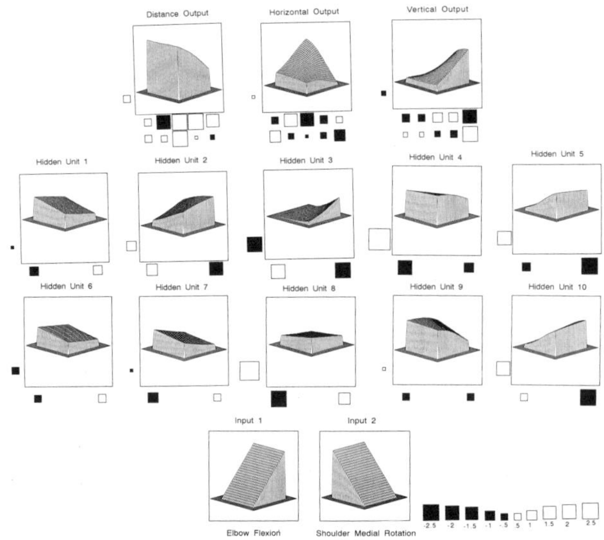

Figure 7: An exhaustive portrait of a ten-hidden-unit network that solves the two-input problem. Each box represents one unit. Within the box there is a surface representing the level of activation of the unit as a function of the two inputs. Each square appended to a box represents the weight of a synapse terminating on the unit represented by the box. The source of the synapse is indicated by the position of the square. The ten synapses beneath each output unit originate from the ten similarly arranged hidden units. The two synapses beneath each hidden unit originate from the two similarly arranged input units. The square to the side of each box represents a biasing synapse. The biasing synapse originates from a unit (not shown) whose level of activation is always 1. The shading of each square indicates the sign of the synapse (black=inhibitory; white=excitatory) and the area of the square indicates the magnitude of the synaptic weight.

surfaces of the hidden units. Each hidden-unit activation surface is a sigmoidal monotonic function of the input variables. Thus each output surface is a weighted sum of sigmoids. Surfaces that are weighted sums of sigmoids differ from sigmoidal surfaces in the crucial respect that they can be nonmonotonic. Nonmonotonicity is required for the solution of some problems. For example, in the two-inputs problem, with the shoulder in a state of full lateral rotation, the eye must move to the right as the elbow flexes (bringing the hand up toward the ear), whereas, with the shoulder in a state of full medial rotation, the eye must move to the left as the elbow flexes (bringing the hand in toward the chest). This requirement is reflected in the local changes of slope of the horizontal-gaze target surface (figure 6, bottom row, midddle surface). The transition from linear input-unit activation surfaces through sigmoidal hidden-unit activation surfaces to sum-of-sigmoidal output-unit activation surfaces can be seen in the representative network of figure 7.

It would be of considerable theoretical interest to know whether networks trained on the same problem contain similar kinds of hidden units. The claim that this is so has been made on subjective grounds (Lehky and Sejnowski, 1988; Zipser and Anderson, 1988) but supporting it objectively has proven difficult. Two reservations should be noted at the outset. (1) To expect that networks trained on the same problem will necessarily contain identical hidden units is unrealistic because any given input-output function can be embodied in different networks. A network functionally equivalent to a trained network can be produced simply by inverting the sign of every input and output synapse on an arbitrary set of hidden units. The resulting "anti-hidden units" produce the same effect at the output layer as the hidden units that they have replaced, but their dependence on the input variables is exactly reversed. (2) To expect that a trained network will contain multiple examples of a given "type" of hidden unit is unrealistic because duplication is incompatible with an efficient solution to the problem. Any effect produced by two hidden units with identical activation surfaces can also be produced by just one. Consequently, the efficient use of the second hidden unit would be to capture a distinct inflection of the target function. Although this is true regardless of the error metric that is used, it is worth noting that different error metrics can give rise to different definitions of efficiency and consequently to different patterns of unit allocation (Hanson & Burr, 1987).

With these qualifications in mind, we now consider how to determine whether networks trained on the same problem contain reidentifiable units - hidden units that recur across networks. To test the reidentifiability hypothesis requires devising a measure of similarity between hidden units. The measure that we have used is the absolute value of the correlation coefficient of the two activation surfaces. This index indicates just how closely two hidden units could be brought, by an appropriate adjustment of the weights of their output synapses, to producing an identical effect on an output unit.

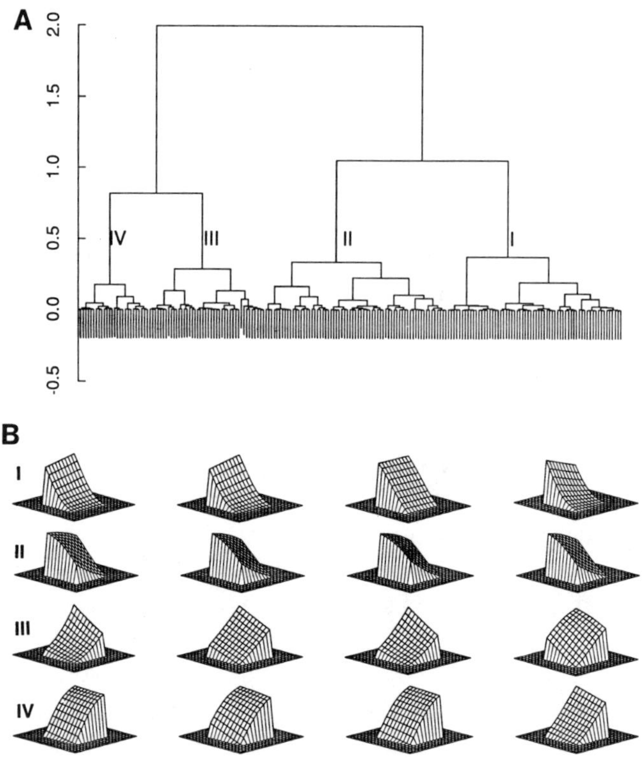

Figure 8: Cluster analysis of hidden units from 20 ten-hidden-unit networks trained on the two-input problem. A. The dendrogram representing successively finer parcellations of the unit population was cut at a level yielding four clusters (I, II, III and IV). B. Activation surfaces of representative hidden units from each of the four clusters. The surfaces have been normalized for ease of comparison by a linear conversion: H'=(H-Hmin)/(Hmax-Hmin), where H' is the normalized height at a given point, H is the unnormalized height, Hmin is the height of the lowest point of the unnormalized surface, and Hmax is the height of the highest point. The conversion does not affect the correlational similarity measure used in clustering analysis.

It throws away all other information about the activation surfaces. A hidden unit and its anti-unit would receive a maximal similarity rating of one because, although they respond inversely to every input, nevertheless, when acting on an output unit through synapses of opposite sign and equal magnitude, they mediate identical effects.

We tested the reidentifiability hypothesis by analyzing a population of 200 hidden units drawn from 20 ten-hidden-unit

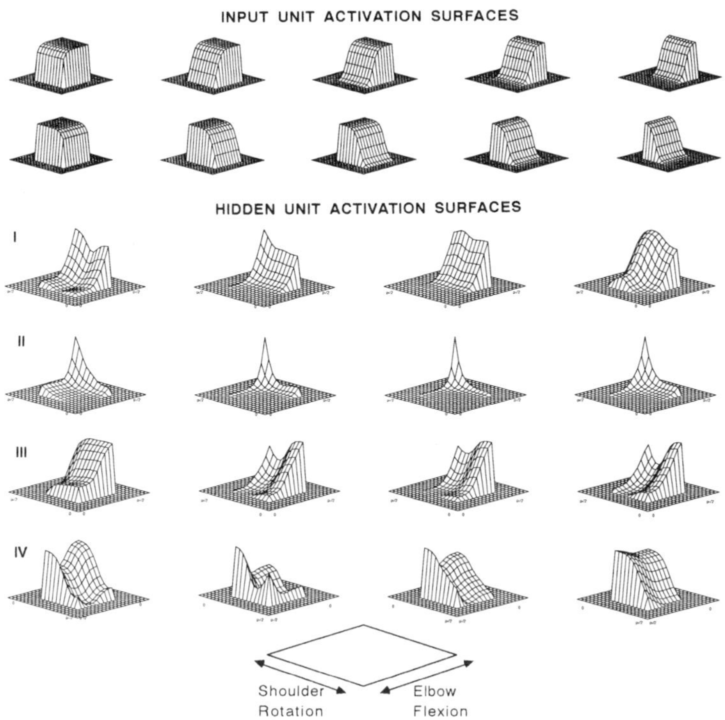

Figure 9: Activation surfaces from ten ten-hidden-unit networks trained on the two-input problem with the two input variables represented by a recruitment code. Above: Each input variable is represented by the activity of five input units. When it is at a minimum, no input units are active; when it is at a maximum, all input units are active. Below: The hidden units were divided into ten clusters as described in the text. Normalized activation surfaces of hidden units from representative clusters (I, II, III and IV) are shown. Attributes associated uniquely with the use of a recruitment code are: very sharp cut-offs (II) and nonmonotonicity (III and IV).

networks trained on the same problem. The hidden units were represented as points in a multidimensional space with the interpoint distance set equal to one minus the pairwise similarity index. Complete linkage clustering procedures were then applied. The results of the clustering analysis are illustrated in figure 8A. Passing down the dendrogram from top to bottom is equivalent to cutting the population of hidden units into progressively finer clusters. The activation surfaces in figure 8B belong to representative hidden units drawn from the four subpopulations formed when the cut is made at the level of four clusters. On the basis of the reidentifiability

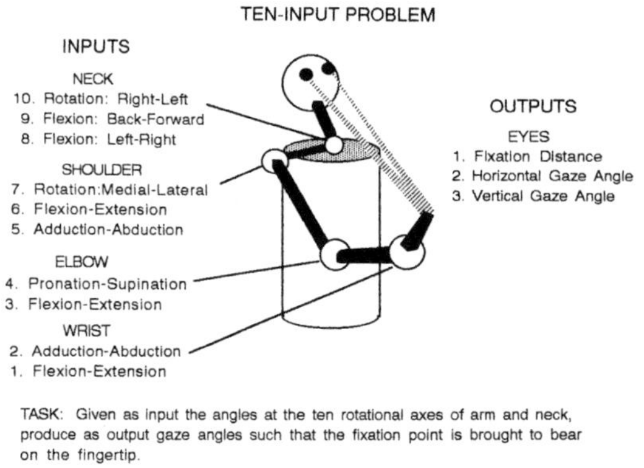

Figure 10: The ten-input problem. Solving this problem is equivalent to looking by means of an eye movement toward something touched in darkness.

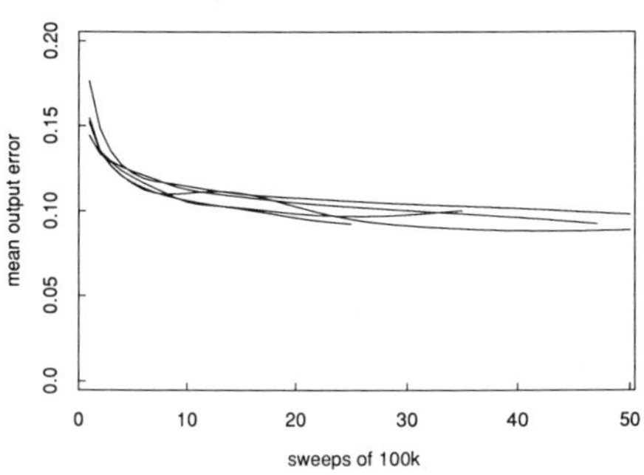

Figure 11: Learning curves of five 100-hidden-unit networks trained on the ten-input problem. The output error was computed by expressing the distance between the fixation point and the fingertip as a fraction of the length of the arm from shoulder to fingertip. The mean error was computed across all 100,000 randomly generated arm-and-neck postures that made up the training set. By the end of the first sweep through the training set, considerable learning had already occurred. The initial mean error (around 15%) is thus lower than the mean obtained from an untrained network (around 160%). The mean error after 100 presentations of the training set was consistently around 10% of the length of the arm.

hypothesis, one would predict that when a cut is made at around ten clusters there should be a tendency for the clusters to be of uniform size and for hidden units from the same network to be dispersed among clusters. Both tendencies were present and significant.

In the cases we have considered up to this point, the angle of

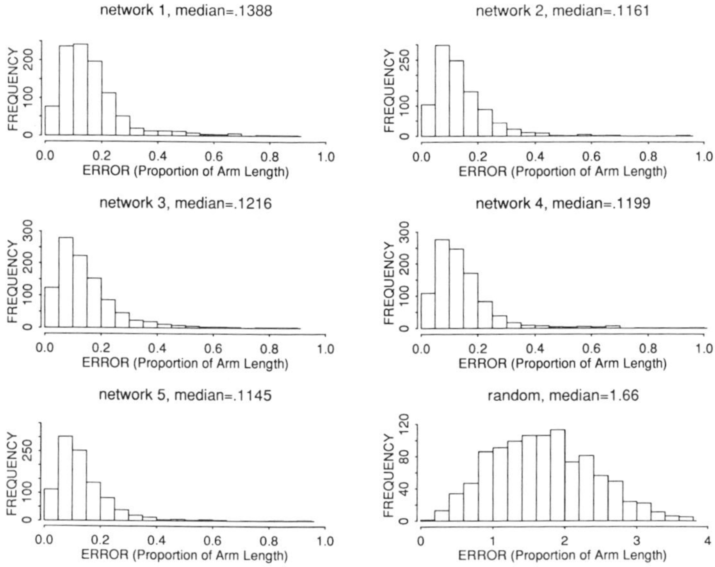

Figure 12: Networks trained on the ten-input problem showed good generalization to a randomly generated transfer set of 1000 arm-and-neck postures. The frequency histograms of error magnitude were obtained from five trained networks and from one untrained network with randomly generated weights (lower right). Note the difference in horizontal scale between the histogram at lower right and the others.

each joint is represented by the level of activation of one input unit. To find out whether the solubility of the two-input problem and the form of the solution depend on the nature of the input representation, we have experimented with other, more biological, input codes including a recruitment code in which each angle is represented by activity in a population of input units with different thresholds (figure 9, top two rows). Networks presented with recruitment-coded joint angles solve the two-input problem and actually perform marginally better at asymptote than networks with an activation-coded input (mean error around 1% as opposed to 2%). However the hidden-unit representation by means of which they solve the problem is qualitatively different. They contain many hidden units with sharply inflected and nonmonotonic activation surfaces (figure 9, bottom rows). Nonmonotonic dependence on joint-angle is possible because a hidden unit may be excited by an input recruited over one range of joint angles and inhibited by an input recruited over a separate range.

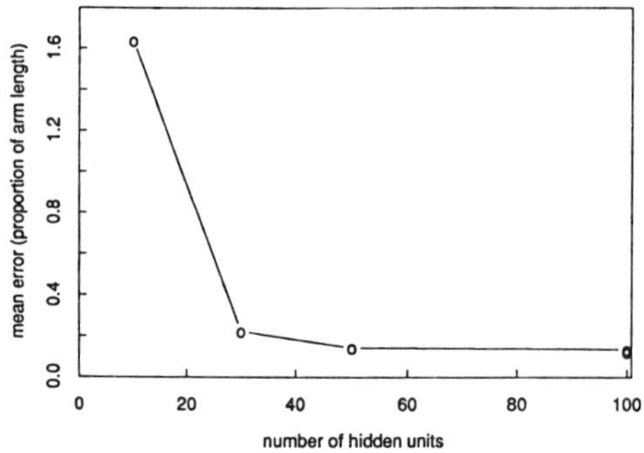

Figure 13: The effect of number of hidden units on ability to master the ten-input problem. Performance was measured after twenty presentations of the 100,000 item training set.

The examples described up to this point have afforded some insight into how multilayered feedforward networks handle postural information but they have not provided an adequate indication of how powerful the networks are. The complexity of postural problems increases rapidly as degrees of freedom are added to an articulated system. Can networks of the type that we have described master problems with a biologically realistic level of complexity? We have investigated this issue by presenting networks with a postural problem as complex as any that the brain must solve. In the "ten-input" task, the network is provided, as input, the angles at ten rotational axes of the arm and neck, and must produce, as output, values of fixation distance, horizontal gaze angle and vertical gaze angle that will bring the eye to bear on the fingertip (figure 10). The training set consists of 100,000 randomly generated postures paired with corresponding eye positions. The set is presented 20-50 times for a total of 2,000,000-5,000,000 trials. The number of trials, larger than necessary for efficient training, was chosen to ensure that network performance would achieve an asymptotic level. Three-layered forward-pass networks do indeed master this problem. They require about 50 times as many trials as networks trained on the two-input problem and their asymptotic performance is worse by a factor of around 10 (figure 11 vs. figure 4). Nevertheless, they achieve a mean error level of around 12% of the length of the arm (figure 12), which is comparable to levels achieved by humans on comparable tasks. To solve the 10-input problem, networks require approximately ten times as many

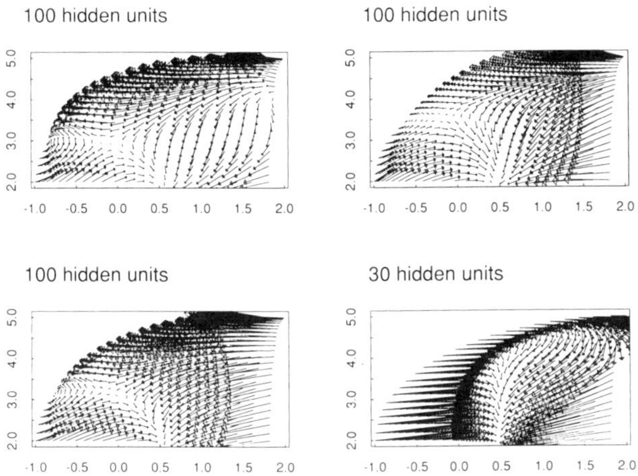

Figure 14: Networks trained on the ten-input problem make similar patterns of errors, as demonstrated by deviation-field plots obtained from four networks. The plots were generated by assessing performance on the restricted set of postures used initially in the two-input problem. With the head pointing straight ahead on the torso, the upper arm pointing straight out from the shoulder and the wrist straight, the network was exposed to a 25x25 raster of elbow flexional angles and shoulder rotational angles. Each vector represents, for one combination of elbow and shoulder values, the location of the fingertip (base of arrow) and the location of the fixation point (head of arrow). Thus the vector represents the error made by the network. All points are projected onto a horizontal plane as viewed from above. Straight ahead is toward the top of the figure. The units on the vertical axis represent distance forward from the head. The units on the horizontal axis represent distance to the right (positive) or left (negative) of the head. The units are defined as in the legend to figure 1 (the length of the arm from shoulder to fingertip is 5 units).

hidden units as for the two-input problem (figure 13).

The fact that these networks make significant errors permits us to ask a question of considerable theoretical interest: do different networks make the same mistakes? The answer seems to be yes. To support this point we gave each network, as input, a stereotyped set of arm-and-neck postures and recorded the outputs that it produced. Then we represented the performance of the network as a set of vectors. For each test item, there was one vector. Its base was at the location of the fingertip (and thus was the same for all networks) but its head was at the location of the fixation point (and thus was determined by the outputs of the individual network). The vectors in figure 14 are projected onto a horizontal plane. Although the performance of each network displays idiosyncratic features, the general consistency of the deviation-field plots is obvious.

Summary of models Although postural coordinate transformations can be solved analytically, at least in the case of an ideal jointed system, connectionist solutions are more plausible as models of brain function and are better able to deal with the complexities of nonideal biological bodies. Linear connectionist networks are unsuited for capturing the complex inflections of coordinate transformation functions and combinatorial connectionist networks capture them in an inefficient manner unlikely to scale up well to problems with a realistic level of complexity. In contrast, function-approximating multilayered networks, with sum-of-sigmoids input-output functions, achieve accurate performance even on very complex problems. We have shown that such networks are able to compute the coordinates of the hand in a head-centered frame of reference even when the arm and neck are permitted a realistic number of degrees of rotational freedom and that their levels of performance are unaffected by changes in the input code that would strain analytic methods (for instance, recruitment-coding of joint angle by a population of nonlinear units). We have developed objective methods for analyzing the representations formed by networks and have found that networks trained on the same problem contain consistent classes of hidden units. The success of this approach underscores the usefulness of simplified neural networks as a means for elucidating principles of postural analysis compatible with the known properties of the brain.

Discussion

We have described behavioral studies of postural perception in humans, biological studies of brain substrates for postural analysis, and modeling studies of connectionist networks that solve postural problems. One point emerging from our presentation is that research in the three areas has been carried out largely independently. This is not surprising in light of the fact that modeling studies are of very recent origin but it is not ideal. In the ideal case, modeling studies should function as a theoretical resource for behavioral and biological studies, obeying constraints from the other disciplines and generating predictions that they can test. For interactions of this type to occur, it will be necessary that the models become more sophisticated in two basic regards: they must achieve greater biological realism and greater functional realism.

Biological realism might be imposed at any or all of the following levels: input codes; neural architecture; functional properties of units; learning rules. With respect to *input codes*, we have shown that network performance is robust in the face of major changes in the way in which joint angles are represented, but that the form of the input representation affects the nature of hidden-unit receptive fields. Thus, for comparison between hidden units and somatosensory neurons to be meaningful, it will be necessary to provide networks with inputs approximating as closely as possible those that impinge on the brain. With respect to *neural architecture*, we have shown that networks consisting of a handful of simply interconnected units are able to carry out coordinate transformations. We are left with the problem of explaining how these spare networks relate logically to a somatosensory system that consists of hundreds of millions of neurons arranged into a series of nuclei and areas with reciprocal, topographic interconnections. With respect to *unit properties*, we have shown that it is possible, making the simplest and most conservative assumptions about how neurons work, to design efficient networks. The fact remains that neurons in the somatosensory system may well possess transfer functions more complicated than the ones we have assumed. Shunting synapses that permit highly occlusive or facilitative interactions between inputs (see Sejnowski et al., Chapter 2) might be particularly useful in capturing sharply inflected coordinate-transformation functions. Finally, with respect to *learning rules*, we have shown that postural input-output functions can be programmed by means of the back-propagation algorithm, but we must deal with the issue of how the brain is programmed: by genetic hardwiring, by a training process more plausible than back-propagation, or (see Tesauro, Chapter 4) by a neuronal implementation of back-propagation.

Functional realism requires that whatever individual problem is under study be placed in a broad systems context. The eye-hand problem, heavily studied though it may be, comprises only one instance of a very large class of related body-spatial problems. We are able to judge the location of any body part relative to any other and to judge the orientation or force or velocity of any point on a limb. These operations rely on the same corpus of information about the structure of the body and it is unrealistic to suppose that they are carried out by altogether independent systems. Future modeling studies will have to address the issue of how multiple related functions might be efficiently represented in a single system.

From the principle of functional realism it also follows that postural analysis should be studied in relation to motor control processes. The motor system is a major user of postural information and it is likely that the demands of the motor system influence the form in which the output of postural analysis systems is expressed. Postural information might be utilized during at least five distinct stages of motor control. *1. Computing the spatial coordinates of the target.* Presumably the motor system formulates the location of the target as a first step toward programming any directed movement. If the target is sensed through touch, then this requires postural analysis. *2. Predefining the muscle-length endstate.* It is an essential feature of most models of targeted reaching that the endpoint activation levels of the arm muscles are explicitly formulated before the beginning of the movement (Polit and Bizzi, 1979; Bizzi et al., 1984; Hinton, 1984; Bullock and Grossberg, 1988), although models that do not predefine the endstate levels of muscle activation also have been proposed (Massone and Bizzi, 1989). A system that takes as input the desired spatial location of the hand and generates as output the required lengths of the arm muscles thereby solves the inverse postural problem. To solve this problem requires the same information about the structure of the body that is utilized in solving the forward postural problem. The motor system might acquire this knowledge through training by the postural system (see Rumelhart et al., 1986b for a discussion of "mental practice" by networks with inverse functions) or might somehow draw on information stored by the postural system without forming an independent representation of the inverse transformation. *3. Predefining the muscle-force endstate.* For any movement at the end of which force is applied against an external resistance, it is necessary not merely to establish the lengths of the muscles at endstate but also to establish the torque to be applied at each joint. There are two basic ways in which torque might be computed. It could be calculated along a series of joints one at a time or it could be calculated simultaneously for all the joints. The latter approach, with its desirable traits of rapidity and simplicity, necessitates representing the spatial coordinates of the joints and the point of force application in a common framework. *4. Programming the trajectory.* During human reaching movements, the hand tends to move in a straight line with a bell-shaped velocity profile (Morasso, 1981; Flash and Hogan, 1985). It is generally assumed that these properties reflect an underlying tendency for cost variables associated with the movement to be minimized, but there is no agreement about the identity of the variables or about whether they are best described in terms of anatomical or spatial coordinates (Flash and Hogan, 1985; Cruse and Bruwer, 1987; Hollerbach and Atkeson, 1987; Uno et al., 1989). It is assumed in some models that trajectories are predefined in spatial coordinates (Bizzi et al., 1984; Flash, 1987), and it follows in these models that an inverse postural computation must be carried out to yield the levels of muscle activation corresponding to successive points along the trajectory. In other models, however, the successive levels of muscular activation are generated by a process that takes account of the starting and final positions but does not explicitly compute the spatial coordinates of the hand during the movement (Polit and Bizzi, 1979; Hinton, 1984; Bullock and Grossberg, 1988; Uno et al., 1989). *5. Correcting the trajectory.* Trajectories are subject to midflight correction on the basis of error messages conveyed by sensory feedback or corollary discharge. The computation of corrective torques is greatly simplified by representing the moving elements in an inertial framework (Hinton, 1984).

Over the short term, the demands of functional realism probably should be given greater weight than the demands of biological realism. By designing connectionist systems that deal with entire sets of related postural functions, and by requiring that postural analysis be geared to the needs of the motor system, it should be possible to move toward a realistic formulation of the process and its possible implementations. Rapprochement with brain studies will come in due course as questions arising from modeling are tested in recording and lesion experiments and as knowledge of brain anatomy and phsyiology approaches the level of detail at which meaningful guidance of network design is possible.

Acknowledgements

C.R.O. was supported by NIH grant 1 RO1 NS27287-01.

References

Adams, R.D., and M. Victor (1981) Principles of Neurology (Second Edition). McGraw-Hill Book Co., New York, 1981.

Albus, J.S. (1975) A new approach to manipulator control: the cerebellar rticulation controller (CMAC). J. Dyn. Sys. Meas., Contr. 97:228-233.

Attneave, F., and B. Benson (1969) Spatial coding of tactual stimulation. J. Exp. Psychol. 81:216-222.

Barker, D. (1974) The Morphology of Muscle Receptors. In D. Barker, C.C. Hunt, and A.K. McIntyre (Eds.), Handbook of Sensory Physiology, Vol. II, Part 2, Muscle Receptors (pp. 1-190).

Basmajian, J.V. (1967) Muscles Alive (Second Edition). Baltimore: The Williams and Wilkins Company.

Benati, M., S. Gaglio, P. Morasso, V. Tagliasco, and R. Zaccaria (1980a) Anthropomorphic robots: I. Representing mechanical complexity. Biol. Cybernetics 38:125-140.

Benati, M., S. Gaglio, P. Morasso, V. Tagliasco, and R. Zaccaria (1980b) Anthropomorphic robots: II. Analysis of manipulator dynamics and the output motor impedance. Biol. Cybernetics 38:141-150.

Binder, M.D., J.S. Kroin, G.P. Moore, E.K. Stauffer, and D.G. Stuart (1976) Correlation analysis of muscle spindle responses to single motor unit contractions. J. Physiol. (Lond.) 257:325-336.

Bioulac, B., and Y. Lamarre (1979) Activity of postcentral cortical neurons of the monkey during conditioned movements of a deafferented limb. Brain Res. 172:427-437.

Bisiach, E., and C. Luzzatti (1978) Unilateral neglect of representational space. Cortex 14:129-133.

Bisiach, E., D. Perani, G. Vallar, and A. Berti (1986) Unilateral neglect: Personal and extrapersonal. Neuropsychologia 24:759-767.

Bizzi, E., N. Accornero, W. Chapple, and N. Hogan (1984) Posture control and trajectory formation during arm movement. J. Neurosci. 4: 2738-2744.

Bock, O. (1986) Contribution of retinal vs. extraretinal signals towards visual localization in goal-directed movements. Exp. Brain Res. 64:476-482.

Bowker, R.M., and J.D. Coulter (1981) Intracortical connectivities of somatic sensory and motor areas: Multiple cortical pathways in monkeys. In C.N. Woolsey (Ed.), Cortical Sensory Organization, Vol. 1, Multiple Somatic Areas (pp.205- 242).

Boyd, I.A. (1980) The isolated mammalian muscle spindle. Trends in Neurosci. 3:258-265.

Bruce, C., R. Desimone, and C.G. Gross (1981) Visual properties of neurons in a polysensory area in superior temporal sulcus of the macaque. J. Neurophysiol. 46:369-384.

Bullock, D., and S. Grossberg (1988) Neural dynamics of planned arm movements: Emergent invariants and speed-accuracy properties during trajectory formation. Psych. Review 95:49-90.

Burbaud, P., C. Gross, and B. Bioulac (1985) Peripheral inputs and early unit activity in area 5 of the monkey during a trained forelimb movement. Brain Res. 337:341-346.

Burgess, P.R., J.Y. Wei, F.J. Clark, and J. Simon (1982) Signalling of kinesthetic information by peripheral sensory receptors. Ann. Rev. Neurosci. 5:171-187.

Burke, D. (1980) Muscle spindle function during movement. Trends in Neurosci. 3:251-253.

Burke, D. (1981a) The activity of human muscle spindle endings in normal motor behavior. In R. Porter (Ed.), Neurophysiology IV, International Review of Physiology, Vol. 25 (pp. 91-126). Baltimore: University Park Press.

Burke, D., S.C. Gandevia, and G. Macefield (1988) Responses to passive movement of receptors in joint, skin and muscle of the human hand. J. Physiol. (Lond.) 402:347-361.

Burke, D., K.-E. Hagbarth, L. Lofstedt, and B.G. Wallin (1976a) The responses of human muscle spindle endings to vibration of noncontracting muscles. J. Physiol. (Lond.) 261:673-693.

Burke, D., K.-E. Hagbarth, L. Lofstedt, and B.G. Wallin (1976b) The responses of human muscle spindle endings to vibration during isometric contraction. J. Physiol. (Lond.) 261:695-711.

Burke, D., K.-E. Hagbarth, and N.F. Skuse (1978) Recruitment order of human spindle in isometric voluntary contractions. J. Physiol. (Lond.) 285:101-112.

Burke, R.E. (1981b) Motor units: anatomy, physiology, and functional organization. In J.M. Brookhart, V.B. Mountcastle, V.B. Brooks, and S.R. Geiger (Eds.), Handbook of Physiology, The Nervous System, Vol. II, Part 1, Motor Control (pp. 345-422). Bethesda, MD: American Physiological Society.

Chan, Y.S., J. Kaspar, and V.J. Wilson (1987) Dynamics and directional sensitivity of neck muscle spindle responses to head rotation. J. Neurophysiol. 57:1716-1729.

Chapman, C.E., G. Spaldieri, and Y. Lamarre (1984) Discharge properties of Area 5 neurones during arm movements triggered by sensory stimuli in the monkey. Brain Res. 309:63-77.

Cheney, P.D., and J.B. Preston (1976) Effects of fusimotor stimulation on dynamic and position sensitivities of spindle afferents in the primate. J. Neurophysiol. 39:20-30.

Clark, F.J., and P.R. Burgess (1975) Slowly adapting receptors in cat knee joint: Can they signal joint angle? J. Neurophysiol. 38:1448- 1463.

Clark, F.J., R.C. Burgess, J.W. Chapin, and W.T. Lipscomb (1985) Role of intramuscular receptors in the awareness of limb position. J. Neurophysiol. 54:1529-1540.

Clark, F.J., and K.W. Horch (1986) Kinesthesia. In L. Kaufman and J.P. Thomas (Eds.), Handbook of Perception and Human Performance, Vol. 1 (pp.13.1-13.62). New York: Wiley.

Corkin, S., B. Milner, and T. Rasmussen (1970) Somatosensory thresholds: contrasting effects of postcentral-gyrus and posterior parietal-lobe excisions. Arch. Neurol. 23:41-58.

Costanzo, R.M., and E.P. Gardner (1981) Multiple-joint neurons in somatosensory cortex of awake monkeys. Brain Res. 214:321-333.

Craig, J.J. (1989) Introduction to Robotics. New York: Addison-Wesley Publishing Company.

Critchley, M. (1953) Disorders of the Parietal Lobe. New York: Hafner Publishing Company.

Cross, M.J., and D.I. McCloskey (1973) Position sense following surgical removal of joints in man. Brain Res. 55:443-445.

Crowe, A., W. Keessen, W. Kuus, R. Van Vliet, and A. Zegeling

(1987) Proprioceptive accuracy in two dimensions. Percept. Mot. Skills 64:831-846.
Crowe, A., T. de Ruyter, M. Blaauw, and B. Oosthoek (1988) Information transmission in nonvisual fingertip matching along a horizontal track in the median plane. Biol. Cybern. 58:141-148.
Cruse, H., and M. Bruwer (1987) The human arm as a redundant manipulator: The control of path and joint angles. Biol. Cybern. 57:137-144.
Damasio, A.R., and A.L. Benton (1979) Impairment of hand movements under visual guidance. Neurology 29:170-178.
Denes, G., F. Caviezel, and C. Semenza (1982) Difficulty in reaching objects and body parts. Cortex 18:165-173.
De Renzi, E. (1982) Disorders of Space Exploration and Cognition. New York: John Wiley and Sons.
De Renzi, E., P. Faglioni, and G. Scotti (1970) Hemispheric contribution to exploration of space through the visual and tactile modality. Cortex 6:191-203.
De Renzi, E., and G. Scotti (1970) Autotopagnosia: Fiction or reality? Arch. Neurol. 23:221-227.
Dreyer, D.A., P.R. Loe, C.B. Metz, and B.L. Whitsel (1975) Representation of head and face in postcentral gyrus of the macaque. J. Neurophysiol. 38:714-733.
Duffy, F.H., and J.L. Burchfiel (1971) Somatosensory system: Organizational hierarchy from single units in monkey Area 5. Science 172:273-275.
Eriksen, C.W., M.G.H. Coles, L.R. Morris, and W.P. O'Hara (1985) An electromyographic examination of response competition. Bull. Psychonom. Soc. 23:165-168.
Ettlinger, G. (1963) Defective identification of fingers. Neuropsychologia 1:39-45.
Flash, T. (1987) The control of hand equilibrium trajectories in multi-joint arm movements. Biol. Cybern. 57:257-274.
Flash, T., and N. Hogan (1985) The coordination of arm movements: An experimentally confirmed mathematical model. J. Neurosci. 5:1688-1703.
Foley, J.M., and R. Held (1972) Visually directed pointing as a function of target distance, direction, and available cues. Percept. Psychophys. 12:263-268.

Fromm, C., and E.V. Evarts (1982) Pyramidal tract neurons in somatosensory cortex: central and peripheral inputs during voluntary movement. Brain Res. 238:186-191.

Gainotti, G., C. Cianchetti, and C. Tiacci (1972) The influence of the hemispheric side of lesion on non verbal tasks of finger localization. Cortex 8:364-381.

Gandevia, S.C., and D.I. McCloskey (1976) Joint sense, muscle sense, and their combination as position sense, measured at the distal interphalangeal joint of the middle finger. J. Physiol. (Lond.) 260:387-407.

Gardner, E.P., and R.M. Costanzo (1981) Properties of kinesthetic neurons in somatosensory cortex of awake monkeys. Brain Res. 214:301-319.

Georgopoulos, A.P., R. Caminiti, and J.F. Kalaska (1984) Static spatial effects in motor cortex and Area 5: Quantitative relations in a two-dimensional space. Exp. Brain Res. 54:446-454.

Georgopoulos, A.P., R.E. Kettner, and A.B. Schwartz (1988) Primate motor cortex and free arm movements to visual targets in three-dimensional space. II. Coding of the direction of movement by a neuronal population. J. Neurosci. 8:2928-2937.

Georgopoulos, A.P., J.T. Lurito, M. Petrides, A.B. Schwartz, and J.T. Massey (1989) Mental rotation of the neuronal population vector. Science 243:234-236.

Georgopoulos, A.P., and J.T. Massey (1985) Static versus dynamic effects in motor cortex and Area 5: Comparison during movement time. Behav. Brain Res. 18:159-166.

Georgopoulos, A.P., and J.T. Massey (1988) Cognitive spatial-motor processes. 2. Information transmitted by the direction of two-dimensional arm movements and by neuronal populations in primate motor cortex and Area 5. Exp. Brain Res. 69:315-326.

Gielen, C.C.A.M., and E. J. van Zuylen (1986) Coordination of arm muscles during flexion and supination: application of the tensor analysis approach. Neuroscience 3:527-539.

Goodwin, G.M., D.I. McCloskey, and P.B.C. Matthews (1972) The contribution of muscle afferents to kinesthesia shown by vibration induced illusions of movement and by the effects of paralyzing joint afferents. Brain 95:705-748.

Grigg, P. (1975) mechanical factors influencing response of joint afferent neurons from cat knee. J. Neurophysiol. 38:1473-1484.

Grigg, P., and B.J. Greenspan (1977) Response of primate joint afferent neurons to mechanical stimulation of the knee joint. J. Neurophysiol. 40:1-8.

Grigg, P., and A.H. Hoffman (1982) Properties of Ruffini afferents revealed by stress analysis of isolated sections of cat knee capsule. J. Neurophysiol. 47:41-54.

Gross, Y., R. Webb, and R. Melzack (1974) Central and peripheral contributions to localization of body parts: Evidence for a central body schema. Exp. Neurol. 44:346-362.

Hanson, S. J. and Burr, D. J., (1987) Minkowski-r Backpropagation: Learning in Connectionist Models with Non-Euclidean Error Signals, In the Proceedings of Neural Networks: Natural and Synthetic, D. Anderson (Ed.) , American Institute of Physics.

Hasan, Z. (1983) A model of spindle afferent response to muscle stretch. J. Neurophysiol. 49:989-1006.

Hatze, H. (1977) A myocybernetic control model of muscle. Biol. Cybernetics 25:103-119.

Hatze, H. (1978) A general myocybernetic control model of muscle. Biol. Cybernetics 28:143-157.

Head, H., and G. Holmes (1911) Sensory disturbances from cerebral lesions. Brain 34:102-254.

Hecaen, H., and M.L. Albert (1978) Human Neuropsychology. New York: John Wiley and Sons.

Heilman, K.M., and E. Valenstein (1979) Clinical Neuropsychology. New York: Oxford University Press.

Heister, G., W.H. Ehrenstein, and P. Schroeder-Heister (1986) Spatial S-R compatibility effects with unimanual two-finger choice reactions for prone and supine hand positions. Perception and Psychophysics 40:271-278.

Heister, G., W.H. Ehrenstein, and P. Schroeder-Heister (1987) Spatial S-R compatibility with unimanual two-finger choice reactions: Effects of irrelevant stimulus location. Perception and Psychophysics 42:195-201.

Henneman, E., and L.M. Mendell (1981) Functional organization of motoneuron pool and its inputs. In J.M. Brookhart, V.B. Mountcastle, V.B. Brooks, and S.R. Geiger (Eds.), Handbook of Physiology, The Nervous System, Vol. II, Part 1, Motor Control (pp. 423-507). Bethesda, MD: American Physiological Society.

Hikosaka, K., E. Iwai, H.-A. Saito, and K. Tanaka (1988) Polysensory properties of neurons in the anterior bank of the caudal superior temporal sulcus of the macaque monkey. J. Neurophysiol. 60:1615- 1637.

Hinton, G. (1984) Parallel computations for controlling an arm. J. Motor Behav. 16:171- 194.

Hollerbach, J.M., and C.G. Atkeson (1987) Deducing planning variables from experimental arm trajectories: Pitfalls and possibilities. Biol. Cybern. 56:279-292.

Horch, K.W., F.J. Clark, and P.R. Burgess (1975) Awareness of knee joint angle under static conditions. J. Neurophysiol. 38:1436-1447.

Houk, J., and E. Henneman (1967) Responses of Golgi tendon organs to active contractions of the soleus muscle of the cat. J. Neurophysiol. 30:466-481.

Houk, J.C., and W. Z. Rymer (1981) Functional organization of motoneuron pool and its inputs. In J.M. Brookhart, V.B. Mountcastle, V.B. Brooks, and S.R. Geiger (Eds.), Handbook of Physiology, The Nervous System, Vol. II, Part 1, Motor Control (pp. 257-323). Bethesda, MD: American Physiological Society.

Howard, I.P. (1982) Human Visual Orientation. New York: John Wiley and Sons.

Howard, I.P., and W.B. Templeton (1966) Human Spatial Orientation. New York: John Wiley and Sons.

Hulliger, M. (1984) The mammalian muscle spindle and its central control. Rev. Physiol. Biochem. Pharmacol. 101:1-110.

Hyvarinen, J. (1981) Regional distribution of functions in parietal association area 7 of the monkey. Brain Res. 206:287-303.

Hyvarinen, J., and A. Poranen (1978) Receptive field integration and submodality convergence in the hand area of the post-central gyrus of the alert monkey. J. Physiol. (Lond.) 283:539-556.

Hyvarinen, J., and Y. Shelepin (1979) Distribution of visual and somatic functions in the parietal associative area of the monkey. Brain Res. 169:561-564.

Iwamura, Y., M. Tanaka, and O. Hikosaka (1980) Overlapping representation of the fingers in the somatosensory cortex (Area 2) of the conscious monkey. Brain Res. 197:516-520.

Jami, L., and J. Petit (1978) Fusimotor actions on sensitivity of spindle secondary endings to slow muscle stretch in cat peroneus tertius. J. Neurophysiol. 41:860-869.

Jennings, V.A., Y. Lamour, H. Solis, and C. Fromm. Somatosensory cortex activity related to position and force. J. Neurophysiol. 49:1216-1229.

Jones, E.G., and H. Burton (1976) Areal differences in the laminar distribution of thalamic afferents in cortical fields of the insular, parietal and temporal regions of primates. J. Comp. Neurol. 168:197-247.

Jones, E.G., and R. Porter (1980) What is Area 3a? Brain Res. Rev. 2:1-43.

Jongen, H.A.H., J.J.D. van der Gon, and C.C.A.M. Gielen (1989) Activation of human arm muscles during flexion/extension and pronation/supination tasks: A theory on muscle coordination.

Kaas, J.H. (1983) What, if anything, is SI? Organization of first somatosensory area of cortex. Physiol. Rev. 63:206-231

Kaas, J.H., R.J. Nelson, M. Sur, C.-S. Lin, and M.M. Merzenich (1979) Multiple representation of the body within the primary somatosensory cortex of primates. Science 264:521-523.

Kalaska, J.F., R. Caminiti, and A.P. Georgopoulos (1983) Cortical mechanisms related to the direction of two-dimensional arm movements: Relations in parietal Area 5 and comparison with motor cortex. Exp. Brain Res. 51:247-260.

Kalaska, J.F., and M.L. Hyde (1985) Area 4 and Area 5: Differences between the load direction-dependent discharge variability of cells during active postural fixation. Exp. Brain Res. 59:197-202.

Kettner, R.E., A.B. Schwartz, and A.P. Georgopoulos (1988) Primate motor cortex and free arm movements to visual targets in three-dimensional space. III. Positional gradients and population coding of movement direction from various movement origins. J. Neurosci. 8:2938-2947.

Kinsbourne, M., and E.K. Warrington (1962) A study of finger agnosia. Brain 85:47-66.

Kohonen, T. (1988) Self-Organization and Associative Memory (Second Edition). New York: Springer-Verlag.

Kuperstein, M. (1987) Adaptive visual- motor coordination in multijoint robots using parallel architecture. In Proceedings of the IEEE International Conference on Robotics and Automation, March 31-April 3 (pp. 1595-1602).

Kuperstein, M. (1988) Neural model of adaptive hand-eye

coordination for single postures. Science 239:1308-1311.

Ladavas, E. (1987) Is the hemispatial deficit produced by right parietal lobe damage associated with retinal or gravitational coordinates? Brain 110:167-180.

Leinonen, L., J. Hyvarinen, G. Nyman, and I. Linnankoski (1979) I. Functional properties of neurons in lateral part of associative area 7 in awake monkeys. Exp. Brain Res. 34:299-320.

Leinonen, L., J. Hyvarinen, and A.R.A. Sovijarvi (1980) Functional properties of neurons in the temporo-parietal association cortex of awake monkey. Exp. Brain Res. 39:203-215.

Lewis, D.M., and U. Proske (1972) The effect of muscle length and rate of fusimotor stimulation on the frequency of discharge in primary endings from muscle spindles in the cat. J. Physiol. (Lond.) 222:511-535.

Loemker, K.K. (1930) Certain factors determining the accuracy of a response to the direction of a visual object. J. Exp. Psychol. 30:500-518.

Matthews, P.B.C. (1977) Muscle Afferents and Kinesthesia. Brit. Med. Bull. 33:137-142.

Matthews, P.B.C. (1981) Muscle spindles: their messages and their fusimotor supply. In J.M. Brookhart, V.B. Mountcastle, V.B. Brooks, and S.R. Geiger (Eds.), Handbook of Physiology, The Nervous System, Vol. II, Part 1, Motor Control (pp. 1415-1447). Bethesda, MD: American Physiological Society.

Matthews, P.B.C. (1982) Where does Sherrington's "muscular sense" originate? Muscles? Joints? Corollary discharges? Ann. Rev. Neurosci. 5:189-218.

McCloskey, D.I. (1973) Differences between the senses of movement and position shown by the effects of loading and vibration of muscles in man. Brain Res. 63:119-131.

McCloskey, D.I. (1978) Kinesthetic Sensibility. Physiol. Rev. 58:763-820.

McCloskey, D.I. (1981) Corollary discharges: Motor commands and perception. In J.M. Brookhart, V.B. Mountcastle, V.B. Brooks, and S.R. Geiger (Eds.), Handbook of Physiology, The Nervous System, Vol. II, Part 2, Motor Control (pp. 1415-1447). Bethesda, MD: American Physiological Society.

McKay, W.A., M.C. Kwan, J.T. Murphy, and Y.C. Wong (1978) Responses to active and passive wrist rotation in Area 5 of

awake monkeys. Neurosci. Lett. 10:235-239.
McKenna, T.M., B.L. Whitsel, and D.A. Dreyer (1982) Anterior Parietal Cortical Topographic Organization in Macaque Monkey: A Reevaluation. J. Neurophysiol. 48:289-317.
Mel, B. (1988) Building and using mental models in a sensory-motor domain: A connectionist approach. Proceedings of the 4th International Conference on Machine Learning, 207-213
Milner-Brown, H.S., R.B. Stein, and R. Yemm (1973) Changes in firing rate of human motor units during linearly changing voluntary contractions. J. Physiol. (Lond.) 230:371-390.
Moberg, E. (1983) The role of cutaneous afferents in position sense, kinaesthesia, and motor function of the hand. Brain 106:1-19.
Morasso, P. (1981) Spatial control of arm movements. Exp. Brain Res. 42:223-227.
Mountcastle, V.B. (1984) Central nervous mechanisms in mechanoreceptive sensibility. In J.M. Brookhart, V.B. Mountcastle, I. Darian-Smith, and S.R. Geiger (Eds.), Handbook of Physiology, The Nervous System, Vol. III, Part 2, Sensory Processes (pp. 789-878). Bethesda, MD: American Physiological Society.
Mountcastle, V.B., J.C. Lynch, A. Georgopoulos, H. Sakata, and C. Acuna (1975) Posterior parietal association cortex of the monkey: Command functions for operations within extrapersonal space. J. Neurophysiol. 38:871-908.
Murray, E.A., and J.D. Coulter (1981) Supplementary Sensory Area: The Medial Parietal Cortex in the Monkey. In C.N. Woolsey (Ed.), Cortical Sensory Organization, Vol. 1, Multiple Somatic Areas (pp. 167-195).
Murthy, K.S.K. (1978) Vertebrate fusimotor neurones and their influences on motor behavior. Prog. in Neurobiol. 11:249-307.
Nicoletti, R., G.P. Anzola, G. Luppino, G. Rizzolatti, and C. Umilta (1982) Spatial compatibility effects on the same side of the body midline. J. Exp. Psychol.: Human Percep. and Perform. 8:664-673.
Nicoletti, R., and C. Umilta (1985) Responding with hand and foot: The right/left prevalence in spatial compatibility is still present. Perception and Psychophysics 38:211-216.
Ogden, J.A. (1985a) Autotopagnosia: Occurrence in a patient without nominal aphasia and with an intact ability to point to parts of

annimals and objects. Brain 108:1009-1022.

Ogden, J.A. (1985b) Contralesional neglect of constructed visual images in right and left brain-damaged patients. Neuropsychologia 23:273-277.

Paillard, J., and M. Brouchon (1968) Active and passive movements in the calibration of position sense. In S.J. Freedman (Ed.), The Neuropsychology of Spatially Oriented Behavior (pp. 37-55). Homewood, Illinois: The Dorsey Press.

Paillard, J. and M. Brouchon (1974) A proprioceptive contribution to the spatial encoding of position cues for ballistic movements. Brain Res. 71:273- 284.

Partridge, L.D., and L.A. Benton (1981) Muscle, the motor. In J.M. Brookhart, V.B. Mountcastle, V.B. Brooks, and S.R. Geiger (Eds.), Handbook of Physiology, The Nervous System, Vol. II, Part 1, Motor Control (pp. 43-106). Bethesda, MD: American Physiological Society.

Pellionisz, A., and R. Llinas (1982) Space-time representation in the brain: The cerebellum as a predictive space-time metric tensor. Neuroscience 7:2949-2970.

Pellionisz, A., and W. Graf (1987) Tensor network model of the "three-neuron vestibulo-ocular reflex-arc" in the cat. J. Theoret. Neurobiol. 5:127-151.

Perenin, M.-T., and A. Vighetto (1988) Optic ataxia: A specific disruption in visuomotor mechanisms. Brain 111: 643-674.

Phillips, C.G., T.P.S. Powell, and M. Wiesendanger (1971) Projection from low-threshold muscle afferents of hand and forearm to Area 3a of Baboon's cortex. J. Physiol. (Lond.) 217:419-446.

Poeck, K., and B. Orgass (1969) An experimental investigation of finger agnosia. Neurology 19:801-807.

Poeck, K., and B. Orgass (1971) The concept of the body schema: A critical review and some experimental results. Cortex 7:254-277.

Polit, A., and E. Bizzi (1979) Characteristics of motor programs underlying reaching in monkeys. J. Neurophysiol. 42:183-194.

Pons, T.P., P.E. Garraghty, C.G. Cusick, and J.H. Kaas (1985) A sequential representation of the occiput, arm, forearm annd hand across the rostrocaudal dimension of areas 1, 2 and 5 in macaque monkeys. Brain Res. 335:350-353.

Prablanc, C., J.F. Echallier, E. Komilis and M Jeannerod (1979)

Optimal response of eye and hand motor systems in pointing at a visual target. Biol. Cybern. 35:113-124.

Prochazka, A. (1981) Muscle spindle function during normal movement. In R. Porter (Ed.) Neurophysiology IV, International Review of Physiology, Vol. 25 (pp. 47-90).

Proske, U. (1981) The Golgi tendon organ. In R. Porter (Ed.), Neurophysiology IV, International Review of Physiology, Vol. 25 (pp. 127-171).

Proske, U., H.-G. Schaible, and R.F. Schmidt (1988) Joint receptors and kinesthesia. Exp. Brain Res. 72:219-224.

Rack, P.M.H., and D.R. Westbury (1969) The effects of length and stimulus rate on tension in the isometric cat soleus muscle. J. Physiol. (Lond.) 204:443-460.

Raibert, M.H. (1978) A model for sensorimotor control and learning. Biol. Cybernetics 29:29-36.

Richmond, F.J.R., and D.A. Bakker (1982) Anatomical organization and sensory receptor content of soft tissues surrounding upper cervical vertebrae in the cat. J. Neurophysiol. 48:49-61.

Richmond, F.J., and P.P. Vidal (1988) The motor system: Joints and muscles of the neck. In B.W. Peterson and F.J. Richmond (Eds.), Control of Head Movement (pp. 1-21). New York: Oxford University Press.

Ritter, H., and K. Schulten (1988) Extending Kohonen's self-organizing mapping algorithm to learn ballistic movements. In R. Eckmiller and C. Marlsburg (Eds.), NATO ASI Series, Vol. F41, Neural Computers. Berlin: Springer-Verlag.

Robinson, C.J., and H. Burton (1980) Somatic submodality distribution within the second somatosensory (SII), 7b, retroinsular, postauditory, annd granular insular cortical areas of M. fascicularis. J. comp. Neurol. 192:93-108.

Robinson, D.A. (1982) The use of matrices in analyzing the three-dimensional behavior of the vestibulo-ocular reflex. Biol. Cybernetics 46:53-66.

Roland, E. (1976) Astereognosis: tactile discrimination after localized hemispheric lesions in man. Arch. Neurol. 33:551-558.

Roland, P.E., and H. Ladegaard-Pedersen (1977) A quantitative analysis of sensations of tension and of kinaesthesia in man: Evidence for a peripherally originating muscular sense and for a sense of effort. Brain 100:671-692.

Rondot, P., J. de Recondo, and J.L.R. Dumas (1977) Visuomotor ataxia. Brain 100:355-376.

Rondot, P. (1989) Visuomotor ataxia. In J.W. Brown (Ed.), Neuropsychology of Visual Perception. Hillsdale, NJ: Lawrence Erlbaum Associates.

Rothwell, J.C., M.M. Traub, B.L. Day, J.A. Obeso, P.K. Thomas, and C.D. Marsden (1982) Manual motor performance in a deafferented man. Brain 105:515-542.

Rumelhart, D.E., G.E. Hinton, and R.J. Williams (1986a) Learning internal representations by error propagation. In D.E. Rumelhart and J.L. McClelland (Eds.), Parallel Distributed Processing, Explorations in the Microstructure of Cognition, Vol. 1: Foundations (pp. 318-364). Cambridge, Massachusetts: The M.I.T. Press.

Rumelhart, D.E., P. Smolensky, J.L. McClelland and G.E. Hinton (1986b) Schemata and sequential thought processes in PDP models. In J.L. McClelland and D.E. Rumelhart (Eds.), Parallel Distributed Processing, Explorations in the Microstructure of Cognition, Vol. 2: Psychological and Biological Models (pp. 7-57). Cambridge, Massachusetts: The M.I.T. Press.

Rymer, W.Z., and A. D'Almeida (1980) Joint position sense: the effects of muscle contraction. Brain 103:1-22.

Sakata, H., Y. Takaoka, A, Kawarasaki, and H. Shibutani (1973) Somatosensory properties of neurons in the superior parietal cortex (Area 5) of the rhesus monkey. Brain Res. 64:85-102.

Sauguet, J., A.L. Benton, and H. Hecaen (1971) Disturbances of the body schema in relation to language impairment and hemispheric locus of lesion. J. Neurol. Neurosurg. Psychiat. 34:496-501.

Schwartz, A.B., R.E. Kettner, and A.P. Georgopoulos (1988) Primate motor cortex and free arm movements to visual targets in three-dimensional space: I. Relations between single cell discharge and direction of movement. J. Neurosci. 8:2913-2927.

Schwarz, D.W.F., L. Deeke, and J.M. Frederickson (1973) Cortical projection of group I muscle afferents to areas 2, 3a, and the vestibular field in the rhesus monkey. Exp. Brain Res. 17:516-526.

Seal, J., C. Gross, and B. Bioulac (1982) Activity of neurons in Area 5 during a simple arm movement in monkeys before and after

deafferentation of the trained limb. Brain Res. 250:229-243.
Semenza, C., and H. Goodglass (1985) Localization of body parts in brain injured subjects. Neuropsychologia 23:161-175.
Semmes, J. (1965) A nontactual factor in astereognosis. Neuropsychologia 3:295-315.
Semmes, J., S. Weinstein, L. Ghent, and H.-L. Teuber (1963) Correlates of impaired orientation in personal and extrapersonal space. Brain 86:747-772.
Simon, J.R., and J.L. Craft (1970) Effects of an irrelevant auditory stimulus on visual choice reaction time. J. Exp. Psychol. 86:272-274.
Sittig, A.C., J.J.D. van der Gon, and C.C.A.M. Gielen (1985) Separate control of arm position and velocity demonstrated by vibration of muscle tendon in man. Exp. Brain Res. 60:445-453.
Skoglund, S. (1973) Joint receptors and kinesthaesis. In A. Iggo (Ed.), Handbook of Sensory Physiology, Vol. 2, The Somatosensory System (pp. 111-136). New York: Springer-Verlag.
Slinger, R.T., and V. Horsley (1906) Upon the orientation of points in space by the muscular, arthrodial, and tactile senses of the upper limbs in normal individuals and in blind persons. Brain 29:1-27.
Soechting, J.F. (1982) Does position sense at the elbow reflect a sense of elbow joint angle or one of limb orientation? Brain Res. 248:392-395.
Soechting, J.F., and B. Ross (1984) Psychophysical determination of coordinate representation of human arm orientation. Neuroscience 13:595-604.
Soso, M.J., and E.E. Fetz (1980) Responses of identified cells in postcentral cortex of awake monkeys during comparable active and passive joint movements. J. Neurophysiol. 43:1090-1110.
Stephens, J.A., R.M. Reinking, and D.G. Stuart (1975) Tendon organs of cat medial gastrocnemius: Responses to active and passive forces as a function of muscle length. J. Neurophysiol. 38:1217-1231.
Tanji, J. (1975) Activity of neurons in cortical area 3a during maintenance of steady postures by the monkey. Brain Res. 88:549- 553.
Tanji, J., and S.P. Wise (1981) Submodality distribution in sensorimotor cortex of the unanesthetized monkey. J.

Neurophysiol. 45:467-481.

Uno, Y., M. Kawato, and R. Suzuki (1989) Formation and control of optimal trajectory in human multijoint arm movement. Biol. Cybern. 61:89-101.

Vallbo, A.B. (1970a) Slowly adapting muscle receptors in man. Acta Physiol. Scand. 78:315-330.

Vallbo, A.B. (1970b) Discharge patterns of human muscle spindle afferents during voluntary isometric contractions. Acta Physiol. Scand. 78:315- 330.

Vallbo, A.B. (1979) Somatosensory, proprioceptive, and sympathetic activity in human peripheral nerves. Physiol. Rev. 59:919-957.

Vallbo, A.B., M. Hulliger, and E. Nordh (1981) Do spindle afferents monitor joint position in man? A study with active position holding. Brain Res. 204:209-213.

Villis, T., and D. Tweed (1988) A matrix analysis for a conjugate vestibulo-ocular reflex. Biol. Cybernetics 59:237-245.

Wallace, R.J. (1972) Spatial S-R compatibility effects involving kinesthetic cues. J. Exp. Psychol. 1:163-168.

Warfel, J.H. (1974) The Extremities (Fourth Edition). Philadelphia: Lea and Fibiger.

Warfel, J.H. (1985) The Head, Neck and Trunk (Fifth Edition). Philadelphia: Lea and Febiger.

Wei, J.Y., J. Simon, M. Randic, and P.R. Burgess (1986) Joint angle signalling by muscle spindle receptors. Brain Res. 370:108-118.

Worringham, C.J., and G.E. Stelmach (1985) The contribution of gravitational torques to limb position sense. Exp. Brain Res. 61:38-42.

Worringham, C.J., G.E. Stelmach, and Z.E. Martin (1987) Limb segment inclination sense in proprioception. Exp. Brain Res. 66:653-658.

Yumiya, H., K. Kubota, and H. Asanuma (1974) Activities of neurons in Area 3a of the cerebral cortex during voluntary movements in the monkey. Brain Res. 78:169-177.

CHAPTER 9

The Development of Ocular Dominance Columns: Mechanisms and Models

K. D. Miller and M. P. Stryker

Introduction

The function of the visual cortex depends upon the precision with which its cells are connected to their inputs. Such precise connections are formed during normal development in large part by the rearrangement of initial connections whose pattern is much more diffuse. The cortical network reorganizes itself under the influence of its own neural activity.

The notion that the visual cortex is a self-organizing biological system has prompted much experimental work and theoretical interest over the past 30 years. One approach toward understanding this self-organization is to construct mathematical models of the system that incorporate the microscopic cellular properties that are believed to exist on experimental grounds. The models may then be followed, from an initial state of diffuse connectivity through a stage of refinement to an ultimate stage of precise connectivity, by computer simulation or with analytic methods. This approach can be used to determine the range of microscopic properties that can give rise to the macroscopic organization observed experimentally.

In this chapter we first present some of the experimental background to studies of the development of ocular dominance columns, a striking feature of organization of the visual cortex. We then present a formal model of the system, the elements of which correspond to identified neural structures whose relevant properties may be measured experimentally (and have been measured in adult animals). We show that several different biologically reasonable mechanisms of synaptic plasticity may be described by the same mathematics. These mechanisms are distinguished by the different biological features that are summarized in each term of the mathematical model.

The model develops patterns like those observed in biological development. Analysis of the model allows prediction of the size of ocular dominance columns from potential experimental measurements. This analysis illuminates the relative importance of the properties of the cortex and its inputs in determining this size. The determinants of development in the model are illustrated through simulation and mathematical analysis. Finally, we outline ways in which a combination of theoretical and experimental approaches can be used to further our understanding of neural development.

Development of Ocular Dominance Columns in Visual Cortex

The primary visual cortex contains a single map of the world as seen through the two eyes. At a scale of half a millimeter, however, this map is not continuous. Instead, in adult humans, monkeys, cats, and many other species, inputs serving the two eyes are largely (or in some species completely) segregated into alternate patches serving the left and right eyes.

"Ocular dominance patches" or "columns" are the names given to these alternate patches or stripes of input serving the two eyes (figure 1). The profound influence of neural activity on their development and plasticity makes them an excellent model system for studies of the organization of connections in the central nervous system. This influence was first noted in the clinic more than 100 years ago (von Senden, 1960). At that time, it became possible to remove cataracts, which obscure vision by purely optical means, without any direct effect on the neurons of the retina. In patients who acquired cataracts in adulthood, this operation "miraculously" restored sight. In patients with congenital cataracts in one eye, that had occluded vision from the

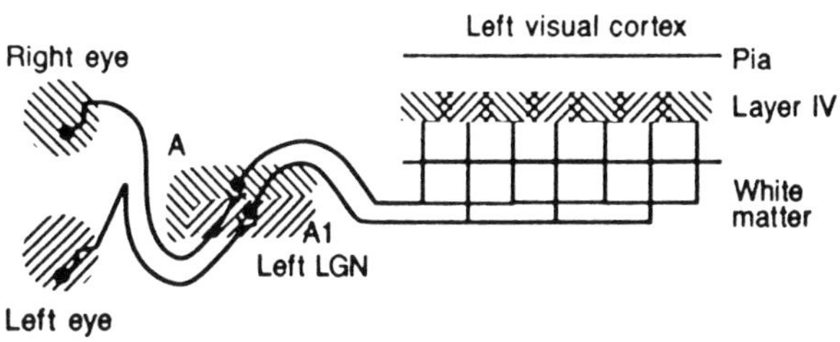

Figure 1: Cartoon showing part of the organization of the mammalian visual system. The two eyes are illustrated to the left, from which emerge the optic nerves. At the optic chiasm the fibers from the nasal side of each retina cross to the opposite optic tract en route to the lateral geniculate nucleus (LGN). The LGN receives these inputs from the two eyes in separate layers (cartooned as layers A and A1 as in the cat) and relays this visual information to the visual cortex, also known as striate cortex or area 17. In the visual cortex, terminations from the lateral geniculate nucleus are found principally in layers IV and VI, with the major concentration in layer IV. In adult animals, as illustrated, these terminals are made in eye-specific patches with a repeat distance (from the middle of one left-eye patch through the right-eye patch back to the middle of the next left-eye patch) of about a millimeter (850 um in the cat). Above and below each ocular dominance patch in layer IV, cortical neurons tend to respond better to the eye that provides input to the patch than to the other eye, revealing a columnar (vertical) organization of ocular dominance throughout all layers of the cortex; hence the name "ocular dominance columns".

time of birth, the same operation did not restore useful vision to that eye. This was true even though no serious histological damage was evident in the retina or in visual structures in the brain.

The changes in the visual system that underly this puzzling phenomenon of *amblyopia ex anopsia* (poor vision in one eye resulting from not seeing through that eye during early life), confirmed in further clinical studies, were finally explained in the early 1960s by the work of Hubel and Wiesel. They showed that most neurons in the cat's visual cortex ordinarily respond to stimulation through either eye, although perhaps one quarter of neurons respond only to a single eye (Hubel and Wiesel, 1962). Such binocular responses in the visual cortex were unchanged by even years of unilateral visual deprivation in adult animals; but as little as a few days or weeks of monocular visual deprivation during a sensitive period in early life left most cortical neurons unresponsive to the eye whose vision had been occluded (Hubel and Wiesel, 1970; Wiesel and Hubel, 1963a). In these animals, the two eyes were entirely normal, and neurons in the lateral geniculate nucleus (LGN, see figure 1), the nucleus that is the major source of input to visual cortex, appeared to be nearly normal. The geniculate neurons in the layer driven by the deprived eye responded almost identically to those in the layer driven by the seeing eye, although they were a bit smaller and stained somewhat more palely (Wiesel and Hubel, 1963b). Thus, it appeared that neonatal monocular visual deprivation had produced a change in the visual cortex, where inputs from the two eyes first had the opportunity to interact on single neurons, but not in earlier, more peripheral stages of the visual system. Binocular visual deprivation (closure of both eyes) in early life produced no ill effects, suggesting that the changes produced by unilateral visual deprivation were due to a competitive interaction between the geniculocortical afferents serving the two eyes, rather than merely to disuse of the occluded eye's afferents (Wiesel and Hubel, 1965). This conclusion was reinforced by failure of monocular deprivation to produce changes either in the most peripheral portion of the visual field, which is viewed through only one eye, or in a region of LGN and visual cortex in which input from the seeing eye was experimentally removed (Guillery, 1972; Sherman et al., 1974).

The changes of binocular connections in the developing visual cortex appeared to be due entirely to alterations of the spatial and temporal patterns of neural discharge in geniculocortical afferents, rather than to metabolic or surgical complications of the deprivation procedure. The monocular deprivation effects did not depend on light deprivation, since similar effects were produced when the image seen by one eye was merely blurred (Wiesel and Hubel, 1963a). Perhaps the most striking finding was that when neural activity during early life was made equal but asynchronous in the two eyes by occlusion of each eye on alternate days, or by surgically or optically misaligning images in the two eyes, the partial segregation of visual responses into ocular dominance columns was made nearly complete (Hubel and Wiesel, 1965; Van Sluyters and Levitt, 1985). In such a cortex nearly all cells in each cortical column were driven exclusively by a single eye, and the eye that dominated alternated from column to column. This is in contrast to the normal case in which most neurons respond to stimulation through either eye, although cells in each column tend

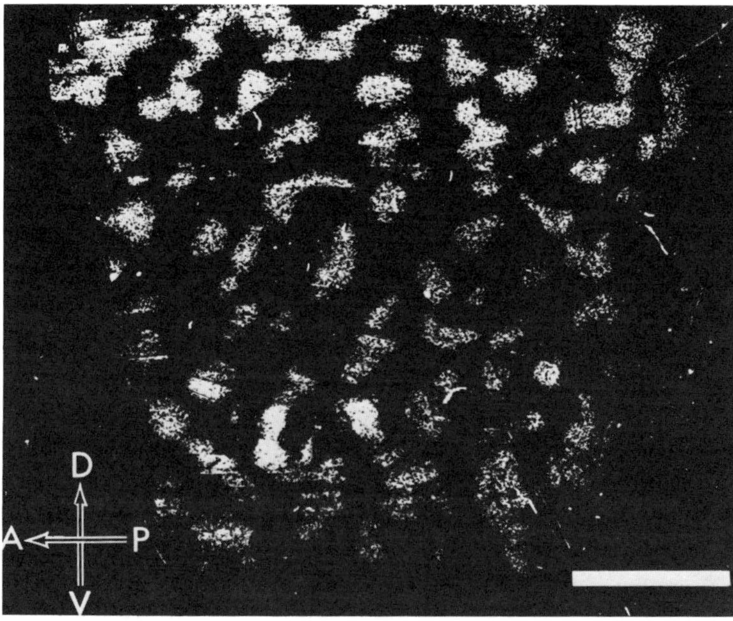

Figure 2: Surface view of ocular dominance patches in a normal cat. This is a reconstruction made from serial autoradiographically labeled parasagittal sections of layer IV of visual cortex, photographed in dark field following an injection of [^{3}H]-amino acid into one eye. Silver grains indicating the presence of geniculocortical afferent terminals serving the injected eye appear as white dots. This picture shows what the ocular dominance patches of one eye would look like if one were to flatten the cortex, remove the superficial layers, and look down on the flattened layer IV from above the cortical surface. The injected eye was ipsilateral to the cortical hemisphere shown. Scale bar = 2 mm. From LeVay, Stryker, and Shatz, 1978. Reprinted by permission of the Journal of Comparative Neurology.

to respond more strongly to stimulation through one eye than through the other (Hubel and Wiesel, 1965). This complete segregation of responses appeared not to result from alterations in levels of activity but merely from alterations in the relative timing of activity in the two eyes.

The structural basis of ocular dominance columns was revealed in experiments in which the population of geniculocortical afferent terminals serving one eye was labelled, either by degeneration methods or autoradiographically by transneuronal axonal transport of [^{3}H]-sugars or amino acids injected into the vitreous humor of one eye

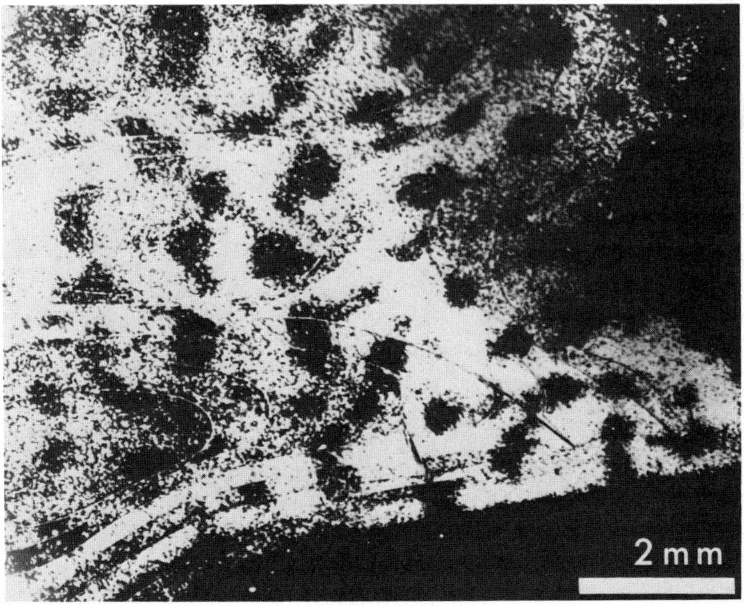

Figure 3: Surface view of ocular dominance patches in a one-year old cat that had experienced monocular visual occlusion beginning before the time of natural eye opening. The eye that had remained open during early life received an injection of [^{3}H]-amino acid. This reconstruction of the cortical hemisphere ipsilateral to the injected eye was made like that of figure 2, to which it should be compared. In these deprived animals, white dots of label indicate terminals serving the eye that had remained open. These open-eye terminals occupy more than 80 percent of layer IV. Physiological experiments in this and similar animals revealed that input from the deprived eye was confined to the holes in this sea of label. Unpublished data from experiments reported in Shatz and Stryker, 1978.

(figure 2; Hubel and Wiesel, 1972; Shatz, Lindstrom, and Wiesel, 1977; Wiesel, Hubel, and Lam, 1974). These methods demonstrated that the geniculate terminals from each eye in normal adult animals were segregated into alternate patches in layer 4 of visual cortex. These methods also revealed that an effect of early monocular deprivation was to change the relative sizes of the patches of inputs to visual cortex serving the two eyes, while keeping the repeat distance the same (figure 3) (Hubel, Wiesel, and LeVay, 1977; Shatz and Stryker, 1978). But the reason for such a sensitive period in early life, in which such small alterations in visual experience as closure of one

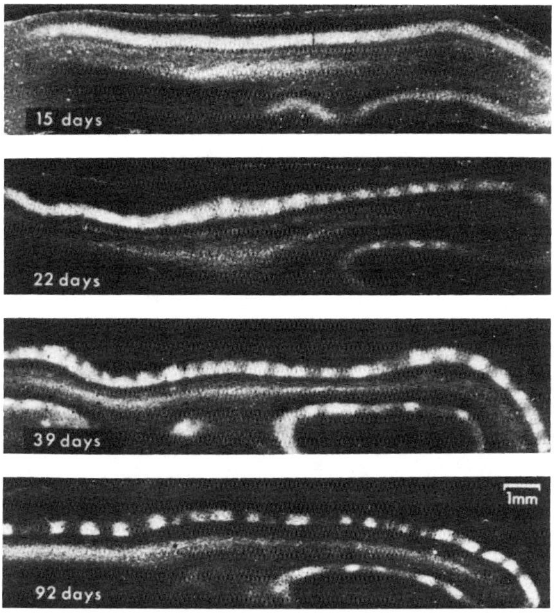

Figure 4: Progressive changes in the transneuronal labelling pattern of visual cortex following an injection of [^{3}H]-amino acid into one eye. Before 15 days of age, label from each eye is found uniformly throughout layer IV. By 21 days, periodicity in the labelling is first evident as afferents begin to segregate. The segregation proceeds rapidly until 5-6 weeks of age and more slowly thereafter, attaining nearly adult levels by 2 months and fully adult levels by 3 months of age. From LeVay and Stryker, 1979. Reprinted with permission from the Society for Neuroscience.

eye for a few days lead to permanent alterations in connectivity, remained obscure.

The reason for this early plasticity was clarified only when it was appreciated that the initial growth of eye-specific inputs into the visual cortex does not take place in the form of ocular dominance patches. Instead, geniculocortical relay cells serving the two eyes initially make connections to the cortex in a uniform, continuous, and completely overlapping fashion (Hubel, Wiesel, and LeVay, 1977; LeVay, Wiesel, and Hubel, 1980; LeVay, Stryker, and Shatz, 1978; Rakic, 1977). Ocular dominance patches then develop by the progressive segregation of these initially overlapping inputs. This development was most clearly revealed by the progressive changes in the transneuronal labelling pattern of visual cortex following an

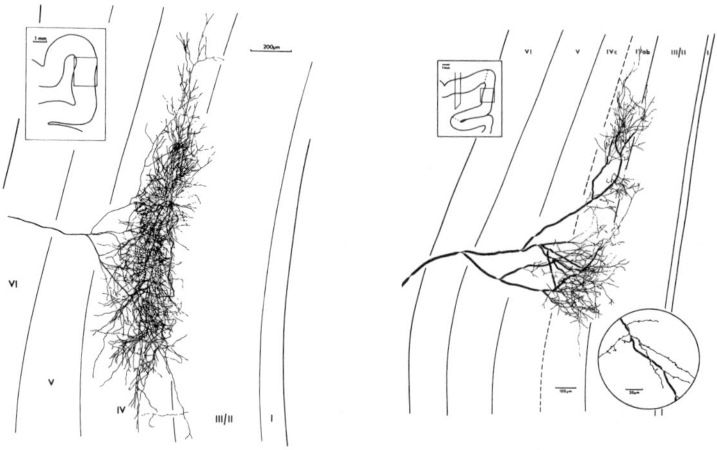

Figure 5: Presumed geniculocortical afferent terminal arbors labelled (left) by a Golgi method in an 8-day-old kitten, before the segregation of ocular dominance patches, and (right) in a normal adult cat by horseradish peroxidase injected into the optic radiations below visual cortex. Each picture shows a large afferent arbor ramifying dorsally in layer IV, presumed to be a Y-type geniculocortical afferent. Note that the arbor in the young animal is continuous over more than 2 mm, while terminals of the arbor from the adult animal are largely confined to two dense, half-millimeter patches separated by a half-millimeter space. Terminal arbors like the one shown from the young animal are not found in adults, and presumably must be "pruned" (or lost) during normal development. Left: from LeVay and Stryker, 1979. Reprinted with permission from the Society for Neuroscience. Right: from Ferster and LeVay, 1978. Reprinted by permission of the Journal of Comparative Neurology.

injection into one eye of animals at different ages (figure 4).

At least some of the early geniculocortical connections are from afferents that arborize over more than 2 mm in cortex, a distance large enough to span two pairs of ocular dominance columns (LeVay and Stryker, 1979) (figure 5). Beginning prenatally in humans and monkeys, but postnatally in cats, these afferents reorganize their terminal arbors so as to partition the cortex into the alternate, half-millimeter ocular dominance patches. It is only during and slightly after this period in which geniculocortical connections are normally reorganizing that they are sensitive to monocular deprivation. Thus, we may interpret the plasticity produced by monocular deprivation not as a bizarre pathological response, but as the abnormal outcome, produced by abnormal patterns of activity, of a normal developmental process.

A mechanism similar to that described by Hebb (1949) was proposed to account for these phenomena produced by visual deprivation during early life (Changeux and Danchin, 1976; Stent, 1973). The Hebb rule postulates that synapses are strengthened to the extent that the activities of pre- and postsynaptic neurons are correlated and that synapses are weakened otherwise. Thus, correlated patterns of inputs would compete for the ability to activate a postsynaptic cell: the synapses corresponding to the pattern that best activated the cell would be strengthened, and other synapses correspondingly weakened. If afferents representing a single eye are better correlated with one another than with afferents of the opposite eye, inputs representing the two eyes would compete with one another in cortex. The outcome of such competition could be altered by alteration of the activity patterns of the two eyes. The experiments

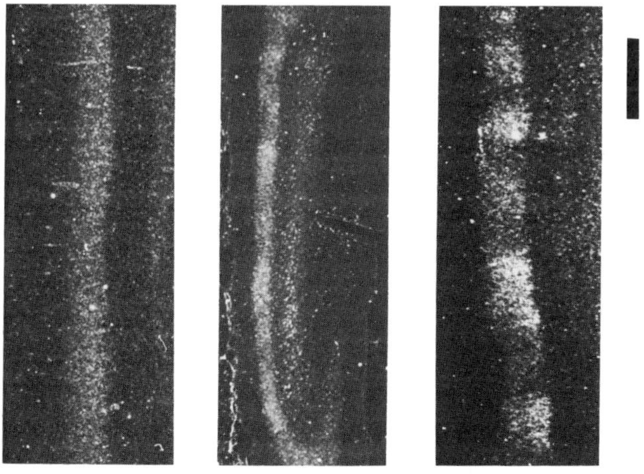

Figure 6: Ocular dominance columns in 7-10 week old TTX-treated (l), dark-reared (c), and control animals (r) labelled autoradiographically as in figures 2-4. Note that ocular dominance patches are absent in the animal in which all retinal activity had been blocked with TTX, so that this animal resembles the youngest animal shown in figure 4. Note also that ocular dominance patches are present in the animals that were reared in darkness. Although these dark-reared animals lacked visual experience, spontaneous neuronal discharge in the retinae was unblocked. Scale bar = 1 mm. Unpublished data from experiments reported in Stryker and Harris, 1986.

described thus far demonstrate that alterations in visual experience interfere with the normal process of geniculocortical afferent segregation. They do not, however, reveal whether neural activity is necessary for normal segregation, as would be the case if the Hebb mechanism were responsible. To investigate this question, Stryker and Harris (1986) blocked neural activity in the two eyes by repeated intraocular injections of the voltage-sensitive sodium channel ligand, tetrodotoxin (TTX), during the period in which ocular dominance columns normally develop. In such animals, neural activity was dramatically reduced in LGN and visual cortex, and geniculocortical afferents did not form ocular dominance patches; they remained instead in their infantile state of complete overlap (figure 6). This lack of segregation was also apparent physiologically in neuronal response properties. In normal animals, many neurons are driven exclusively through one eye or the other, as shown in figure 7 (left). In contrast, in TTX-treated animals nearly all neurons in the cortex were driven well through both eyes, as shown in figure 7 (right).

These experiments suggested that the normal developmental rearrangement of geniculocortical synaptic connections to form ocular dominance columns required neural activity. Since ocular dominance columns form, to a considerable extent, in utero in the monkey (DesRosiers et al., 1978; LeVay, Wiesel, and Hubel, 1980; Rakic, 1977), and in cats reared with bilateral lid suture or in total darkness, it appears that the so-called "spontaneous" or maintained activity of retinal ganglion cells in darkness is sufficient for segregation and that visually driven activity is not required.

The maintained activity of retinal ganglion and geniculate cells in darkness has been investigated in adult cats and in other species. Neighboring ganglion cells of the same center type tend to fire together over time periods of a millisecond to a few tens of milliseconds, and this correlation of activity decreases with increasing distance across the retina (Mastronarde, 1983a,b). In addition, correlations over longer time scales of activities within each eye or each lamina of the LGN are also present (Levick and Williams, 1964; Rodieck and Smith, 1966). With such correlated activity, a Hebb rule for the adjustment of geniculocortical synaptic strengths would be expected to allow the geniculocortical afferents serving each eye to remain together, while lack of correlation between activity in the two eyes causes the two eyes' afferents to segregate from one another.

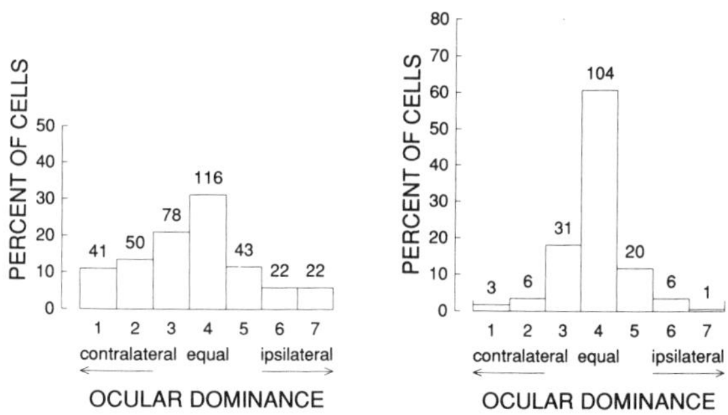

Figure 7: Ocular dominance histograms from visually responsive single units in the visual cortex of 6 normal kittens, 36 to 51 days of age (left) or of four kittens subject to bilateral retinal activity blockade by injection of TTX (right). Histograms plot percentage of units in each of 7 ocular dominance groups defined by Hubel and Wiesel (1962): group 1 (7) for units driven exclusively by contralateral (ipsilateral) eye; group 2 (6) for units strongly dominated by contralateral (ipsilateral) eye; group 3 (5) for units weakly dominated by contralateral (ipsilateral) eye; and group 4 for units driven nearly equally through the 2 eyes. Number of units in each bar of histograms is indicated above bar. Left: Normal kittens. Note that many neurons are driven well by both eyes but that a large fraction are strongly dominated or exclusively driven by one eye or the other (groups 1-2 and 6-7). Right: Kittens subject to bilateral retinal activity blockade. Retinal activity was blockaded beginning at age 14-16 days, and continuing through age 39-57 days. Kittens were allowed to recover from blockade for 2-4 days before microelectrode recording. Note the increased percentage of cells driven well by either eye (groups 3-5). These and the following histograms include cells from the cortical layers above and below layer IV, as well as from within layer IV. In normal animals there is a greater degree of binocular mixing in the layers other than layer IV. Data replotted from experiments reported in Stryker and Harris, 1986.

We have carried out several tests of assumptions implicit in the Hebb-synapse explanation of ocular dominance column development. One basic assumption is that the neural activity relevant to plasticity was the activity in the visual cortex. The experiments described thus far had all interfered with activity at earlier stages of the visual system as well. We tested this assumption by infusing TTX into a region of cortex to block the discharge of cortical cells and their geniculocortical afferent terminals (Reiter, Waitzman and Stryker, 1986). We then instituted a period of monocular deprivation and studied whether this deprivation caused a

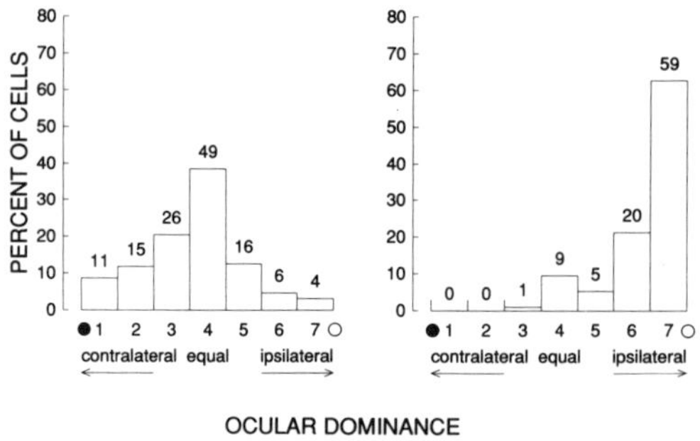

Figure 8: Ocular dominance histograms from visually responsive single units in visual cortex of 38-day old kittens, after 1 week of monocular deprivation during cortical infusion of TTX (left, 4 animals) or of vehicle solution (right, 3 animals). Conventions for histograms as in figure 7. Black dot indicates closed eye, white dot indicates open eye. Left, results from a cortical region in which all neuronal activity was blocked by cortical infusion of TTX throughout the period of monocular deprivation. Ocular dominance distribution was completely normal (compare left of figure 7), including the normal slight bias in favor of the contralateral eye, although the contralateral eye was the visually deprived eye in these animals. Right, results from a cortical region in which vehicle solution was infused throughout the period of deprivation. Most units were dominated by the open, ipsilateral eye. Data replotted from experiments reported in Reiter, Waitzman, and Stryker, 1986.

shift in ocular dominance. As expected, and consistent with the predictions of a Hebb-synapse model, the cortical activity blockade completely prevented plasticity, as is shown from the comparison of figures 7 (left) and 8.

A second assumption implicit in the Hebb-synapse explanation of development is that the statistics of neural activity are sufficient to account for ocular dominance plasticity. In nearly all of the experiments above, animals received visual experience. One might maintain that behaviorally significant visual stimulation, acting via some high-level perceptual mechanism, is crucial to plasticity in this system. In support of this notion, there is considerable evidence that behavioral state and diffuse neuromodulatory systems can influence

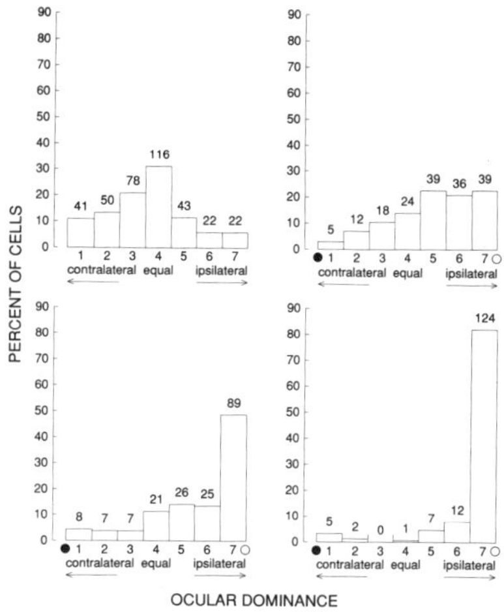

Figure 9: Ocular dominance histograms from visually responsive single units in the visual cortex of normal kittens (upper right) and of kittens subject to three different regimes by which the contralateral eye had decreased activity compared to the ipsilateral eye. Conventions for histograms as in figure 7. Black dot indicates relatively deprived eye, white dot indicates relatively more active eye. Upper left: results from six normal kittens, 36-51 days of age. Upper right: results from five kittens, 37-44 days of age, in which all retinal activity in the contralateral eye was blocked by injection of TTX, while the ipsilateral eye remained in total darkness. Lower left: results from six kittens, 36-40 days of age, in which all retinal activity in the contralateral eye was blocked by injection of TTX, and the ipsilateral eye was lid sutured. Lower right: results from four kittens, 29-34 days of age, subject to normal monocular deprivation by lid suture of the contralateral eye. Ocular dominance plasticity, resulting in a bias in favor of the ipsilateral eye, is seen in all deprivation regimes, although that at upper right and lower left included no behaviorally meaningful visual stimulation. Data replotted from experiments reported in Chapman et al., 1986.

plasticity in the visual system. These findings have led to Singer's "gating hypothesis" that plasticity occurs only when some combination of adrenergic, cholinergic, or glutamatergic "gates" are opened in visual cortex by the activity of various diffuse projection systems (Singer, 1985). To test this assumption, we attempted to produce ocular dominance plasticity in a situation in which neither eye received behaviorally significant visual stimulation (Chapman et al., 1986). Activity in one eye was blocked with TTX, while maintained but not visual activity was present in the other eye, which either remained in total darkness or received diffuse-light stimulation.

The maintained activity of the occluded eye, without any meaningful visual input, was sufficient to cause ocular dominance plasticity, as shown in figure 9. We conclude that at least much of the plasticity that takes place in the development of ocular dominance columns does not require meaningful visual input, so that it is reasonable to attempt

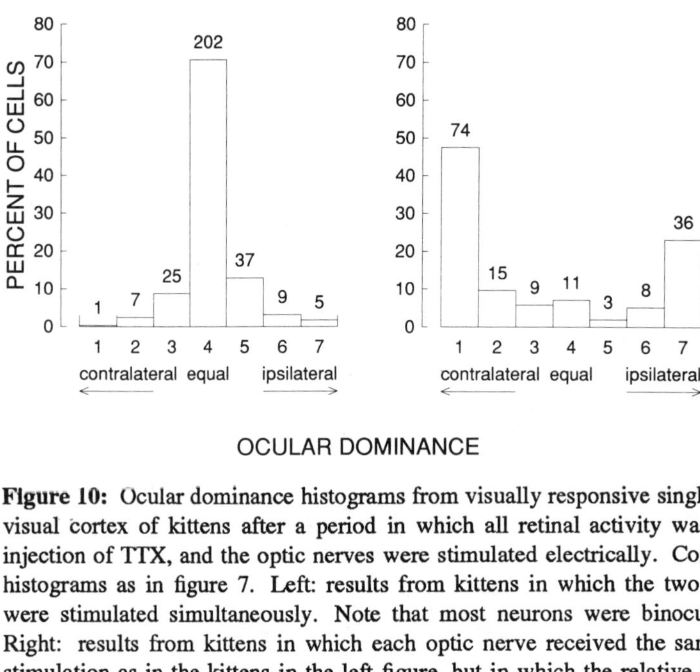

Figure 10: Ocular dominance histograms from visually responsive single units in the visual cortex of kittens after a period in which all retinal activity was blocked by injection of TTX, and the optic nerves were stimulated electrically. Conventions for histograms as in figure 7. Left: results from kittens in which the two optic nerves were stimulated simultaneously. Note that most neurons were binocularly driven. Right: results from kittens in which each optic nerve received the same pattern of stimulation as in the kittens in the left figure, but in which the relative timing of the pattern between the two eyes was such that the two eyes were stimulated alternately rather than simultaneously. Note that most neurons were driven monocularly. Unpublished data from experiments described in Stryker, 1986.

to understand its dependence on the statistics of neural discharge, as our model does.

A third assumption of the Hebb-synapse model of ocular

dominance column development is that ocular dominance segregation is governed by the relative timing of neural activity, the greater correlation in the patterns of neural discharge within each eyes' afferents than between the afferents of the two eyes. This assumption is only weakly supported by experiments showing that complete retinal activity blockade prevents ocular dominance column formation (Stryker and Harris, 1986). It is supported more strongly, but still indirectly, by the experiments showing that ocular dominance columns form more rapidly and more nearly completely when animals are reared with experimental strabismus. A direct test of the role of correlated activity was carried out by eliminating the natural maintained retinal discharge and introducing controlled patterns of activity into the two optic nerves by electrical stimulation (Stryker, 1986). These experiments showed that ocular dominance columns did not form when the the two optic nerves were stimulated simultaneously, but that an equal amount of activity delivered alternately to the two nerves did allow ocular dominance segregation. Figure 10 shows that most neurons were binocularly driven in the former case but monocularly driven in the latter.

A final assumption of the Hebb-synapse model is that not only are patterns of activity in the synaptic inputs important, but so are the responses of postsynaptic cortical cells. In numerous earlier experiments in which the responses of cortical cells were perturbed by substances infused into the cortex, ocular dominance plasticity was disrupted to a greater or lesser extent, but the presynaptic effects of these substances on afferent terminals were not known. We tested this assumption by infusing into visual cortex muscimol, a substance that powerfully inhibited cortical cells but appeared (and is thought to have) no effect on activation of or synaptic release from afferent terminals (Reiter and Stryker, 1988). In the region of cortex in which cortical discharge was completely inhibited, inputs from the less-active, occluded eye came to dominate over those from the more active, non-deprived eye, as shown in figure 11. This is a form of synaptic plasticity in the reverse direction from normal, but it is exactly what would be predicted by the Hebb-synapse model. In this

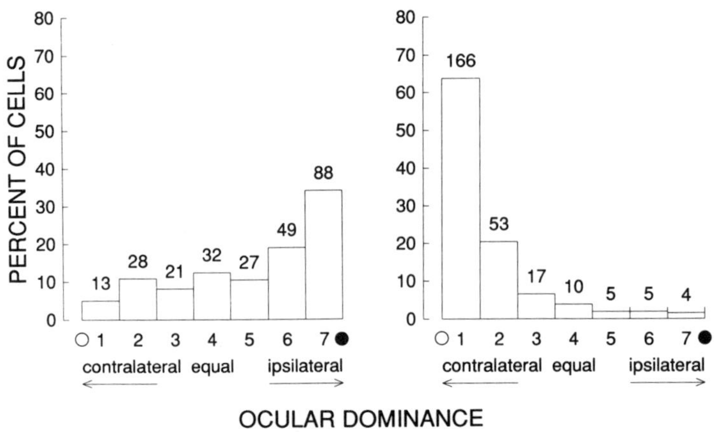

Figure 11: Ocular dominance histograms from visually responsive single units in the visual cortex of 8 kittens, 35-43 days of age, after 5-7 days of monocular deprivation during cortical infusion of muscimol. Left, results from regions in which all cortical cell neuronal activity was blocked by muscimol infusion. Right, results from the same kittens in cortical regions outside the muscimol-induced blocked. Conventions for histograms as in figure 7. Black dot indicates closed eye, white dot indicates open eye. In the muscimol-treated region (left), the less active, closed eye came to dominate cortical cell responses. In untreated regions, the normal domination by the more active, open eye was seen. Data replotted from experiments reported in Reiter and Stryker, 1988.

case, the activity of the less-active, occluded eye is better correlated with that of the inhibited postsynaptic cortical cells than is the activity of the more-active open eye. This finding confirmed the crucial role of the postsynaptic cells, as postulated in the Hebb model. Furthermore, it suggested that a process coupled to postsynaptic membrane voltage or conductance controls the direction of synaptic plasticity, which favors more-active inputs when the postsynaptic cell can be depolarized by them but less-active inputs when the postsynaptic cell is inhibited. Finally, it demonstrates that, at least for this form of plasticity, postsynaptic discharge is not essential, since the reverse plasticity took place while spikes were blocked in the cortical cells.

These experiments reveal a minimal set of features that must be incorporated in a model of ocular dominance column development in mammalian visual cortex. First, the afferents in such a model should initially make widespread overlapping connections, some of which become ineffective or are removed in development. Second, the correlation of activity among afferents serving one eye, and the lack of correlation between the eyes, clearly plays a role. Third, postsynaptic activity in the cortex is also crucial, and, therefore, intracortical synaptic connections by which cortical cells influence one another's activity will play a role.

The model described below incorporates these three features, together with a Hebbian or other rule to govern changes in synaptic strength. These features may all be measured accurately by feasible experiments; in fact, measurements in the literature allow us to make estimates of the values of the parameters describing each. Thus, a model incorporating these features may make genuine predictions about normal development and the outcomes of experiments.

A Model of Ocular Dominance Segregation

We have developed a simple model of the segregation of geniculocortical afferents into ocular dominance columns (Miller, Keller, and Stryker, 1986, 1989). This model summarizes the properties of the initial geniculocortical anatomy and physiology and of proposed plasticity rules in three functions that describe the three features just discussed:

- An arbor function, telling the strength of the initial connection – the number of synapses – between an afferent and a cortical cell, as a function of the retinotopic distance between their receptive field centers;
- A set of afferent correlation functions, summarizing the correlations in firing between afferents as a function of the distance between their receptive field centers and of their eyes of origin;
- A cortical interaction function, telling the influence that two simultaneously active synapses have upon one another's growth, as a function of the distance between them across cortex.

This model can be used to analyze a number of proposed mechanisms of synaptic plasticity, including a Hebbian mechanism. The features that distinguish one proposed mechanism from another will be found in the different forms given by each mechanism to the three functions that summarize the initial anatomy and physiology and the plasticity rule: afferent correlations, arbor/retinotopy, and cortical interactions.

The cortical interaction function is particularly dependent upon the proposed biological mechanism of plasticity. In the case of a Hebbian synapse mechanism, the growth of a synapse is promoted if it is active simultaneously with its postsynaptic cell. A synapse, by being active, increases the chance that its postsynaptic cell will be simultaneously active. Consider two simultaneously active synapses onto two cortical cells. If the two cortical cells excite one another, then the two simultaneously active synapses tend to aid one another's growth by increasing the probability of simultaneous excitation of one another's cortical cell. If the two cortical cells inhibit one another, the two synapses tend to suppress one another's growth. Hence for a Hebb synapse, the cortical interaction function, which describes the influence of two simultaneously active synapses upon one another's growth, is determined by the spread of intracortical synaptic influences. For a mechanism involving release of a diffusible modification factor, the function involves the lateral spread of influences by diffusion as well.

A number of simplifying assumptions are embodied in this model. We assume that, in some averaged sense, it makes sense to talk about afferent correlations and about geniculocortical and corticocortical connectivity as simple functions of the distance between two cells. Afferents actually comprise several classes: on- and off-center cells, and X and Y or parvo and magno cells. Afferent correlations and arbor sizes depend upon the type of afferent. The cortex has many attributes besides position: a cortical location in layer 4 may have associated with it a specificity for orientation or other properties, and there are a variety of cortical cell types. Geniculocortical and corticocortical connections are likely to have specificity with respect to these attributes, even in the young animal before ocular dominance segregation occurs (Lund, 1988; Martin, 1988). By ignoring these details, we reduce the system to the elements that seem essential to columnar development. We similarly assume that every synapse from a given afferent to a given cortical location experiences and exerts identical influences, ignoring neuronal microstructure. We also assume that the process controlling synaptic plasticity can be well approximated by instantaneous interactions rather than by following the detailed temporal structure of neuronal activations. We hypothesize that the tendency of two inputs to have correlated activity on a time scale determined by the plasticity mechanism is the key feature to be abstracted, and that the finer details of activity patterns can be ignored[1]. It is also assumed that the time scale associated with the activity-dependent mechanism is much smaller than the time scale over which synapses are appreciably changing their strengths; this allows averaging of the equations over all afferent activity patterns. Finally, the three functions are considered static during the initial development of a pattern of ocular dominance segregation. This simplification is potentially most problematic for the cortical interaction function, since if cortical interactions are themselves changing by an activity-dependent mechanism, their development would be coupled to the development of the geniculocortical projection.

The model need not depend on other features that might be thought critical. The measure of post-synaptic activity for purposes of plasticity may be action potentials or, as suggested by a number of recent studies (Katz and Constantine-Paton, 1988; Reiter and Stryker, 1988), local membrane depolarization. In principle, it is not even critical that the post-synaptic activity be involved at all. What is critical is that modification depend upon paired activities, either paired presynaptic activities or paired pre- and post-synaptic activities (the latter can be reduced to paired presynaptic activities via an equation like equation 4 below).

---

1. The time within which inputs must be correlated in order to influence one another's growth appears to be on the order of 10^{-1} second (Altmann et al., 1987; Blasdel and Pettigrew, 1979; see also references re LTP in hippocampus: Gustafsson et al., 1987; Larson and Lynch, 1986; Rose and Dunwiddie, 1986) while membrane times and interspike correlation times are of order 10^{-2} or 10^{-3} seconds.

Biological Characterization of the Three Functions Measurements made in cats allow estimation of the three functions that characterize the cortex in the model. As shown in figure 5, Y-cell afferents in visual cortex of young kittens appear to be capable of arborizing over an area more than 2 mm in diameter, or 1 mm in radius. This result, based on the study of putative afferents (LeVay and Stryker, 1979), is consistent with the final extent of patchy Y-cell arborizations seen in adults (Humphrey et al., 1985a,b). X-cell afferents have not been filled in kittens before columns develop, but based on the extent of patchy arborizations in adults (Humphrey et al., 1985a,b), it appears that X-cells might initially arborize over a region 0.5–0.75 mm in radius.

The ocular dominance patches of adult cats have a periodicity of about 850 µm (width of right-eye plus left-eye patches) (Anderson, Olavarria, and Van Sluyters, 1988; LeVay, Stryker, and Shatz, 1978; Shatz, Lindstrom, and Wiesel, 1977; Swindale, 1988). This period appears to be smaller than, or perhaps about the same size as, the diameter of initial X-cell afferent arborizations and seems clearly smaller than the diameter of initial Y-cell afferent arborizations.

Measurements of maintained activity of retinal ganglion cells in darkness in adult cats demonstrate that nearby ganglion cells have correlated activities, due to their common inputs (Mastronarde, 1983a,b). No indications of anticorrelations at further distances were seen[2]. Converting the retinotopic distance between correlated ganglion cells to a retinotopic distance across cortex (Tusa, Palmer, and Rosenquist, 1978), it appears that incoming afferents representing a single eye are correlated across cortical distances of from 1/2 (X-cells) to 3/2 (Y-cells) of a geniculocortical arbor radius. As previously noted, there may also be more widespread correlations within each eye on longer time scales (Levick and Williams, 1964; Rodieck and Smith, 1966). These estimates are very crude, as they are based upon

2. ON-cells are correlated with ON-cells, and OFF-cells with OFF cells. ON-cells are anticorrelated with OFF-cells over similar distances; no indication of correlation with increasing distance is seen in this case. Some possible implications of the anticorrelations between ON and OFF cells are briefly discussed later in this chapter.

measurements in the retina rather than the LGN and in adults rather than kittens. Further measurements are needed.

Horizontal intracortical synaptic interactions are the feature in the model that is least well characterized experimentally. In adult cat, these connections are likely to be excitatory at short range, and appear to be inhibitory to distances of perhaps 400-500 μm (Hata, et al., 1988; Hess, Negishi and Creutzfeldt, 1975; Toyama, Kimura, and Tanaka, 1981a,b; Worgotter and Eysel, 1989). Long-range synaptic interactions exist over distances of 1 mm or more (Gilbert and Wiesel, 1983). These long-range interactions connect only discrete patches of cortex and so may quantitatively have less impact than the shorter-range interactions. In the adult, the long-range connections may connect cells of similar orientation specificity by excitatory connections, and may also make inhibitory contributions to direction-selectivity (Tso, Gilbert, and Wiesel, 1986; Worgotter and Eysel, 1989). In the kitten, even before patch development but after the development of orientation selectivity, the long-range connections appear to connect patches with a periodicity consistent with that of orientation columns, as well as with that of the subsequent ocular dominance columns (Luhmann, Martinez Millan and Singer, 1986).

In summary, knowledge of the three functions involved in the model is still rudimentary, but better measurements can and will be made. In the adult, incoming afferents from a single eye appear correlated over $1/2-3/2$ of an afferent arbor radius, while anticorrelations between more distant inputs thus far have not been seen. The ocular dominance patches themselves appear to have a period less than, or perhaps equal to, an afferent arbor diameter. Intracortical synaptic interactions appear to be excitatory at short distances and inhibitory at greater horizontal distances in the adult, over distances that are within an arbor radius. In addition, there may be longer range connections with a definite periodicity. We will return to these points after examination of the model and its behavior.

In general, the model and its behavior will first be described intuitively, with a minimum of mathematics. Then more mathematical detail may be presented, keyed by a section heading or lead sentence referring to "mathematical" results. The reader with little mathematical interest may wish to skim or skip these sections; results in other sections do not depend upon them.

General Formulation of the Model We model the cortex as a 2-dimensional lamina, representing layer 4, and the LGN as two 2-dimensional laminae, one representing each eye. Let Roman letters such as x designate a 2-dimensional position on the cortex, and Greek letters such as α designate a 2-dimensional retinotopic position in the LGN. We assume retinotopic location can be directly interconverted with cortical location, via the retinotopic map onto cortex[3].

Let $S^L(x,\alpha,t)$, $S^R(x,\alpha,t)$ represent total synaptic strength at time t of the afferents of left or right eye, respectively, from $\alpha \rightarrow x$. Since an afferent may make many synapses onto a cortical cell, S may represent a sum of strength over many synapses. The number of such synapses is defined by the arbor function, $A(x-\alpha)$. Let $s_j^L(x,\alpha,t)$ represent the strength of the jth individual synapse of the left-eye from α to x. Then S^L is defined by $S^L(x,\alpha,t) \equiv \sum_{j=1}^{A(x-\alpha)} s_j^L(x,\alpha,t)$. We take A to be a single static function for all cells, while s and S vary in time.

The afferent correlation and cortical interaction functions are defined as follows (figure 12). The correlation functions $C^{LL}(\alpha-\beta)$ and $C^{LR}(\alpha-\beta)$ describe the correlation in firing between the left-eye afferent with retinotopic position α, and the left- or right-eye afferent, respectively, with retinotopic position β. $C^{RL}(\alpha-\beta)$ and $C^{RR}(\alpha-\beta)$ are defined similarly. The cortical interaction function $I(x-y)$ describes the influence of simultaneously active synapses at x and y upon one another's growth.

Our model equation for cortical synaptic plasticity is

$$\frac{dS^L(x,\alpha,t)}{dt} = \lambda A(x-\alpha)\sum_{y,\beta} I(x-y)[C^{LL}(\alpha-\beta)S^L(y,\beta,t) + C^{LR}(\alpha-\beta)S^R(y,\beta,t)] - DECAY^L(x,\alpha,t) \tag{1}$$

3. The retinotopic map onto cortex is linear in a local region of cortex. We assume it to be isotropic, ignoring differences in magnification with direction across cortex or LGN. Hence it is assumed to be a simple identity map between the cortical and geniculate grids.

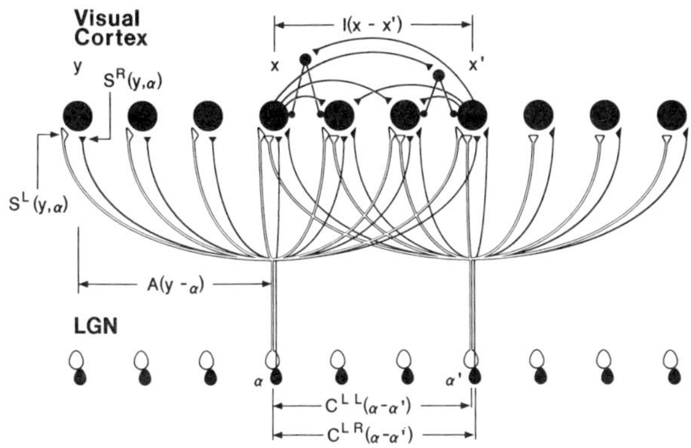

Figure 12: Notation. Afferents from left-eye (white) and right-eye (black) layers of the LGN innervate layer 4 of the visual cortex. The LGN is modeled as consisting of two layers, one serving each eye. Each layer is two-dimensional, though only one dimension is illustrated. The cortex is modeled as a single, two-dimensional layer. In the figure, α and α' label two-dimensional positions in the LGN, and x and x' label the retinotopically corresponding points in the cortex; y labels an additional position in the cortex. The afferent correlation functions C^{LL} (correlation in activity between two left-eye afferents) and C^{LR} (correlation in activity between a left-eye and a right-eye afferent) are functions of separation across the LGN. The arbor function A measures anatomical connectivity (number of synapses) from a geniculate point to a cortical point, as a function of the retinotopic distance between them. The cortical interaction function I depends on a distance across cortex. The left-eye and right-eye synaptic strengths, S^L and S^R, from a geniculate location to a cortical location, depend upon both locations. S^L and S^R change during development in the model, while the arbor function A is held fixed; the assumption is made that anatomical changes occur late in development, after a pattern of physiological synaptic strengths is established. From Miller, Keller, And Stryker, 1989. Copyright 1989 by the AAAS.

where $\text{DECAY}^L(x,\alpha,t) = \gamma S^L(x,\alpha,t) + \varepsilon A(x-\alpha)$ for some constants γ and ε. The equation for S^R is identical with interchange of L and R everywhere. The DECAY term involves changes in each synapse independent of all other synapses, while the non-DECAY term describes changes due to interactions between synapses.

The synaptic interaction term can be expressed in words as follows. First, the change in time of any synapse's strength is a sum of the influences exerted on it by all other synapses, so that the equation is linear in the variables s and thus in S. Second, the influence of any one synapse upon another is a product of the correlation between their activities, which gives a measure of the likelihood that the two synapses fire together; the strength of the influencing synapse, which tells the amount of influence exerted by that synapse when it is active; and the strength of the intracortical interaction between them, which tells the attenuation or change in sign with cortical distance of the influence exerted. Third, the change in the total synaptic strength S between an afferent and a cortical cell is given by multiplying the total influence on one such synapse by the number of synapses from the afferent to the cortical cell. That number is given by the arbor function. This multiplication occurs because each such synapse receives identical influences.

In studying this equation we will ignore boundary effects. Stripe width is at least an order of magnitude smaller than the overall width of striate cortex. We assume that the three functions C,A, and I are relatively local, and hence most stripes develop without any direct influence of the boundaries. Thus, the boundaries should not influence the occurrence and width of periodic segregation, and it is on these features that our analysis will focus. The boundaries may influence the overall form or layout of the stripes, which we will not attempt to analyze[4]. Thus, we do away with the boundaries, either by using periodic boundary conditions (in simulations) or by assuming an infinitely wide cortex.

Equation 1 for S^L and the corresponding equation for S^R constitute our basic mathematical model of synaptic strength development. The data used in the model are the arbor function $A(x-\alpha)$, the cortical interaction function $I(x-y)$, and the correlation functions such as $C^{LL}(\alpha-\beta)$. When the initial values of S^L and S^R are given at time t=0, the equations determine S^L and S^R at any later time t.

Geniculocortical synapses are exclusively excitatory. Furthermore, synapses are limited in strength. Therefore, this basic model must be modified to prevent synaptic strength from becoming negative or from becoming too large. Nonlinearities must be included

---

4. The occurrence and width of segregation are determined by simple linear interactions, as will be described, and hence are robust features determined by very general features of a model. In contrast, the overall form of the pattern is determined by nonlinear interactions.

in the equation to enforce these conditions.

In addition, the total synaptic strength supported by a cortical cell or afferent may be limited beyond the extent implicit in the limits on the strengths of individual synapses. There is no direct evidence for such limits, but the phenomena associated with ocular dominance plasticity are suggestive of a competition for finite resources between afferents serving the two eyes. Such limits may also be suggested by biological evidence for limits to the total number of synapses supported by a cell. In the goldfish optic tectum, if presynaptic cells are forced to innervate only half of a normal tectum, the number of synapses formed per postsynaptic cell remains the same as in the normal case, so that on average each input forms only half its normal number of synapses (Hayes and Meyer, 1988b; Murray, Sharma, and Edwards, 1982). This suggests that the presynaptic inputs may be competing for a fixed number of postsynaptic sites. Similarly, if only a fraction of the normal presynaptic cells are allowed to innervate the tectum, the number of synapses formed per postsynaptic cell is smaller than in the normal case, suggesting intrinsic limits to the number of synapses that each presynaptic cell can make (Hayes and Mayer, 1988a). Presynaptic limits have also been observed in other systems (Fladby and Jansen, 1987; Brown, Jansen and Van Essen, 1976; Schneider, 1973). We refer to such limits as *constraints* on the total synaptic strength supported by a postsynaptic or presynaptic cell. The use of such constraints in modeling was first suggested by von der Malsburg (1973). To test the possible role of constraints in development, the model can be modified to limit the total synaptic strength supported by a cortical or afferent cell. Terms must be added to the equations to enforce such limits.

Mathematical Derivation of the Model Equation From a Hebbian Mechanism Equation 1 can be simply derived from a linear Hebb synapse mechanism. The Hebb synapse rule can be expressed, for individual synapses, as

$$\frac{ds_j^L(x,\alpha,t)}{dt} = \lambda[\mathrm{POST}(x)\mathrm{PRE}^L(\alpha)] - \mathrm{decay} \qquad (2a)$$

where POST(x) is some function of postsynaptic activity at x, $\mathrm{PRE}^L(\alpha)$

is some function of presynaptic activity from α in the left eye, and λ is a constant. Let $c(x,t)$ be the activity of the cortical cell at position x at time t, and similarly let $a^L(\alpha,t), a^R(\alpha,t)$ represent the activity of left- and right-eye afferents. Taking post to be a linear function of c (see the Appendix for the nonlinear case), this is

$$\frac{ds_j^L(x,\alpha,t)}{dt} = \lambda[c(x,t) - c_1]f_1[a^L(\alpha,t)] - \gamma s_j^L(x,\alpha,t) - \varepsilon' \qquad (2b)$$

where c_1 is a constant, f_1 is a function that may incorporate threshold or saturation effects, and γ and ε' are intrinsic decay (or growth) factors. Summing over j on both sides, this can be reexpressed as

$$\frac{dS^L(x,\alpha,t)}{dt} = \lambda A(x-\alpha)[c(x,t) - c_1]f_1[a^L(\alpha,t)] - \gamma S^L(x,\alpha,t) - \varepsilon' A(x-\alpha) \qquad (3)$$

The cortical activities in this equation can be replaced by a function of afferent activities and synaptic strengths. Define the net geniculate input to a cortical cell by

$$NET_{LGN}(x) \equiv \sum_{\alpha} \left\{ S^L(x,\alpha,t)f_2[a^L(\alpha,t)] + S^R(x,\alpha,t)f_2[a^R(\alpha,t)] \right\}$$

f_2 incorporates thresholds and saturations, like f_1 in equation 2. If cortical activity is a linear function of NET_{LGN} (again, the nonlinear case is considered in the Appendix), then

$$c(x,t) = \sum_{y} I(x-y)NET_{LGN}(y) + c_2. \qquad (4)$$

$I(x-y)$ describes the total influence on cortical point x of geniculate excitation of the cortical point at y, including direct excitation (when $y=x$), as well as indirect effects via intracortical synaptic connections by which the activity[5] of the cortical cell at y influences activity at x.

c_2 represents intrinsic activity of cortical cells in the absence of geniculate input. Substituting equation 4 for c(x,t) into equation 3, and averaging over input activity patterns[6], we obtain the model equation, equation 1. Let < > denote the average. Then[7]

$$C^{LL}(\alpha-\beta) \equiv \langle f_1[a^L(\alpha,t)]f_2[a^L(\beta,t)]\rangle, \quad C^{LR}(\alpha-\beta) \equiv \langle f_1[a^L(\alpha,t)]f_2[a^R(\beta,t)]\rangle$$

and

$$\varepsilon \equiv \varepsilon' - \lambda[c_2 - c_1]\langle f_1[a^L(\alpha,t)]\rangle$$

Which Mechanisms Can Be Studied Within This Framework? We have shown elsewhere (Miller, Keller, and Stryker, 1989) that equation 1 may be derived from a number of proposed biological plasticity mechanisms. These include mechanisms involving activity-dependent release by either cortical cells or afferents of a diffusible modification factor, combined with activity-dependent

---

5. The nature of I can be further understood if we postulate, instead of eq 4, $c(x,t) = NET_{LGN}(x) + \sum_y B(x-y)c(y,t) + c'$ where $B(x-y)$ summarizes the effects of cortico-cortical synaptic interconnections, and c' is intrinsic activity of the cortical cell in the absence of all input. Assume that cortical activity is determined by afferent activity, so that the matrix $(\mathbf{1}-\mathbf{B})$ is invertible (where $\mathbf{1}$ is the identity matrix). Then equation 4 follows, with, as matrix equations, $\mathbf{I}=(\mathbf{1}-\mathbf{B})^{-1} = \mathbf{1}+\mathbf{B}+\mathbf{B}^2 + \cdots$ and $\mathbf{c_2} = \mathbf{I}\mathbf{c}'$.

6. We have continued to use the notation S for $\langle S\rangle$ after averaging. Averaging produces an infinite series of terms, of which we keep only the first. Higher order terms involve the tendency of three or more afferents to be active conjointly beyond the extent predicted by pairwise correlations. These terms are small either for small fluctuations or for small $\lambda,\gamma,\varepsilon$. Averaging is done by the smoothing procedure, described in Keller (1977) and in Miller (1989c).

7. As $\alpha-\beta \to \infty$, $C^{IJ} \to (\alpha-\beta) \langle f_1[a^I(\alpha,t)]\rangle \langle f_2[a^J(\beta,t)]\rangle$. Hence if $\langle f_1[a]\rangle \neq 0$ and $\langle f_2[a]\rangle \neq 0$, the various C's converge to a constant at large distances. This constant is the constant $-k_2$ of Linsker (1986a-c). Because it disappears from C^D and hence from the equation for S^D, described later in this chapter, it has no influence on the initial development of ocular dominance segregation.

of chemospecific adhesion such that retinotopically matched afferents and cortical cells adhere best to one another.

The mechanisms that lead to equation 1 have three points in common. They all include some arboreal and/or retinotopic factor $A(x-\alpha)$, and some lateral interconnection, either synaptic or via chemical diffusion or transport, between points in cortex. They also depend for modification upon paired activity which, via equation 4, reduces to dependence on paired afferent activities. The mathematical model common to these mechanisms, excluding the decay terms, can be summed up by the following heuristic for the interactions between synapses:

$$\frac{dS^L(x,\alpha,t)}{dt} = \sum_{E=L,R}\sum_{y,\beta}[\text{Influence Felt by } S^L(x,\alpha,t)][\text{Influence Exerted by } S^E(y,\beta,t)]$$

Here the influence felt by $S^L(x,\alpha,t)$ is $f_2[a^L(\alpha,t)]A(x-\alpha)$, that is, some measure of the presynaptic activity of the synapse, multiplied by the number of synapses from α innervating x (perhaps scaled by the retinotopic affinity of synapses from α for x). The influence exerted by $S^E(y,\beta,t)$ is $f_1[a^E(\beta,t)]S^E(y,\beta,t)I(x-y)$, that is, some measure of the presynaptic activity of these synapses, multiplied by the total synaptic strength of such synapses, times a factor indicating the influence felt at a distance across cortex from the influencing synapse.

The mechanisms have one additional point in common. Synapses causing modification act in proportion to their total strength, S^L or S^R, while synapses being modified respond independently of their strength, but in proportion to their total number A. Biologically, the first feature seems natural, but the second is arbitrary given our current biological knowledge. Choosing to let synapses exert influence in proportion to A results in a trivial theory. With this choice, synaptic strengths do not interact. Instead, all synapses simply decay or grow uniformly. Thus, the only interesting alternative choice is to let both influence exerted and influence felt be proportional to S. In this case, the lowest-order term embracing synaptic interactions is intrinsically nonlinear. However, in certain limits this case, also, approaches equations like those we study[8].

Because several different biologically reasonable mechanisms of synaptic plasticity are described by the same mathematical formalism, agreement between the predictions from a particular mechanism and the biological reality can not by itself be taken as strong evidence in favor of the mechanism. Such agreement appears at first sight only to distinguish among very large classes of mechanisms, where the members of a class make similar predictions. Therefore, one is led to question the extent to which such models can be informative to biologists. It is important to note, however, that different mechanisms are distinguished by the different biological features that are summarized in the three functions characterizing the mathematical model. That is, these functions will have different biological interpretations under different mechanisms. Furthermore, as we will show, measurements of these functions can allow prediction of whether ocular dominance segregation should occur, and the

8. Let $S^S = S^L + S^R$, $S^D = S^L - S^R$. Assume equivalence of the two eyes, and ignore decay. Then for this case, before averaging over input activity patterns,

$$\frac{dS^D(x,\alpha,t)}{dt} = \tfrac{1}{2} S^S(x,\alpha,t) \sum_{y,\beta} I(x-y)\, C^D(\alpha-\beta,t)\, S^D(y,\beta,t) \qquad \text{(N1)}$$

$$+ \tfrac{1}{2} S^D(x,\alpha,t) \sum_{y,\beta} I(x-y)\, C^S(\alpha-\beta,t)\, S^S(y,\beta,t)$$

Here,

$$C^S(\alpha-\beta,t) = f_1[a^L(\alpha,t)] \left\{ f_2[a^L(\beta,t)] + f_2[a^R(\beta,t)] \right\},$$

$$C^D(\alpha-\beta,t) = f_1[a^L(\alpha,t)] \left\{ f_2[a^L(\beta,t)] - f_2[a^R(\beta,t)] \right\}.$$

Note that $\langle C^S \rangle = C^{LL} + C^{LR}$, $\langle C^D \rangle = C^{LL} - C^{LR}$. Suppose there are constraints on the sum S^S such that receptive fields are relatively uniform across the cortex, except for ocular dominance. Then $S^S(x,\alpha,t) \approx f(x-\alpha,t)$ for some function f. Suppose that S^S and S^D can be taken to be statistically independent. Then, after averaging, the first term in equation N1 becomes the non-decay term in equation 5, with f playing the role of an arbor function. The second term in equation N1 contributes to the decay term in equation 5.

periodicity of such segregation if it does occur. Therefore, experimental measurements of the biological features that define the functions under the assumption that one or another mechanism is active, can serve to distinguish among the mechanisms of synaptic plasticity.

Studying the Model Through Simulations

Simulations of Development and Deprivation We have studied the behavior of the model through computer simulation of development, by carrying equation 1 forward in time from an initial state in which synapses of the two eyes have nearly equal strengths. For this purpose, we model cortex and the left- and right-eye layers of the LGN as three 25×25 layers of neurons. To eliminate boundary effects, we use periodic boundary conditions, so that the leftmost and rightmost columns of each grid are adjacent, as are the bottom and top rows. Each geniculate cell is connected to a 7×7 square of cortical cells, centered about the retinotopically corresponding cell in the cortical grid.

Initially, we consider the following choice of functions. The arbor function is equal to 1 over the 7 x 7 set of connections and 0 outside. The correlation function includes positive correlation within each eye, falling to zero over about an arbor radius; there is neither correlation nor anticorrelation between the eyes ("same-eye correlations", gaussian parameter 0.3, in figure 13).

The cortical interaction function is excitatory among nearest neighbors on the grid, and weakly inhibitory more peripherally for several gridpoints (Mexican hat cortical interaction in figure 13). Other choices of functions will be examined subsequently. Each of the 2×7×7×25×25=61,250 synapses is assigned an initial weight randomly drawn from a distribution uniform between 0.8 and 1.2. Synaptic weights grow or decay according to the model equation until they reach 8.0 or 0, at which point no further change is allowed.

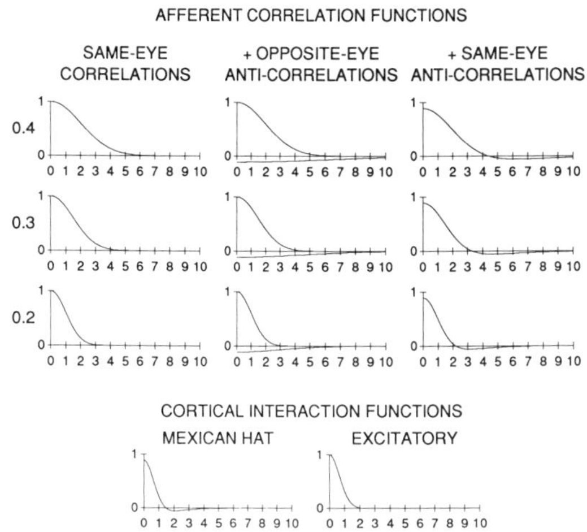

Figure 13: Functions used for simulations of full geniculocortical innervation. Function value, vertical axis, versus distance in grid intervals, horizontal axis. All functions are circularly symmetric in two dimensions. **Top**: Correlation functions $C(\alpha)$. These summarize the correlation in activity between two afferents separated by the retinotopic distance α. We assume $C^{LL}(\alpha)=C^{RR}(\alpha)$, $C^{LR}(\alpha)=C^{RL}(\alpha)$. The correlation functions in the left column (same-eye correlations) are positive within each eye, and zero between the two eyes: $C^{LL}(\alpha)=e^{-\alpha^2/(\kappa D)^2}$, where $D=7$ is the arbor diameter and $\kappa=0.4$, 0.3 or 0.2 for the top, middle, or bottom row respectively, and $C^{LR}=0$. The positive correlations within each eye are shown. The correlation functions in the middle column (+ opposite-eye anticorrelations) are positive within each eye as in the left column, as shown by the curves above the horizontal axes; but in addition there are weaker, more broadly ranging negative correlations between the two eyes, shown by the curves below the horizontal axes. These negative correlations are given by $C^{LR}(\alpha)=-1/9e^{-\alpha^2/(3\kappa D)^2}$. The correlation functions in the right row (+ same-eye anticorrelations) have these same negative correlations added to the positive correlations within each eye, and have zero correlation between the two eyes. This creates a "Mexican hat" function within each eye, so that inputs are correlated at shorter distances and anticorrelated at longer distances, as illustrated. **Bottom**: Cortical interaction functions $I(x)$. Left, Mexican hat function, given by gaussians as for the rightmost correlation functions but with $\kappa=0.1333$. Right, a purely excitatory cortical interaction function, given by the excitatory gaussian of the Mexican hat function. From Miller, 1989a.

Constraints are added, fixing the total synaptic strength on a cortical cell and fixing or limiting the total synaptic strength coming from each afferent. We will subsequently discuss the effect of these constraints.

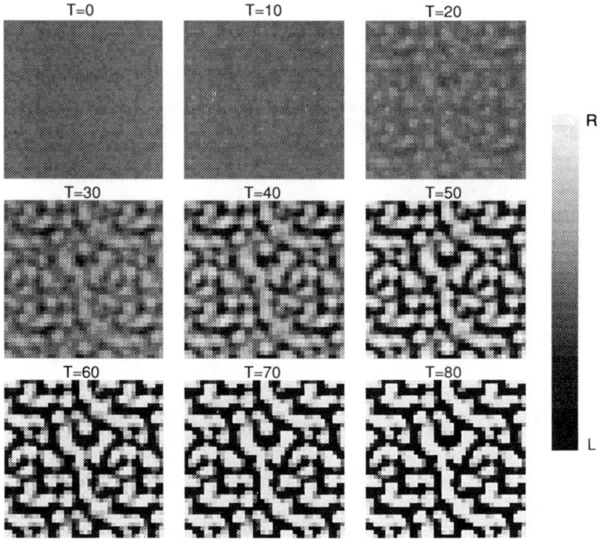

Figure 14: Development of ocular dominance of cortex in the model. Cortex is shown at nine times, from time 0 to the 80th iteration. Each pixel represents a single cortical cell. The greyscale represents ocular dominance D of each cell, the difference between the total strength of right-eye and of left-eye geniculate inputs to the cell: $D(x) \equiv \sum_{\alpha}[S^R(x,\alpha) - S^L(x,\alpha)]$. The greyscale runs linearly from monocular for the right eye (white) to monocular for the left eye (black). Final (timestep 200) cortex is the lower right of figure 15. This development used the following functions (figure 13): The correlation functions have same-eye correlations only, with parameter 0.3. The intracortical interaction function is Mexican hat. The arbor function is taken to be 1 over a 7×7 arbor, 0 elsewhere. Constraints were used to conserve total synaptic strength over each cortical cell and over each afferent arbor (Miller, 1989c; Miller, Keller, and Stryker, 1989). *Convention for all simulations*: Illustrations of cortex show 40×40 grids, although the model cortex is 25×25. Periodic boundary conditions were used, so this display shows continuity of the pattern across what would otherwise appear to be a boundary. Thus, the top 15 and bottom 15 rows within each square are identical, as are the left 15 and right 15 columns. From Miller, 1989c.

The randomly assigned synaptic strengths result in an initially nearly uniform innervation by the two eyes, much like that seen by autoradiography in the young kitten before segregation of left-eye from right-eye afferents. Segregation and development of an overall pattern of ocular dominance in the model cortex occurs even while the

Figure 15: Cortex, timestep 200, resulting from nine different random initial conditions. Cortical interaction, arbor, and correlation functions and conventions as in figure 14. Results are qualititatively and quantitatively similar for all initial conditions that have been tried; that is, the 2-dimensional Fourier transforms yield similar power spectra. From Miller, 1989c.

amplitude of the pattern is still quite small, perhaps smaller than could be detected experimentally. This is illustrated in figure 14. In this figure, each pixel represents the ocular dominance of a single cortical cell. Black or white represent dominance by left or right eyes, respectively, while intermediate greys represents equality or varying degrees of dominance of the two eyes. The pattern continues to develop as it grows nearly to saturation, so that most cortical cells eventually become fully monocular, completely dominated by one eye or the other. The resulting development and final pattern of ocular dominance closely resembles the patterns of periodic segregation seen in cat or monkey, although cat layer 4 has more binocular cells than this model cortex. This development is completely robust across randomly generated initial conditions: every initial condition leads, given this same choice of functions, to a qualitatively similar, though distinct, final outcome. Figure 15 illustrates the final cortices resulting from nine different sets of initial synaptic strengths.

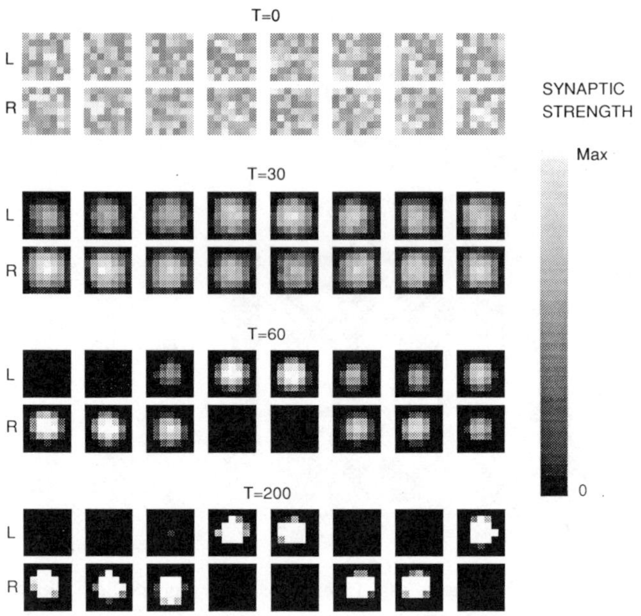

Figure 16: Receptive fields (ignoring the contribution of corticocortical connections) of eight cortical cells at timesteps 0, 30, 60, 200. Each vertical L,R pair of large squares shows strengths of the 7×7 left-eye and 7×7 right-eye synapses onto a cortical cell at one time. Synaptic strengths onto eight adjacent cortical cells are shown. Cortical cells shown are the eight leftmost cells in the bottom row of the cortices of figure 14. The greyscale varies linearly in synaptic strength from 0 (black) to the maximum strength present at the given timestep (white). These maximum strengths are: timestep 0, 1.2; timestep 30, 3.3; timesteps 60 and 200, 8.0. Receptive fields first refine in size, concentrating their strength centrally. They then become monocular with synaptic strength confined to left- or right-eye inputs. Adjacent groups of cells tend to become dominated by the same eye, providing the basis for ocular dominance segregation across the cortex as a whole. From Miller, 1989c.

The pictures of cortex just presented collapse information about the 98 synapses onto each cortical cell into a single pixel representing net ocular dominance. More can be learned by examining in detail the development of each individual synapse onto a given cortical cell. The geniculate synapses onto the cortical cell represent the cell's receptive field, discounting the effects of cortico-cortical synaptic connections. We illustrate (figure 16) the development of these receptive fields for 8 cortical cells. These are the 8 leftmost cortical

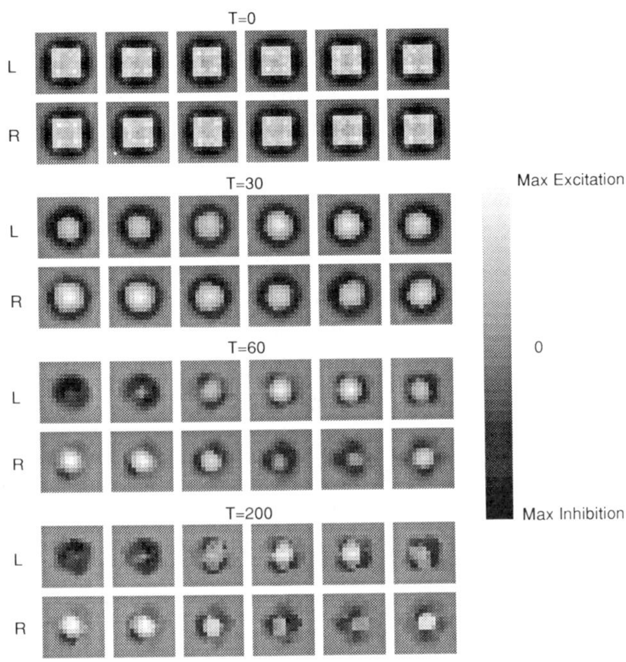

Figure 17: Physiological receptive fields at timesteps 0, 30, 60, 200, for the six leftmost cortical cells of the 8 shown in figure 16. Each vertical L,R pair shows physiological input to the cortical cell from 15×15 left-eye or right-eye geniculate positions. Physiological input from a geniculate cell is defined as the linear sum of the cell's direct synaptic input onto the cortical cell, and its synaptic input onto other cortical cells weighted by the synaptic influence of those cortical cells on the cortical cell observed. Thus, physiological input to a cortical cell at x from the left-eye geniculate position α is defined by $P^L(x,\alpha) = \sum_y I(x-y)S^L(y,\alpha)$. Input may be excitatory (grey, strength 0, to white, maximum excitatory strength present at a timestep) or inhibitory (grey, strength 0, to black, maximum inhibitory strength present at a timestep). Maximum excitatory strengths at the four times, respectively, are 1.4, 4.1, 13.6, 14.4; maximum inhibitory strengths are -0.6, -0.9, -2.5, -2.9. At time 0, the 7×7 arbor of direct synaptic connections provides excitatory input, while surrounding geniculate positions give inhibition. Development of physiological receptive fields parallels the development of the direct geniculocortical input shown in figure 16, but in addition the development of sidebands of inhibitory input from the dominant eye and of inhibitory input from the nondominant eye can be observed. From Miller, 1989c.

cells in the bottom row of the cortices of figure 14. Each vertical pair of 7×7 "L" and "R" squares displays the strengths of the 49 left-eye and 49 right-eye synapses onto one cortical cell at one developmental time. The receptive fields first refine in size, while developing only weak biases of ocular dominance. Subsequently, the receptive fields become monocular as they continue to refine in size.

Intracortical synaptic interconnections modify cortical receptive fields. Assuming local excitation and more distant inhibition between cortical cells as given by the excitatory/inhibitory cortical interaction function, and using a simple linear model of cortical activation[9], we can compute the resulting receptive fields. This is illustrated in figure 17, for the six leftmost cells of the eight illustrated in figure 16. Two features are added to cortical receptive fields by intracortical connectivity. First, the nondominant eye gains input to the cortical cell, but it is inhibitory input, as occurs physiologically in some cases (Bishop, Henry, and Smith, 1971; Kato, Bishop, and Orban, 1981; Ohzawa and Freeman, 1986). Second, though the receptive fields initially develop a circularly symmetric structure as they refine, this circular symmetry of individual receptive fields is broken as each eye's inputs become segregated into its eye-specific stripes. While this is reminiscent of the development of orientation selectivity in cortical cells, it cannot explain that phenomenon, because orientation selectivity develops biologically even in regions of cortex where there is innervation by only a single eye.

Just as we can study the receptive fields of individual cortical cells, we can study the terminal arbors – the projection onto cortex – of individual geniculate afferents. These arbors are illustrated in the same format used for the receptive fields in the absence of intracortical connectivity (figure 18). Vertical pairs of "L" and "R"

9. Our crude, linear model of cortical activation can be sufficient to study the periodic segregation of ocular dominance, as we will discuss, but it lacks the detail to make precise predictions about the effects of intracortical synapses on physiological receptive field structure. Thus, results with this linear model are only intended to be suggestive of some of the additional features that intracortical connectivity might add to receptive fields; results might be altered considerably by nonlinearities.

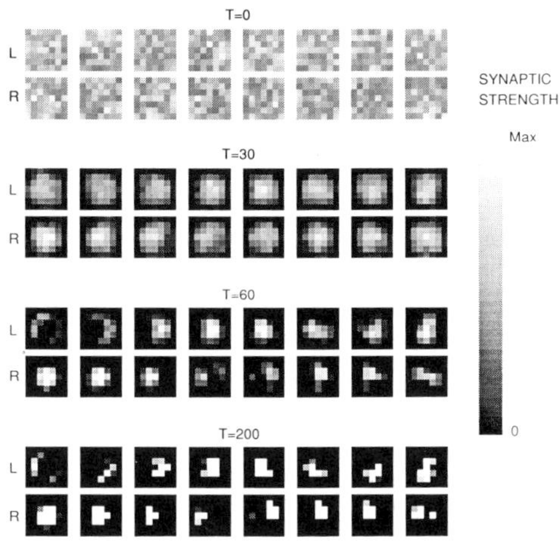

Figure 18: Afferent arbors at timesteps 0, 30, 60, 200. Conventions as in figure 16, except that here the strengths of the synaptic arbors projected by eight adjacent left-eye and eight adjacent right-eye LGN afferents onto cortex are pictured, whereas in figure 16 the strengths of the left-eye and right-eye synapses received by 8 cortical cells were pictured. The afferents shown are the eight leftmost cells in the bottom row of the respective geniculate grids. Thus, each vertical L, R pair illustrates the arbors of a left- and a right-eye geniculate cell from identical geniculate grid positions at one time. Greyscale shows synaptic strength of connection to each of the 7 by 7 cortical cells contacted by the arbor. Arbors first concentrate their strength centrally, then break up into patches confined to complementary cortical ocular dominance patches. From Miller, 1989c.

7×7 squares here display the strengths of the synapses projected to cortex by the left-eye and right-eye afferents representing a single receptive field location. The arbors initially refine in size much like the receptive fields, but then segregate into complementary cortical regions as each eye's input becomes confined to the appropriate eye's patches. This is reminiscent of the terminal arbors in the cat (figure 5). Thus, many features of normal biological development emerge in the model. These include periodic ocular dominance segregation across cortex, refinement and development of monocularity in individual cortical receptive fields, and refinement and confinement to patches of afferent arbors.

Using the same choice of functions and the same initial conditions as in the simulation just presented, we now proceed to study the effects of visual deprivation.

Monocular visual deprivation is modelled as a decrease in the amount of activity within one eye, and hence, as a decrease in the amplitude of the correlation function within that eye. Possible disruption of the correlational structure within the deprived eye, which would only increase the effects of deprivation, is ignored. We consider the effects on the final cortex of initiating monocular deprivation at increasingly later times in development (figure 19). With early initiation of deprivation, the "open" eye takes over much

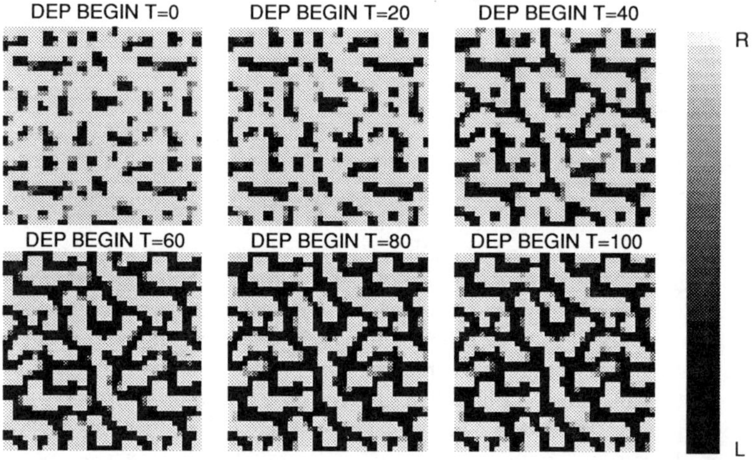

Figure 19: Cortical patterns of ocular dominance resulting from sustained monocular deprivation initiated at various times. Results of initiating the deprivation at iteration 0, 20, 40, 60, 80, or 100 are shown. All parameters, functions, and initial conditions are identical to those used in figure 14, except that (1) the deprived eye's correlation function was multiplied by 0.7 during the deprivation, and (2) Only partial constraints on afferent arbors were used, so that afferent arbors were able to vary their total synaptic strength by ±50 percent from the original value. Final cortices (T=200) are shown for each case. Synaptic strengths were frozen upon reaching the limiting values of 0 or 8; this makes very little difference in final results for these functions (Miller, 1989c). From Miller, 1989c.

more than its normal share of cortex. As in biological development, there is a critical period for this phenomenon in the model: the later in development deprivation is initiated, the less effect deprivation has, and with sufficiently late initiation there is no effect. The precise degree of ocular dominance shift produced by deprivation initiated at a particular time in development depends upon model parameters. These include the amount by which activity in the deprived eye is reduced and the degree of change allowed in the total synaptic strength supported by an afferent arbor[10]. It is very robust, however, that early onset of deprivation leads to strong ocular dominance shifts, and that initiation of deprivation progressively later in development leads to progressively smaller effects.

This critical period occurs in the model for two reasons. First, there is a strictly dynamical or organizational component. As ocular dominance segregation proceeds, each eye attains dominance in synaptic strength within its stripes. Since the influence of synapses involves a product of the presynaptic activity times the synaptic strength, this dominance in synaptic strength can more than compensate for the decrease in activity caused by deprivation and allow the deprived eye to maintain dominance within its stripes. Second, there is a component due to "stabilization" of synapses: once synapses are fully saturated (strength 8.0 or 0), they may be frozen, so that they are no longer allowed to change in strength. If stabilization of synapses is used, no changes in ocular dominance can occur once most synapses have been saturated.

With correlation functions like that used here, or more broadly ranging, layer 4 normally becomes fully monocular (as will be discussed). Then dynamical or organizational factors are sufficient to entirely account for a critical period. In this case, the critical period is unchanged by the presence or absence of stabilization: it emerges entirely from the fact that a sufficient degree of ocular dominance organization becomes dynamically irreversible. When correlation functions are narrower, the final layer 4 normally includes a greater

---

10. In the simulation shown, we have used constraints allowing progressively less change in the total synaptic strength over an afferent arbor as that total approaches ±50 percent of its initial value. Afferents are not allowed to gain or lose more than 50 percent of their total strength, although they continue to redistribute strength between their synapses even after the 50 percent limit is reached. Deprivation initiated at time 0 leads deprived-eye arbors to shrink to the maximum degree allowable, and to disappear completely if that is allowed. Later initiation of deprivation leads to smaller effects, irrespective of the precise constraint used.

number of binocular cells. Such an outcome may also result if the development of vision tends to correlate the activity of the two eyes and if these changes occur while segregation is incomplete. Binocular cells remain dynamically susceptible to an ocular dominance shift at any stage of development. Hence, if the final layer 4 includes a significant percentage of binocular cells, as in the cat, a complete explanation of a critical period requires stabilization of saturated synapses. In this case, however, dynamical factors will still contribute to the critical period by ensuring that regions sufficiently dominated by one eye will no longer be subject to an ocular dominance shift. The biological critical period may, of course, involve other factors such as molecular changes that switch off the possibility of plasticity, in addition to the factors discussed here.

Functional Dependence of the Simulation Results The results above resulted from simulation using a particular choice of the correlation, cortical interaction, and arbor functions. The shapes of these functions are not well known in developing animals and, furthermore, it will be of experimental interest in the future to perturb these functions to study the effect upon development. Hence, to connect the model to biology, the dependence of developmental results on the choice of the correlation, cortical interaction, and arbor functions must be understood.

To determine this dependence, we have systematically varied each function. In the results that follow, one function at a time is varied, while the other two functions are left identical to those used in the simulation illustrated in figs. 16 and 18-20. The random initial conditions in all cases are identical to those used for that simulation, to facilitate comparison of results. This does not allow a complete exploration of the parameter space, but in combination with our analytical results, discussed subsequently, yields an understanding of the role played by each function in ocular dominance segregation.

We first consider the effects of varying the correlation functions. Figure 13 illustrates a series of correlation functions obtained by broadening or narrowing a gaussian correlation function within each eye ("same-eye correlations"). To such gaussian correlations within each eye, weak, more broadly ranging gaussian anticorrelations may be added either between the two eyes ("+ opposite-eye anticorrelations"), or within each eye ("+ same-eye anticorrelations").

Opposite-eye anticorrelations can be considered a model of strabismus or of alternating monocular deprivation; but such anticorrelations might also be present during normal development as a result of inhibitory interactions within the LGN. Same-eye anticorrelations are of interest both because of their use in other models (Linsker, Chapter 10), and because it is possible that they too might be introduced in the LGN or by the statistics of retinal response to visual input.

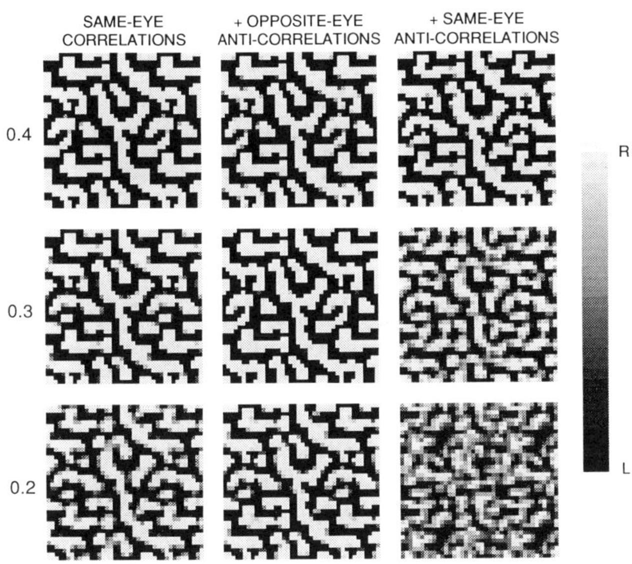

Figure 20: Development of ocular dominance using the nine afferent correlation functions illustrated in figure 13. Initial conditions, conventions, and all functions (cortical interaction, arbor) save the correlation function are as in figure 14. Cortex is shown in its final state (timestep 200). Broader correlations, or opposite-eye anticorrelations, lead to increased monocularity of the final cortex without changing the basic periodicity. Same-eye anticorrelations lead to decreased monocularity. If present within an arbor radius (0.2), same-eye anticorrelations lead to a largely binocular cortex with a weak and irregular pattern of ocular dominance segregation. The periodicity of this pattern is fundamentally altered from the others and is quite noisy (distributed over many wavelengths).

Results of simulations using each of these correlation functions are shown in figure 20. Broader correlations within each eye or addition of opposite-eye anticorrelations leads to a more completely monocular cortex. This agrees with experimental findings in which artificially induced strabismus or alternating monocular deprivation, which induce anticorrelations or reduce correlations between the eyes, lead to an increase in monocularity throughout the cortex (Hubel and Wiesel 1965). Narrower correlations within each eye lead to more binocular cells at the borders between eye stripes without otherwise altering the pattern of stripes. Same-eye anticorrelations also lead to more binocular cells at the borders between eye stripes and, if significant within an arbor radius ("same-eye anticorrelations 0.2"), tend to destroy monocularity and to destroy the regular cortical organization of ocular dominance. In this case, afferents from a single eye giving input to different halves of a single cortical cell's receptive field are anticorrelated with one another. Then development of one eye's strength in one part of the cell's receptive field discourages development of that same eye's strength elsewhere in the same field, leading to binocularity.

The above results on narrowing of the correlation function suggest that even nearest neighbor correlations might be sufficient to yield ocular dominance segregation. To test this, a gaussian correlation function within each eye ("same-eye correlations 0.4") was set equal to zero outside squares of various sizes (Miller, 1989C). Correlation only between afferents within one square ring of one another[11] is sufficient to yield monocular cells and a normal periodic pattern of segregation. Correlation over an arbor radius (3 square rings) seems both necessary and sufficient to yield a completely monocular layer 4. Similar conclusions result from the study of correlation functions that are constant over some distance, and equal to zero outside (Miller, 1989C)[12]. An increase in the amplitude of the correlations can largely, but not completely, compensate for a decrease in spatial extent. In all of these cases, when segregation occurs, it appears to occur with an invariant period.

Next, we consider the effects of varying the cortical interaction

11. α and β are within N square rings of each other if $|\alpha_1-\beta_1|\leq N$ and $|\alpha_2-\beta_2|\leq N$.

12. The correlation functions used were $C(0)=1$; $C(\alpha-\beta)=c$ for $\alpha\neq\beta$, α and β within N square rings of one another; $C(\alpha-\beta)=0$ otherwise.

function. Results using a purely excitatory cortical interaction function, without the more peripheral inhibition present in the Mexican hat function studied previously (figure 13), depend critically on the presence of constraints conserving the sum of synaptic strength over an afferent arbor. In contrast, results obtained with the Mexican hat function depended little on such constraints. Figure 21 shows the

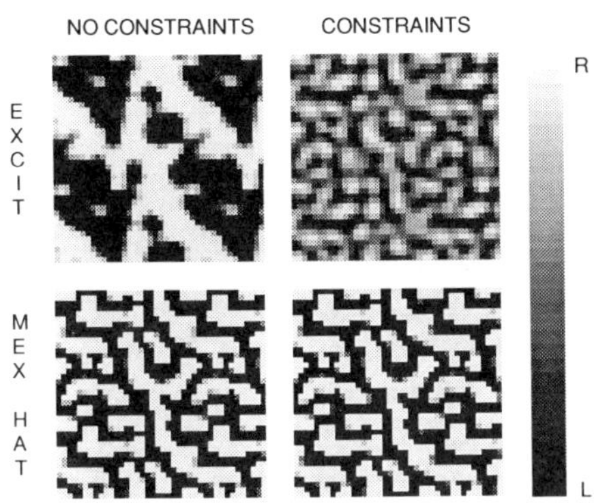

Figure 21: Cortical patterns of ocular dominance resulting from development with the excitatory cortical interaction function of figure 13, with and without constraints fixing the total synaptic strength over each afferent arbor. For comparison, results using the Mexican hat cortical interaction function of figure 13, with identical constraints, are shown below. The final cortex (T=200) is shown in each case. Functions, initial conditions, and conventions otherwise as in figure 14. Constraints on the total synaptic strength over a cortical cell are as in figure 14; only the constraint on total strength over an afferent is being varied. In the absence of constraints, the excitatory function leads to an arbitrarily large scale of ocular dominance segregation; constraints lead to segregation with width of left-eye plus right-eye stripes equal to about an arbor diameter. Much greater monocularity of segregation is seen if a slightly broader correlation function ("same-eye correlations 0.4") is used (Miller, 1989a; Miller, Keller, and Stryker, 1989). The Mexican hat function selects a width of segregation smaller than an arbor diameter, and results are unaffected by the constraints. From Miller, 1989c.

periodic pattern of ocular dominance resulting from the purely excitatory cortical interaction function, and from the Mexican hat cortical interaction function, in the presence or absence of such constraints. In the absence of constraints, a purely excitatory cortical interaction leads one eye to take over large regions of cortex, while the opposite eye loses all input to these regions. Constraints force afferents to redistribute strength, rather than to uniformly gain or lose strength, and thus force the two eyes to retain equal cortical innervation. This leads a periodic pattern of ocular dominance segregation to develop even when the cortical interaction function is entirely excitatory. Close examination reveals that the average width of the periodicity in this case is somewhat greater than with the Mexican hat cortical interaction function. The width of right-eye plus left-eye bands is now about an arbor diameter, as will be explained

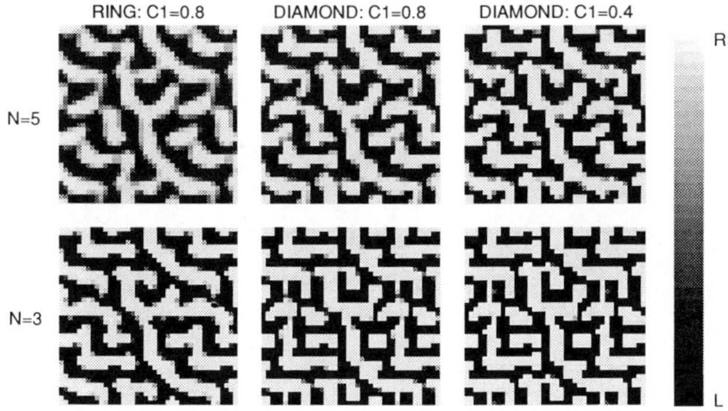

Figure 22: Cortices resulting from development with varying cortical interaction functions. These functions were zero outside of N=5 or 3 square rings. Functions were equal to c_1 (0.8 or 0.4) over either one square ring ("RING") or over the diamond of four nearest neighbors on the grid ("DIAMOND"), and equal to a constant negative value elsewhere, adjusted so that the function sums to zero. The correlation function used was "same-eye correlations 0.4". Arbor function and initial conditions as in figure 14. Cortices shown at T=100; runs were only carried out to that time. From Miller, 1989c.

below.

We have studied a variety of other forms of mixed excitatory/inhibitory cortical interaction functions. The resulting cortices, shown in figure 22, illustrate that the same initial conditions can lead to broader or narrower ocular dominance patches as the intracortical interactions are varied.

Finally, the effect of using an arbor function that tapers over the 7 x 7 region in which the arbor function is nonzero has been studied. Normal periodic segregation occurs, but if there are constraints fixing the total synaptic strength over an arbor, then the width of the periodicity is decreased (Miller, 1989c). The effect is like that of using a smaller arbor.

To summarize, the model robustly reproduces periodic ocular dominance segregation like that seen experimentally, beginning from an essentially uniform initial innervation by the two eyes. Individual cortical receptive fields refine in size and grow monocular; physiological fields develop inhibition from the nondominant eye. Afferent arbors refine in size and restrict their innervation to the appropriate eye's patches. The effects of monocular deprivation are reproduced, and a critical period is seen. Broader correlations within each eye, or anticorrelations between the eyes as in strabismus, enhance monocularity, while narrower correlations or anti-correlations within an eye reduce monocularity. Changes in the correlation function do not obviously modify the width of periodic segregation, provided that segregation occurs. A purely excitatory cortical interaction function is sufficient to yield ocular dominance patches, provided arbors are constrained to conserve their total synaptic strength. The width of the periodic pattern can be altered by varying the cortical interaction or arbor functions.

We now seek to understand more generally what is responsible for periodic segregation. Can we predict the forms of the correlation, cortical interaction, and arbor function that will lead to segregation, and can we predict the width of the stripes that will result when segregation does occur?

Analysis of the Model

To analyze the model, we assume equality of the two eyes (ignoring the case of monocular deprivation, and relatively small asymmetries in normal development). This means $C^{LL} = C^{RR} = C^{SameEye}$, $C^{LR} = C^{RL} = C^{OppEye}$. We analyze the development of the *difference* between the two eyes' synaptic strengths, $S^D(x,\alpha,t) \equiv S^L(x,\alpha,t) - S^R(x,\alpha,t)$. The equation for S^D is

$$\frac{dS^D(x,\alpha,t)}{dt} = \lambda A(x-\alpha)\sum_{y,\beta} I(x-y)C^D(\alpha-\beta)S^D(y,\beta,t) - \gamma S^D(x,\alpha) \qquad (5)$$

where $C^D(\alpha-\beta) \equiv C^{SameEye}(\alpha-\beta) - C^{OppEye}(\alpha-\beta)$. C^D represents the degree to which an afferent is more correlated with another afferent from its own eye than with an afferent from the opposite eye, at a fixed retinotopic separation.

Initially the innervations representing the two eyes are nearly equal. Hence S^D is very nearly zero everywhere. If this difference were initially exactly zero everywhere, it would stay so forever. Each eye's synapses would experience exactly the same environment at each moment in time, and we have assumed that the two eyes obey identical rules. Given the small perturbation in the difference, however, one can imagine two outcomes: either the small perturbation might decay to a state of complete equality; or it might grow, developing into some large pattern of differences between the two eyes. We will analyze whether the condition of equality is stable to small perturbations. If it is unstable, we will determine the pattern that will grow out of the instability.

This analysis, as to the existence and nature of a possible pattern-forming instability, depends only upon the behavior of the model very near the condition of complete equality. In this region, S^D is very small, hence nonlinear terms are negligibly small. Therefore, only the linear terms in S^D play a role in this region. Thus, even quite complicated, nonlinear models can be reduced, near the condition of equality, to a linear equation. If the linear equation we use for S^D approximates the linear equation resulting from a nonlinear model, our results will describe the pattern formation expected in such a fully nonlinear model.

To conduct this analysis, we determine the patterns of ocular dominance that grow or decay exponentially and independently, each at its own rate, in the linear regime. We call these the characteristic patterns of ocular dominance. Each choice of functions yields a distinct set of characteristic patterns and growth rates. If some of the patterns have a positive growth rate for a given choice of functions, so that they will grow in time rather than decay, then the condition of equality will be unstable to small perturbations for that choice of functions. In this case, we determine the characteristic patterns of ocular dominance that grow most quickly. These patterns are spatially periodic. Their wavelength determines the *width* of the final periodic pattern of ocular dominance across cortex that will result for that choice of functions. This is the width of a complete cycle of right-eye plus left-eye ocular dominance patches or stripes. This width is determined by the wavelengths of the fastest-growing patterns, because in the linear regime these patterns quickly dominate. The *form* that this periodic pattern takes – blobs, patches, stripes, hexagons – depends on the nonlinearities, and cannot be predicted from such a simple analysis.

This discussion can be expressed mathematically. Write equation 5 as $\frac{dS^D(x,\alpha,t)}{dt} = LS^D(x,\alpha,t)$ where L stands for the linear operator on the right side of equation 5. Suppose we can find eigenfunctions of the operator L, that is, a set of functions $S_i^D(x,\alpha)$ such that $LS_i^D(x,\alpha) = \omega_i S_i^D(x,\alpha)$ for constants ω_i, where i is an index enumerating the eigenfunctions. The constants ω_i are the *eigenvalues* of the operator L. Suppose further that these eigenfunctions are *complete* in the sense that any initial condition $S_0^D(x,\alpha) \equiv S^D(x,\alpha,t{=}0)$ can be expressed as a linear sum of these eigenfunctions: $S_0^D(x,\alpha) = \sum_i c_i S_i^D(x,\alpha)$ for some constants c_i[13]. Then the solution to equation 5 is $S^D(x,\alpha,t) = \sum_i c_i e^{\omega_i t} S_i^D(x,\alpha)$. If the real equation has nonlinearities, this solution only holds near the initial condition, where the c_i are small and the equation remains effectively linear.

From this mathematical solution we can conclude several things about development of S^D near the initial condition. First, in this region the eigenfunctions each grow independently of one another, with the rate of growth of each determined by its eigenvalue ω_i. The eigenfunctions are thus the characteristic patterns of ocular dominance, defined above; while the eigenvalues (or, in the case of complex eigenvalues, the real parts of the eigenvalues) are the growth rates of the characteristic patterns. Second, if any of the growth rates are positive, the corresponding characteristic patterns will grow rather than decay from the initial condition (this assumes that, in the perturbation representing the initial condition, none of the c_i are exactly zero). Hence, the condition for a pattern-forming instability is

---

13. If $I(x)=I(-x)$ and $C^D(x)=C^D(-x)$, the substitution $S'^D(x,\alpha,t) = \frac{S^D(x,\alpha,t)}{\sqrt{A(x-\alpha)}}$ transforms L into a symmetric operator; hence in this case L will have a complete set of eigenfunctions, with real eigenvalues.

the presence of at least one positive growth rate. Third, if there is such an instability, the exponential growth in time of each pattern ensures that the fastest growing patterns will quickly outgrow the others and dominate the final solution. This is the basis for the statement that the fastest-growing pattern in the linear regime sets the wavelength, even in the final fully non-linear conditions.[14] If there are a variety of fastest-growing patterns, for example, ones involving parallel ocular dominance stripes of a fixed wavelength but each with stripes oriented in a different direction across cortex, then the final form of the overall pattern will depend on the initial conditions and upon dynamical interactions among these individual oriented patterns. Such interactions are exclusively nonlinear; hence, the form but not the wavelength of the final pattern is determined by the nonlinearities in the process.

The Characteristic Patterns of Ocular Dominance

The characteristic patterns of ocular dominance are the patterns of S^D that grow independently and exponentially, each at its own rate, from an initial condition of near equality of the two eyes. Each characteristic pattern of ocular dominance consists of a characteristic receptive field of ocular dominance, and an oscillation of ocular dominance across cortex. Figure 23 shows the fastest growing such pattern for the functions used in the simulation of figure 14. The characteristic receptive field is the pattern of differences between left- and right-eye synaptic strengths in the input to a cortical cell. Where it is positive one eye is dominant, and where it is negative the opposite

14. It is possible for the initial pattern that develops out of an instability to be only metastable, so that eventually it may reorganize into a pattern with a different period determined by nonlinearities. This seems unlikely in the ocular dominance system, as studies of development of ocular dominance columns (LeVay, Stryker, and Shatz, 1978) show no obvious change in periodicity between its earliest detection and the final adult pattern. With the use of voltage sensitive dyes (Blasdel and Salama, 1986), it may be possible to test this by following development of the columns within a single animal.

eye is dominant. A monocular characteristic receptive field, like that of figure 23, is one dominated by a single eye throughout, so that the

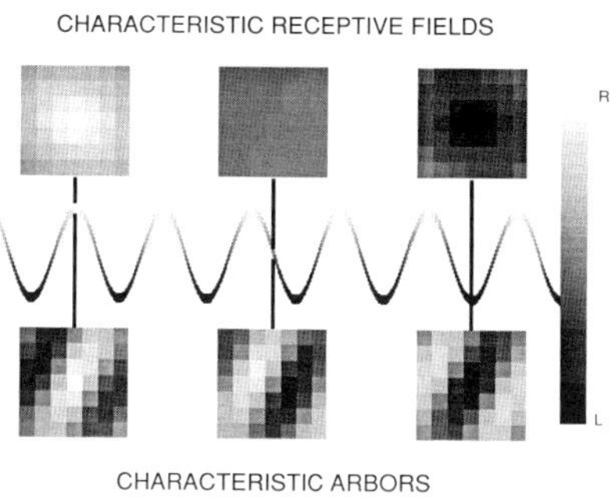

Figure 23: A monocular characteristic pattern of ocular dominance. The characteristic receptive field, and associated characteristic arbor, are illustrated at three cortical locations. The sinusoid illustrates the oscillation of ocular dominance across cortex associated with the characteristic pattern, correctly scaled to the arbor and receptive field sizes. Greyscale codes S^D, the difference between the synaptic strengths of the two eyes, varying from dominance by one eye to dominance by the other. The pattern shown here is one of the set (identical except for rotations of the direction of the oscillation) of fastest-growing characteristic patterns for functions used in the simulation of figure 14. The characteristic receptive field illustrates the retinotopic positions from which a cortical cell at a given position in cortex receives stronger left-eye or right-eye geniculate input. The characteristic arbor illustrates the cortical positions to which the left eye or right eye geniculate afferents representing a given retinotopic position project more strongly. At the cortical point corresponding to the leftmost receptive field, cortical cell inputs are dominated by the right eye everywhere within the receptive field. Afferents with the corresponding retinotopic position therefore project arbors such that the right eye afferents preferentially project to the central patch of the arbor (cortical right-eye stripe) and the left-eye afferents preferentially project to the peripheral patches in the cortex (cortical left-eye stripe). Similarly, the central receptive field is at the border between left-eye and right-eye stripes, where the two eyes have equal innervation, and the rightmost receptive field is in the center of a left-eye stripe. The oscillation projects in a direction perpendicular to the stripes across the arbors, rather than horizontally as depicted.

characteristic receptive field can be taken positive everywhere. Characteristic receptive fields need not be monocular: they may be binocular, showing division of the receptive field into domains dominated by opposite eyes. The characteristic receptive field will be positive in the domains dominated by one eye, and negative in the domains dominated by the opposite eye.

The characteristic receptive field represents the pattern of ocular dominance *within* a cortical receptive field. The characteristic pattern of ocular dominance also includes an oscillation in the degree of dominance of receptive fields across the cortex. This is an oscillation of ocular dominance *between* receptive fields across the cortex. In figure 23, the leftmost receptive field occurs at a cortical point where the right eye is dominant. The central receptive field occurs at a point that differs by 1/4 cycle of the cortical oscillation, so the two eyes are equal. The rightmost receptive field occurs at a point that differs by 1/2 cycle of oscillation, so the left eye is dominant. At points a distance across the cortex corresponding to an integral number of full cycles of oscillation, the right eye is again dominant. In the case of monocular characteristic fields, it is this oscillation, between ocular dominance by one eye and by the other, that causes organization of monocular cortical cells into ocular dominance patches. We refer to the spatial period or wavelength of this oscillation as the wavelength of the characteristic pattern. This wavelength corresponds to the width of left-eye patch plus right-eye patch, which we refer to simply as the patch width.

As figure 23 indicates, the characteristic receptive fields have associated with them characteristic afferent arbors, given by multiplying the receptive field by the oscillation in ocular dominance. In other words, when characteristic receptive fields are monocular, so that ocular dominance patches arise, the afferent arbors will only innervate the patches from the relevant eye. Thus, characteristic arbors show patches with a periodicity equal to that of the cortical oscillation, as is seen both in the simulations and in actual biological development.

Different characteristic patterns, then, are distinguished by (1) The nature of their characteristic receptive field; (2) The wavelength of the oscillation of ocular dominance across cortex. The characteristic receptive field may be monocular, corresponding to development of ocular dominance segregation, or binocular, corresponding to a cortex in which every cell receives input from both eyes. If it is monocular, the wavelength of the cortical oscillation will determine the patch width of ocular dominance organization associated with the pattern. Many patterns will compete from the initial condition of near equality of the two eyes; those that grow fastest will win out. If the fastest-growing patterns are all monocular and all have a similar wavelength, the cortex will develop ocular dominance organization with that wavelength.

To express these results mathematically, transform variables in equation 5 from cortex and LGN (x,α) to cortex and receptive field (x,r), where $r = x - \alpha$. The resulting equation is a simple convolution in the cortical variable x. By fourier transform, such a convolution has eigenfunctions of the form $S^D_{mj}(x,\alpha) = e^{im\cdot x}R_{mj}(r)$[15], where m is a two-dimensional real vector, the wavevector of cortical oscillation, and j is an additional index of the eigenfunctions for a given m. R is a characteristic receptive field, because it represents the dependence of the eigenfunction upon r for a fixed cortical position x. We can also write this as $S^D_{mj}(x,\alpha) = e^{im\cdot\alpha}B_{mj}(r)$ where $B_{mj}(r) \equiv e^{im\cdot r}R_{mj}(r)$. B represents the dependence of the eigenfunction upon r for a fixed afferent position α, and hence is a characteristic arbor.

These eigenfunctions are complex. If $I(x) = I(-x)$ and $C^D(\alpha) = C^D(-\alpha)$, the real part and imaginary part of a complex eigenfunction are both real eigenfunctions, with the same eigenvalue. These real eigenfunctions can be chosen of the form $\cos k\cdot x\, R^+(r) + \sin k\cdot x\, R^0(r)$, where R^0 has zero net ocular dominance $(\sum_r R(r)=0)$ and R^+ may have net ocular dominance. For the eigenfunction illustrated in figure 23, $R^0 \equiv 0$; the nonmathematical discussion was simplified by neglect of the possibility that R^0 may be nonzero[16].

---

15. We are assuming here, and subsequently, either an infinite grid or a continuum. On the finite grid with periodic boundary conditions of the simulations, we would instead obtain $e^{\frac{2\pi i m\cdot x}{L}} R_{mj}(r)$ where the finite grid is of length L and m is a two-vector of integers.

16. To construct real eigenfunctions, note that, if the operator L can be made symmetric (see note 12), the complex conjugate of an eigenfunction is also an eigenfunction with identical (real) eigenvalue. Hence, given any eigenfunction S^D_{mj}, its real part S^R_{mj} and its imaginary part S^I_{mj} are both real eigenfunctions with the same eigenvalue. Let R^R and R^I be the real and imaginary parts, respectively, of R. Then these real eigenfunctions are

$$\begin{pmatrix} S^R_{mj}(x,\alpha) \\ S^I_{mj}(x,\alpha) \end{pmatrix} = \begin{pmatrix} \cos mx & -\sin mx \\ \sin mx & \cos mx \end{pmatrix} \begin{pmatrix} R^R_{mj}(r) \\ R^I_{mj}(r) \end{pmatrix}$$

Thus real eigenfunctions involve a sinusoidal mixing, across cortex, between R^R

In summary, if the fastest-growing characteristic pattern of ocular dominance is monocular – that is, if its characteristic receptive field is monocular – it will yield segregation across cortex into ocular dominance patches. The width of a cycle of these patches will be given by the wavelength of the pattern's associated cortical

---

and R^I. Note that S^R and S^I are identical except for a cortical phase shift of 90°.

A more convenient representation can be found by shifting phase so that one of the two basis receptive fields has zero dominance. The subscript [mj] will be assumed where not explicitly indicated. Define the (unnormalized) dominance of an eigenfunction by $D \equiv |\sum_r R(r)|$, and define the angle θ by $\sum_r R(r) = De^{i\theta}$. Define new complex eigenfunctions by $\tilde{S}^D = e^{-i\theta} S^D = e^{ik\cdot x}\tilde{R}$ where $\tilde{R} = e^{-i\theta}R$. The imaginary part of $\tilde{R}$ has zero dominance, $\sum_r \tilde{R}^I(r) = 0$, while the real part of $\tilde{R}$ has the maximum dominance for any cortical phase of receptive fields associated with that eigenfunction, $\sum_r \tilde{R}^R(r) = D$. The new real eigenfunctions are given by

$$\begin{pmatrix} \tilde{S}^R_{mj}(x,\alpha) \\ \tilde{S}^I_{mj}(x,\alpha) \end{pmatrix} = \begin{pmatrix} \cos\theta_{mj} & \sin\theta_{mj} \\ -\sin\theta_{mj} & \cos\theta_{mj} \end{pmatrix} \begin{pmatrix} S^R_{mj}(x,\alpha) \\ S^I_{mj}(x,\alpha) \end{pmatrix}$$

Define $R^+ \equiv \tilde{R}^R$, $R^0 \equiv \tilde{R}^I$. Then

$$\begin{pmatrix} \tilde{S}^R_{mj}(x,\alpha) \\ \tilde{S}^I_{mj}(x,\alpha) \end{pmatrix} = \begin{pmatrix} \cos mx & -\sin mx \\ \sin mx & \cos mx \end{pmatrix} \begin{pmatrix} R^+_{mj}(r) \\ R^0_{mj}(r) \end{pmatrix}$$

Hence the eigenfunctions can be described as a sinusoidal mixing, across cortex, between a maximum-dominance (for that eigenfunction) receptive field, and a zero-dominance receptive field. Figures 23-25 illustrate R^+ and R^0 for each eigenfunction.

In the general case, the operator L is not assumed symmetrizable. The fact that L is real ensures that $\tilde{S}^D_j$ and its complex conjugate $\tilde{S}^{D*}_j$ are both eigenfunctions, with complex conjugate eigenvalues. If the leading eigenvalues are not real, the result will be temporal oscillations of ocular dominance. These will combine with the spatial oscillations to yield traveling waves of ocular dominance. This is discussed in more detail in Miller, 1989c. One expects that such an effect should be suppressed in any reasonable theory, so that the leading eigenvalues will always be real, although empirically it is not impossible that such waves could exist early in development.

oscillation. Afferent arbors will become confined to those periodic patches. To understand the development of ocular dominance segregation within the model, then, it is sufficient to answer two questions: (1) Under what circumstances will the fastest growing pattern be monocular? (2) When it is monocular, what will determine the wavelength of its cortical oscillation?

Monocularity and Wavelength of the Fastest Growing Patterns

The results of both analysis and computation lead to the following conclusion: The fastest growing patterns are monocular whenever the correlation function C^D is positive at least between nearest neighbors, and not significantly negative within an arbor radius. In this case, the wavelength of cortical oscillation is the wavelength corresponding to the maximum of the fourier transform of the cortical interaction function.

Intuitively, this can be understood as follows. By a correlation-based mechanism, an individual cortical cell develops two features. First, its receptive field comes to consist of inputs whose activities are mutually correlated. To the extent to which, in the local area sampled by the cortical cell, this represents inputs from a single eye, the cortical cell will become monocular. Alternatively, when there are same-eye anticorrelations within an arbor radius, the most correlated inputs are obtained if one part of the receptive field represents one eye, and an adjacent part of the receptive field represents the opposite eye. This yields a binocular field. Second, due to cortical interactions, a cortical cell's receptive field becomes both maximally correlated with the receptive fields of neighboring cortical cells at distances over which the cortical interactions are excitatory, and minimally correlated (or maximally anti-correlated) with the receptive fields of neighboring cortical cells at distances over which cortical interactions are inhibitory. For a monocular cell, this is best achieved when nearby cells within the region of excitatory cortical interactions are monocular cells that represent the same eye; while more distantly neighboring cells, within the region of inhibitory cortical interaction, are monocular cells from the opposite eye. This cannot be achieved simultaneously for all cells; but it is most nearly achieved, over all cells, by matching the cortical oscillation of the ocular dominance

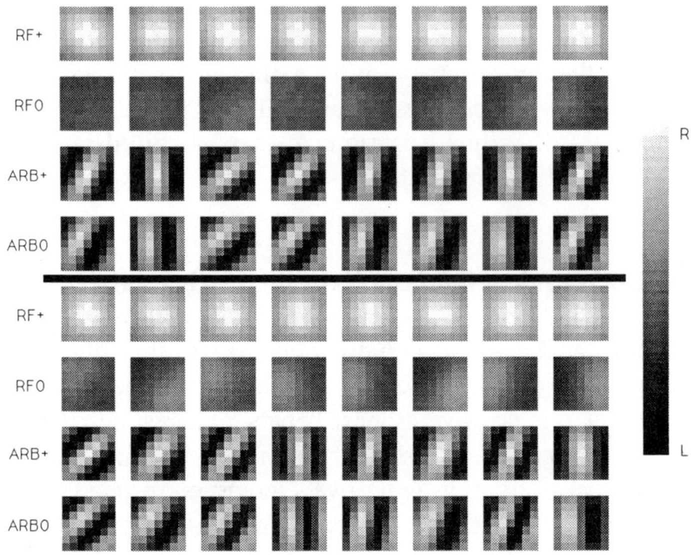

Figure 24: The 16 fastest-growing characteristic patterns of ocular dominance for the functions used in the time development of figure 14. Each vertical set of four squares shows, from top to bottom, the maximum dominance receptive field, the minimum dominance receptive field, the arbor corresponding to the maximum dominance position, and the arbor corresponding to the minimum dominance position. These correspond, respectively, to the leftmost receptive field, central receptive field, leftmost arbor, and central arbor in figure 23. The 16 patterns are arranged in decreasing order of growth rate, so that the fastest growing is at upper left and most slowly growing of the 16 at lower right. For each pattern, additional patterns with identical growthrates exist, differing only by rotations or reflection of the pictured pattern. All the leading patterns have highly monocular receptive fields. They differ significantly only in the precise width and orientation of their cortical oscillations, which is visible as the oscillation across the arbors. From Miller, 1989c.

pattern to the dominant wavelength in the cortical interaction.

We will illustrate these results both through direct computations of the characteristic patterns for varying parameters, and by analytic solutions to the equations in various limiting cases.

Computation of the Characteristic Patterns We can directly compute the characteristic patterns for the model when it is placed on a grid as in the simulations of time development. For the choice of functions used in the time development displayed in figure 14, for which C^D

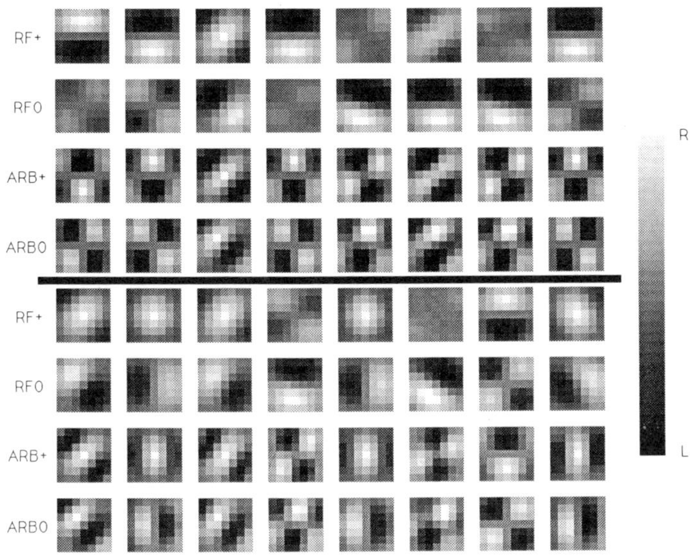

Figure 25: The sixteen fastest-growing patterns for the correlation function "+ same-eye anticorrelations 0.2", with functions otherwise as in figure 24. The fastest-growing patterns are binocular. Conventions as in figure 24. From Miller, 1989c.

satisfied the above criteria for emergence of monocularity, the fastest growing characteristic patterns are indeed monocular (figure 24). There are 625 oscillation frequencies on the 25×25 grid, and for each such frequency there are 49 characteristic patterns due to the 7×7 arbor. Among the 49 characteristic patterns with a single oscillation frequency, for the same choice of functions, only a single pattern – the fastest growing one for that frequency – is monocular (Miller, 1989c). Thus, monocular patterns are rare among all possible patterns, yet they are selected as the fastest-growing for this choice of functions. In contrast, with a different choice of correlation function that includes prominent anticorrelations within an arbor radius ("+ same-eye anticorrelations 0.2"), the fastest growing patterns across all frequencies are binocular (figure 25). For this correlation function, only weak and irregular ocular dominance segregation developed in simulations (figure 20).

Results across all frequencies and for varying correlation functions are shown in figure 26. This figure illustrates the results of computation of the characteristic patterns for each of the nine sets of functions, identical except for varying correlation functions, whose developmental outcome was illustrated in figure 20. The graphs show the growth rate (vertical axis) of the characteristic patterns versus their inverse wavelength (horizontal axis). The graph's grey scale represents the monocularity of patterns: the shading at a point corresponds to the degree of monocularity seen for patterns with the given wavelength and growth rate, from lightest for fully monocular, to black for fully binocular. The curves show analytic predictions for the growth rates as a function of wavelength, derived in limiting cases that will be discussed. In particular, the lower thin lines show the fourier transform of the cortical interaction function, whose peak accurately predicts the wavelength of the fastest-growing patterns.

Increasing the breadth over which the correlation function C^D is positive increases the monocularity of the fastest-growing patterns and increases the dominance in growth rate of monocular patterns over binocular patterns. The breadth of C^D is increased either by broadening of within-eye correlations, or by the addition of more broadly-ranging opposite-eye anticorrelations[17]. The monocularity and the growth rate of monocular patterns is enhanced at wavelengths equal to the peak of the fourier transform of the cortical interaction function. Introduction of anticorrelations within each eye can cause nonmonocular modes to grow fastest; but even in this case, the fastest-growing monocular modes and the most strongly monocular modes are found at the dominant wavelength of the cortical interaction function. Thus, the correlation function C^D provides the precondition for the ascendency of monocular cells. The cortical interaction function selects the wavelength of monocular cells that will grow the fastest, enhancing the growth rate and increasing the monocularity of those fastest-growing monocular patterns.

17. Opposite-eye anticorrelations also enhance the development of monocularity even if they do not extend more broadly than within-eye correlations. This is because their presence increases the rate of growth of S^D relative to the rate of growth of $S^S \equiv S^L + S^R$. This allows the development of ocular dominance segregation to proceed faster relative to the rate of topographic refinement of receptive fields and to the rate of approach of synapses to saturation.

Analysis of Limiting Cases These results can also be derived analytically through solutions of the equations in limiting cases. One simple limit is the case in which the correlation function C^D is constant over at least an arbor diameter plus the maximum distance x for which $I(x) \neq 0$. In this limit the influence between synapses depends only upon their eyes of origin and their cortical locations, and not upon their retinotopic locations. Then there can be no retinotopic refinement within each receptive field, since all synapses from a given eye onto a given cortical cell grow or decay identically. Hence, the only dynamical variable for each cortical cell is its ocular dominance.

The equation can be reduced in this limit to a simpler equation studied previously by Swindale (1980), by performing $\sum_\alpha$ on both sides of equation 5 to obtain

$$\frac{dD(x,t)}{dt} = \sum_y w(x-y)D(y,t) - \gamma D(x,t). \tag{6}$$

$D(x,t) \equiv \sum_\alpha S^D(x,\alpha,t)$ is the ocular dominance of the cortical cell at x, $w(x-y) \equiv \lambda C^D N I(x-y)$ tells the influence between ocular dominance development at two different cortical locations, and $N \equiv \sum_\alpha A(\alpha)$ is the number of synapses in an arbor. Swindale called $w(x-y)$ an "effective interaction", but its biological nature was not clear. In this limit, we may express his effective interaction in terms of correlations, arbors and intracortical interactions. Swindale's equation can be easily solved. It is a convolution. Hence, periodic modes of ocular dominance, $D_m(x) = e^{im \cdot x}$, grow with a rate ω_m determined by the fourier transform $\tilde{w}(m)$ of $w(x)$, $\omega_m = \tilde{w}(m) - \gamma = \lambda C N \tilde{I}(m) - \gamma$. The full eigenfunctions of our original equation can also be found in this limit: for each m, there is one monocular mode, $S^D_m(x,\alpha) = e^{im \cdot x} A(x-\alpha)$, with growth rate ω_m as just given. The Swindale equation describes the growth of this mode. All other modes decay with growth rate $-\gamma$. Thus in this limit, the non-decaying eigenfunctions are completely monocular and the fastest growing mode is selected by the

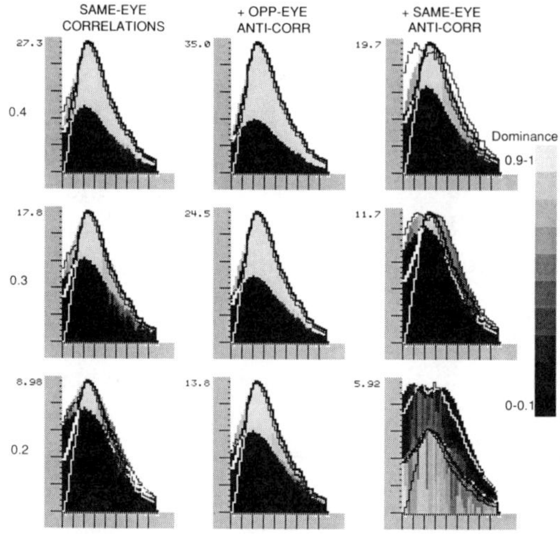

Figure 26: Computed growth rate (vertical axes) of characteristic patterns of ocular dominance, as a function of inverse wavelength of the pattern (horizontal axes). Results are presented for the nine correlation functions of figures 13 and 20, with Mexican hat cortical interaction function and flat arbor function as in those figures. Greyscale indicates the dominance of characteristic patterns: Dominance is a measure of the degree of monocularity of the pattern's characteristic receptive fields, on a scale from 0 for complete binocularity (black) to 1 for complete monocularity (light). Graphs for functions for which monocular patterns are the fastest growing (all but lower right) were constructed by painting lower dominance growthrates on top of higher dominance growthrates, so that the greyscale at any point reflects the lowest dominance patterns with wavelength corresponding to the point and growthrate at least as large as that corresponding to the point. For the case of "+ same-eye anticorrelations 0.2" (lower right), this scheme was reversed: Higher dominance growthrates were painted on top of lower dominance growthrates. Mixed black/white lines show the analytic predictions for growth rate as a function of inverse wavelength derived in limiting cases. Lower thin lines show the fourier transform of the cortical interaction function, which is the prediction in the limit of constant correlations; thick lines show the prediction of the broad correlations limit. These two predictions have been normalized to the maximum growth rate of patterns with dominance ≥0.5. The upper thin lines show the prediction of the limit of full connectivity, normalized to the maximum growth rate of any pattern. Number beside the vertical axis indicates the maximum growth rate of any pattern. The horizontal axis represents wave number; the wavelength in units of grid intervals is 25 divided by the wave number. The first bin on the horizontal axis represents wave numbers 0-0.23; subsequent bins represent increments of 0.4 in wave number, so that the second bin represents wave numbers 0.23 to 0.63, and so forth. From Miller, 1989c.

peak of $\tilde{I}(m)$. This limit is shown as the lower thin lines in figure 26. It accurately predicts the wavelength of the fastest growing monocular pattern, even far from the conditions in which the limit was derived.

In a limit of broad but not constant correlations, we approximate the correlation of one afferent with any afferent synapse onto a cortical cell, by the average correlation of that afferent with all afferent synapses onto that cortical cell. Thus, in equation 5, we set

$$C^D(\alpha-\beta)S^D(\beta,y) \approx \frac{\sum_\delta C^D(\alpha-\delta)A(\delta-y)}{\sum_\delta A(\delta-y)} S^D(\beta,y).$$

In this limit, the influence between synapses depends on the retinotopic location of the influenced synapse, but not of the influencing synapse. This limit also yields a Swindale equation after performing $\sum_\alpha$ on both sides, but now $w(x-y)=\lambda A*C*A(x-y)I(x-y)/N$. The '*' indicates convolution, $X*Y(z)=\sum_k X(z-k)Y(k)$, and $N=\sum_\alpha A(\alpha)$. The growth rates of cortical ocular dominance in this case are $\omega_m=(\lambda/N)\sum_k \tilde{I}(m-k)\tilde{A}^2(k)\tilde{C}(k)-\gamma$. Because C is assumed positive over a broad range, and because biologically A and C are likely to extend without oscillation over a much broader range than I, fourier transforms of A and C will be much narrower than that of I. This expression for ω_m then implies that the peak growth rate is found at or very near the m that maximizes $\tilde{I}(m)$, but the growth rates do not fall to zero as sharply as $\tilde{I}(m)$. This prediction is shown as the thick lines in figure 26. It accurately predicts the growth of monocular patterns of all wavelengths, even far from the limiting conditions in which it was derived.

A third relevant limit is the limit of complete geniculocortical connectivity, $A(x-\alpha)\equiv 1$, in which every geniculate cell is connected to every cortical cell. In this limit the eigenfunctions are $S^D_{k_x k_\alpha}(x,\alpha)=e^{ik_x x}e^{ik_\alpha \alpha}=e^{imx}e^{-ik_\alpha(x-\alpha)}$, with $m=k_x+k_\alpha$. The resulting growth rates are $\omega_{k_x k_\alpha}=\lambda\tilde{I}(k_x)\tilde{C}^D(k_\alpha)-\gamma$. The correlation function drives an oscillation across the receptive field with wavenumber k_α and growth rate proportional to $\tilde{C}^D(k_\alpha)$. For the fastest growing mode to be monocular, the peak of C^D must be at frequency 0. The cortical interaction function drives an oscillation across an arbor with wavenumber k_x and growth rate proportional to $\tilde{I}(k_x)$. The resulting cortical oscillation wavenumber, m, is the sum of k_x and k_α. In the event of monocular modes, $k_\alpha = 0$, the arbor and cortical oscillations are identical. Therefore the peak of $\tilde{I}$ selects the fastest growing cortical wavelength of monocular modes. In general, the intuitive picture found in this limit, that correlations drive oscillations across a receptive field, intracortical interactions drive oscillations across an arbor, and the cortical oscillation is the sum of these two oscillations, is a useful one for understanding a wide variety of cases.

In this limit, the fastest growth rate for a pattern with a given wavenumber m is proportional to the maximum over all k_α of $\tilde{I}(m-k_\alpha)\tilde{C}^D(k_\alpha)$. The resulting predictions for fastest growth rate versus wavenumber m are shown as the upper thin lines in figure 26. When the peak of C^D is at frequency zero, this limit accurately predicts the growth rates of monocular patterns of each frequency. Otherwise, if the peak of C^D is at a wavelength less than about an arbor diameter, binocular patterns become dominant. Then this limit appears qualitatively to predict some aspects of the shape of the curve of fastest growth rates vs. wavelength for these binocular patterns.

We can summarize the results found by computation, and suggested by these limiting cases and other analytical methods, as follows. When same-eye correlations are locally greater than opposite-eye correlations, and not significantly less within an arbor radius, then C^D is locally positive, and nonnegative within an arbor radius. Under these conditions, monocular cells will tend to form. The width of dominance patches will then be determined by the dominant wavelength in the cortical interaction function, which is the peak of its fourier transform. A broader C^D leads to an increased dominance in growth rate of monocular over binocular patterns and thus an increase in the tendency to monocularity. Oscillations within C^D cause a tendency to oscillation of ocular dominance within individual receptive fields, and if present significantly within an arbor radius is inconsistent with monocular segregation.

The Role of Constraints There is one important addition to the above conclusions. For monocular patterns with wavelength longer than an arbor diameter to grow, entire arbors centered within ocular dominance patches must either grow or shrink. If this were the case, some regions of the visual field in each eye would come not to be represented at all in the cortex. Hence, one might expect that the growth of such patterns is somehow prevented in biological systems. This could occur dynamically, if biological parameters select patterns with wavelengths shorter than an arbor diameter. Alternatively, these patterns might otherwise tend to grow, but be suppressed by additional limitations on development. For example, a presynaptic cell whose terminal arbor has a smaller total synaptic strength may compete more effectively for its remaining synapses, while one with more total synaptic strength may become less competitive or reach limits to its ability to support further synaptic strength. This would yield a constraint on the total synaptic strength supported by a presynaptic cell.

Such a constraint can be modelled by fixing the total synaptic strength over a presynaptic arbor. The effect, as suggested, is to prevent the growth of monocular patterns longer than an arbor

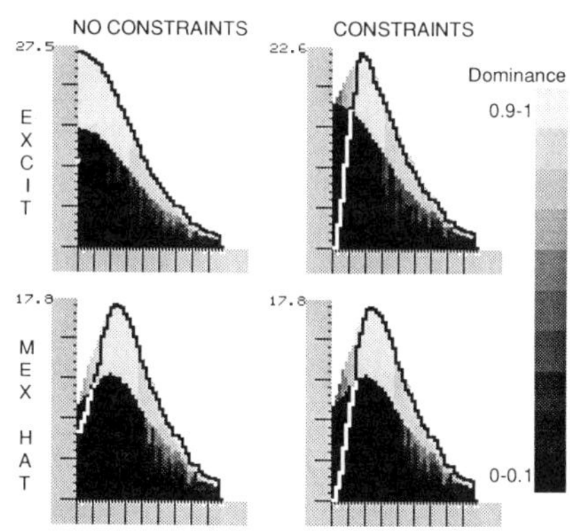

Figure 27: Computed growth rate (vertical axes) of characteristic patterns of ocular dominance, as a function of inverse wavelength of the pattern (horizontal axes), for the two cortical interactions and the two choices of constraints on afferent arbors of figure 21. Mixed black/white lines indicate predictions of the broad correlations limit, described in the text. Conventions otherwise as in figure 26. The constraints suppress the growth of monocular patterns with wavelength longer than an arbor diameter. This has a profound effect on the outcome when the excitatory cortical interaction function is used, selecting a wavelength of about an arbor diameter. There is little effect when the mixed excitatory/inhibitory interaction function is used, because that function normally selects a wavelength shorter than an arbor diameter. Maximum growth rates are at wavelengths of 7.3-8.3 grid intervals (excitatory function, with constraints) or 5.4-5.9 grid intervals (Mexican hat function). From Miller, 1989c.

diameter (figure 27). If the cortical interactions select a pattern with a wavelength shorter than an arbor diameter, the constraints have essentially no effect on the overall pattern that develops. However, if the cortical interactions select a pattern with longer wavelength, as in the case of purely excitatory cortical interactions, the effect is dramatic: the constraints force selection of a wavelength of about an arbor diameter. This was demonstrated in simulations in figure 21.

Mathematically, the constraint on presynaptic cells is modelled as follows. Write equation 5 as $\frac{dS^D(x,\alpha,t)}{dt} = LS^D(x,\alpha,t)$ as previously, and let $\gamma=0$. Now replace this equation with $\frac{dS^D(x,\alpha,t)}{dt} = LS^D(x,\alpha,t) - \kappa\frac{A(x-\alpha)}{N}\sum_y LS^D(y,\alpha,t)$ where $N \equiv \sum_x A(x-\alpha)$, with $0 \leq \kappa \leq 1$. Setting $\kappa=0$ gives the original equation. Setting $\kappa=1$ yields a fully constrained equation for which $\sum_x \frac{dS^D(x,\alpha,t)}{dt} = 0$, as would be the case if total synaptic strength over afferent arbors were fixed. In the constant correlations limit, the result of this extra term is to multiply the $\tilde{m}$-dependent part of the growth rate for monocular modes by $1-\kappa\frac{|\tilde{A}(m)|^2}{|\tilde{A}(0)|^2}$. This suppresses monocular modes whose oscillation frequency m contributes significantly to $\tilde{A}$, that is, modes whose wavelengths are longer than an arbor diameter. In the broad correlations limit, a similar result is obtained: the monocular growth rates become

$$\omega_m = (\lambda/N)\sum_k \tilde{I}(m-k)\tilde{A}^2(k)\tilde{C}(k)\left[1-\kappa\frac{\tilde{A}(m)}{\tilde{A}(0)}\frac{\tilde{A}(k-m)}{\tilde{A}(k)}\right] \qquad (7)$$

Similar results are obtained by modeling the constraint as subtraction of a term $-\kappa'\sum_x S^D(x,\alpha)$, or as combinations of the two terms. Other methods of analysis of these equations also support this conclusion: that constraints serve to suppress the growth of monocular modes whose wavelengths are longer than an arbor width, or more specifically, long enough to contribute significantly to $\tilde{A}$.

Constraints on the total synaptic strength supported by a postsynaptic cell may also play an important role. Biologically, one expects neither that all geniculate synapses onto a cortical cell will decay in strength, resulting in a loss of all visual input, nor that all geniculate synapses onto a cortical cell will grow in strength, leading to hyperresponsiveness and loss of selectivity. Rather, one expects that there will be a competition, by which some patterns of input grow in synaptic strength at the expense of other patterns. Such competition may occur dynamically, or it may be imposed through conditions that limit the total synaptic strength over a cortical cell. Such postsynaptic constraints can be imposed on the equation for the sum of the two eyes' synaptic strengths, $S^S \equiv S^L + S^R$, without effect upon the equation for S^D. Hence, such constraints can be imposed without effect on the development of ocular dominance periodicity.

In summary, given a correlation function that encourages monocularity, the width of a cycle of right-eye plus left-eye ocular dominance patches is determined by the dominant wavelength in the cortical interaction function, up to a limit set by the arbor width and by constraints on total afferent strength. If the cortical interaction function would drive a longer wavelength, constraints on presynaptic cells can select a wavelength set by the arbor diameter.

Comparison of Predictions to Simulated Results We can test the analytical understanding of the model by comparing the period of ocular dominance segregation that emerges in simulations with the predictions made by analysis. To measure the period in the simulated cortex, we examine the relative contributions of each wavelength to the cortical patterns of ocular dominance by measuring the power spectrum of these patterns[18]. Figure 28 shows the time development of the power (vertical axis) at each inverse wavelength (horizontal axis) in the development of a cortical pattern of ocular dominance from a random initial condition. These are the power spectra of the cortical patterns of ocular dominance shown in figure 14. The prediction of growth rate versus inverse wavelength in the broad correlations limit is presented as the grey background to each graph.

---

18. We compute the 2-dimensional fourier spectrum of the pattern of ocular dominance in the simulated cortex. We then determine the average power in small annuli in 2-D fourier space, representing small increments of wavelength.

Early in development, when the equations are linear, the fastest-growing wavelength is predicted perfectly by the broad correlations limit. The nonlinearities can result in small shifts in the final wavelength, in this case by allowing a fast-growing wavelength that dominated the random initial condition to remain dominant.

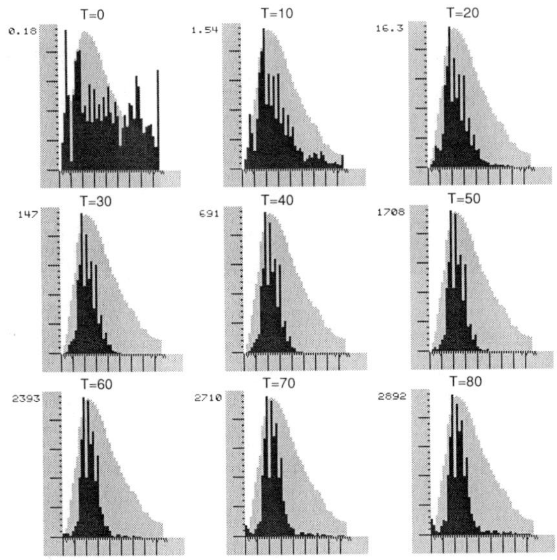

Figure 28: Development of the power (vertical axis) at each inverse wavelength (horizontal axis) in the cortical patterns of ocular dominance that develop in simulations. Power is black, compared to the analytical prediction of the limit of broad correlations, grey. These are the spectra of the cortices shown in figure 14. The number along the vertical axis shows the maximum power present in any wavelength bin at the given time, on a scale which is arbitrary but consistent across the various spectra presented. The binning of the horizontal axis is as described in figure 26. Note that each bin has a width of 0.4 grid intervals in Fourier space, so a difference of two bins represents a difference of less than one Fourier grid interval. From Miller, 1989c.

The power spectra of the final cortices that result from varying the correlation, cortical interaction, and arbor functions yield similar results (Miller, 1989c). The dominant wavelength in each final cortex is well predicted by the analysis, except when anticorrelations within each eye disrupt the overall monocularity.

Though the power spectra resulting from simulations are like those predicted, the precision is not sufficient to closely predict a wavelength. The wavelengths representing peak power are generally found within ±1 grid interval in Fourier space of the predicted peak, while almost all of the power in the final pattern is found within ±1.8 grid intervals in Fourier space of the predicted peak[19]. Using the Mexican hat cortical interaction function and the flat arbor function, these spans of inverse wavelength in Fourier space correspond in real space to wavelengths of 5.4±1.1 grid intervals and 6.05±2.25 grid intervals, respectively. This lack of precision may be related to the coarseness of the grid: wavelengths represent only a small number of grid intervals, so that uncertainty of only a few grid intervals corresponds to a significant variation of wavelength. One can guess that the precision of prediction may be greater when there are many grid intervals to a wavelength. Assuming a horizontal spacing of roughly 20 μm between cortical cells in layer 4, there are perhaps 30-40 cortical cells across one wavelength of right-eye plus left-eye patches in the cat or monkey visual cortex.

On What Do the Model Results Depend? Development of a periodic pattern of ocular dominance is an extremely robust phenomena, dependent only upon the three basic functions of the model, and perhaps on constraints on afferent arbors. However, the full expression of these tendencies depends upon the implementation of the model: on the limits on synaptic strengths, both of individual synapses and of total strengths over an afferent or over a cortical cell; and also on additional nonlinearities.

The analysis of periodicity examined only the development of the *difference* between the two eyes' synaptic strengths. However, the details of receptive field and afferent arbor structure depend also upon the development of the *sum* of the two eyes' synaptic strengths. For example, if the sum is growing everywhere, so that each eye's strength

19. The histograms showing power versus inverse wavelength are binned so that ±2 bins about the peak bin corresponds to a span of ±1 grid intervals in wavenumber, while ±4 bins about the peak bin corresponds to a span of of ±1.8 grid intervals in wavenumber. Details of the binning are found in the legend to figure 26.

is growing everywhere, then although a periodic difference in the strength of the two eyes may develop, this periodic pattern does not result in segregation. If this occured, individual receptive fields could show dominance, but they would not be monocular; similarly, individual afferent arbors could show periodicity, but they would not be confined to periodic patches. It is possible that there would be no final periodicity at all. For example, if all synapses tended to grow, and there were no constraints limiting total synaptic strength either over afferents or cortical cells, all synapses from both eyes would grow to saturation. Then periodicity would be manifest only during development, in the differing rates of reaching saturation. By using constraints that conserve total synaptic strength over each cortical cell, we avoid such problems without affecting the development of a periodic pattern in ocular dominance.

Similarly, receptive fields tend to focus their strength centrally in the model, provided that correlations are positive within each eye and diminish over less than an arbor width. The reason is that central synapses in the receptive field then have a larger number of correlated neighbors than do peripheral synapses, and hence tend to grow more quickly. Without a constraint on the total synaptic strength over a cortical cell, however, this refinement might in the end disappear as the fastest-growing synapses saturate and the slower-growing, more peripheral synapses catch up.

Biologically, there must be some limits on synaptic strengths, but very little is known about what they are or how they are implemented. Many means of achieving such limits have been used in previous models. Two broad classes can be distinguished (Miller, in preparation): multiplicative (von der Malsburg, 1973; Oja, 1982), and subtractive (Bienenstock, Cooper, and Munro, 1982; Linsker, 1986a-c; Miller, Keller, and Stryker, 1989; Sejnowski, 1977). In multiplicative methods, total synaptic strength over a cell is limited or conserved by multiplying each synapse on the cell by a constant determined after each iteration. In subtractive methods, each synapse on the cell has a constant subtracted from it, the constant determined after each iteration. There are important differences between these two methods. In multiplicative methods, ocular dominance segregation cannot occur in the absence of anticorrelations between the eyes[20]. This seems overly restrictive for study of the ocular dominance system. With subtractive methods, such as the one we use, there is no such requirement. With multiplicative methods, final synaptic strengths tend to be proportional to a synapse's average activity when the cortical cell is active, so that for example topographic refinement will manifest in a gradual decrease in synaptic strengths from the center to the periphery of a receptive field. With subtractive methods, final synaptic strengths tend to all be saturated at the maximum or minimum allowed values. In this respect, multiplicative methods might make easier the modeling of systems such as the somatosensory cortex which remain subject to activity-dependent plasticity throughout adulthood.

The robust features of the model are those involving relative rates of growth: the tendency of one eye's synapses to grow faster than another's on a cell, or of central synapses on a cell to grow faster than peripheral ones. These robust features result from interactions among synapses, and the correlations between them. The less robust features of the model involve absolute rates of growth. These are determined by the constraints, and do not depend on specific interactions between synapses. Biologically, robust tendencies to the development of monocularity, the organization of a periodic pattern of ocular dominance, and the refinement of receptive fields do exist, and it appears that sufficient constraints exist to ensure that these tendencies are fully expressed. Except for the differences outlined between subtractive and multiplicative methods, the results of development do not appear to depend greatly upon the method of limiting synaptic strengths. Hence, in the absence of experimental knowledge, we prefer to use the simplest possible model of constraints.

---

20. Multiplicative conservation of the sum of synaptic strengths over a cortical cell reduces that sum at each iteration by an amount proportional to its growth rate. Hence, if the growth rate of the difference of synaptic strengths between the two eyes is smaller than the growth rate of the sum of synaptic strengths in the absence of constraints, the difference of synaptic strengths between the two eyes will end up shrinking after application of multiplicative constraints. The growth of the difference is driven by the correlation function $C^{D}=C^{SameEye}-C^{OppEye}$, while the growth of the sum is driven by $C^{S}=C^{SameEye}+C^{OppEye}$. Therefore, in order for the difference to grow faster than the sum, there must be opposite-eye anticorrelations.

Discussion

Biological Implications of the Model Results We have shown that many phenomena associated with ocular dominance segregation are implicit in a Hebbian synapse mechanism and in other mechanisms involving synaptic reward via correlated activity. These phenomena emerge provided that afferents from a single eye are locally correlated in their activities, and that there are interactions across cortex by which synapses may influence one another's growth. This simple framework can account for the development of monocular cells and their organization into periodic patches or columns, the topographic refinement of receptive fields, and the confinement of afferent arbors to patches. Correlations of inputs over about ½ of an afferent arbor radius, as may be the case for X-cells in the cat (based upon measurements in adult cat: Humphrey et al., 1985a,b; Mastronarde, 1983a,b; Tusa, Palmer, and Rosenquist, 1978), can lead to a layer 4 with many binocular cells at the borders between patches, as observed in the cat (LeVay, Stryker, and Shatz, 1978). Broader correlations can lead to a virtually completely monocular layer 4, as is seen in the monkey (Hubel, Wiesel, and LeVay, 1977). The effects of experimental manipulations of activity, including monocular deprivation (Hubel and Wiesel, 1970; Wiesel and Hubel, 1965), strabismus or alternating monocular deprivation (Hubel and Wiesel, 1965), and cortical infusion of muscimol (Reiter and Stryker, 1988), all follow naturally from the model, and a critical period for these effects emerges.

The model can also account quantitatively for the width of the ocular dominance organization that develops. The observed period of this organization (width of right-eye plus left-eye patch) in the cat is 850 μm (Anderson, Olavarria, and Van Sluyters, 1988; LeVay, Stryker, and Shatz, 1978; Shatz, Lindstrom, and Wiesel, 1977; Swindale, 1988). To understand how a simple intracortical interaction might yield this wavelength as the peak of its fourier transform, it is helpful to keep two thoughts in mind. First, if a function is excitatory over a radius of about 200 μm, and strongly inhibitory more distantly, it can have this desired 850 μm peak. For example, a cosine wave with wavelength 850 μm is positive for ±212 μm from 0, then equally inhibitory over the same distance more peripherally. Second, a function can be excitatory over a much smaller region, and retain the same fourier peak, if the inhibition is made correspondingly weaker. For example, if there is no inhibition at all, the peak wavelength of an excitatory region of any size is at an infinitely long wavelength. Hence, functions excitatory over radii of 50 or 100 μm, with weak inhibition more peripherally, can also yield the desired wavelength.

As previously observed, we do not know enough about the nature of the intracortical interaction in general, and in particular at the time of ocular dominance segregation, to compare it to the 850 μm period in a precise way. We can only say that the distances over which excitation and inhibition are seen in the adult seem consistent with this period (Hata, et al., 1988; Hess, Negishi and Creutzfeldt, 1975; Toyama, Kimura, and Tanaka, 1981a,b; Worgotter and Eysel, 1989). If in addition there are longer-range interactions with a preexisting period on the order of 850 μm (Gilbert and Wiesel, 1983; Luhmann, Martinez Millan and Singer, 1986; Tso, Gilbert, and Wiesel, 1986; Worgotter and Eysel, 1989), this could create, or add to, a peak in the fourier transform at this wavelength.

If the intracortical interaction is purely excitatory, or otherwise selects a wavelength longer than an initial afferent arbor diameter, constraints can lead to wavelength selection by the arbor function. An 850 μm period can be selected by an arbor function that either is uniform over a diameter of 850 μm, or tapers over a somewhat larger diameter. Hence, it seems possible that X-cell afferent arbors could by this limiting mechanism be responsible for the wavelength of ocular dominance (based on maximum extent of patchy adult arbors: Humphreys et al., 1985a,b). However, Y-cell afferent arbors are likely to be be too large to play such a role (Humphreys et al., 1985a,b; LeVay and Stryker, 1979). It is interesting in this context that several experiments suggest that the X-cell projection develops earlier than that of the Y-cells (Sherman and Spear, 1982), and so might play a dominant role at the time of ocular dominance segregation.

The most obvious experiment suggested by the model is to better characterize the nature of the three functions at the onset of ocular dominance segregation. Measurement of initial correlation, cortical interaction, and arbor functions in various brain regions or species can test whether a proposed developmental mechanism is consistent with the periodicity that emerges in each case. Comparisons across such regions or species can enhance insight. For example, ocular dominance patches in area 18 of the cat are more than 1.5 times wider than those in area 17; arbors, and perhaps correlations, are also more widespread (Anderson, Olavarria, and Van Sluyters, 1988; Humphrey et al., 1985a,b; Lowel and Singer, 1987; Shatz, Lindstrom, and Wiesel, 1977; Swindale, 1988). If a Hebb mechanism is responsible for ocular dominance segregation in these two areas, two possibilities arise. First, intracortical connectivities in kittens may be sufficiently different between the two regions to account for the difference in patch width. Alternatively, both regions in kittens might have predominantly excitatory intracortical connections, resulting in arbor-limited patch widths. If measurement were to establish that neither of these possibilities is the case, one would conclude that the patch width is not likely to be determined by a Hebbian mechanism.

Sustained perturbation of the three functions in an experimental preparation beginning before the normal onset of segregation, and comparison of the resulting periodicity to the unperturbed case, can also test mechanisms. Under the hypothesis that a Hebb mechanism underlies ocular dominance development, periodic segregation is driven by intracortical synaptic connections. Local infusion of muscimol, an agonist of $GABA_A$ receptors which inhibits postsynaptic cells, will eliminate activation of such connections. We therefore would predict that no pattern of ocular dominance organization would be seen in a cortical regions into which muscimol was infused, although individual cells might become monocular. Alternatively, intracortical inhibitory connections may be blocked by local infusion of bicuculline, an antagonist of GABA receptors. A resulting increase in patch period would be consistent with the hypothesis that the underlying mechanism is Hebbian and that the patch width is determined by the intracortical interactions. If patch period were unchanged by bicuculline, one would conclude either that the period was normally arbor-limited (which could be tested by measuring whether intracortical interactions were predominantly excitatory during initial column development) or that a non-Hebbian mechanism was involved.

The model predicts that broader correlations within each eye would increase monocularity of layer 4 for mechanisms of the type we study. This could be tested by inducing broader correlations through pharmacological interventions in the retinas. One could also measure whether retinal correlations are broadened in animals deprived of pattern vision. Such animals have increased numbers of monocular cortical neurons (Sherman and Spear, 1982). Finally, one could measure whether geniculate correlations are broader, relative to a geniculocortical arbor radius, in developing monkeys than in kittens, since layer 4 is fully monocular in adult monkeys (Hubel, Wiesel, and LeVay, 1977) but not in cats (LeVay, Stryker, and Shatz, 1978).

In sum, the model suggests that much of the known phenomena of ocular dominance segregation follow from simple dynamics common to a number of proposed mechanims. It does not allow us to specifically distinguish one mechanism from another on the basis of existing experiments. It does, however, provide a framework for experiments capable of making such distinctions. In particular, it seems feasible to distinguish between mechanisms that would determine the period of segregation on the basis of intracortical synaptic connectivity, and mechanisms that determine a period on the basis of either afferent arbor diameter or intracortical interactions that involve diffusion rather than synaptic connections.

Implications for Other Models of Cortical Development and Plasticity
The present work extends previous theoretical studies of ocular dominance column development. Von der Malsburg and Willshaw, in a series of papers in the 1970's, demonstrated through computer simulation that mechanisms like those discussed here were sufficient to yield stripes as well as topographic maps (von der Malsburg, 1979; von der Malsburg and Willshaw, 1976, 1977; Willshaw and von der Malsburg, 1976, 1979). Bienenstock, Cooper and Munro (1982) focused on the development of monocularity in isolated cortical cells, as well as on dynamical means of limiting synaptic strengths, but did not study cortical organization.[21] Legendy (1978) studied a Hebb-like model, and concluded that intracortical synaptic interactions will determine the distances over which cortical cells are similar in their response properties. Swindale (1980) formulated a model in terms of an effective interaction between right-eye and left-eye synapses,

21. The relationship of the present model to the model of Bienenstock, Cooper, and Munro is discussed in detail in Miller, 1989a,c.

which produced stripes like those obtained here. This interaction was not related to measurable biological quantities. In the limit in which correlations among afferents vary slowly over an arbor radius, our model can be reduced to his, allowing us to express his effective interaction in terms of arbors, cortical interactions and afferent correlations.

Two models of cortical development discussed elsewhere in this volume may be compared to the model presented here. These are the models of Linsker (1986a-c, and Chapter 10)[22] and of Pearson, Finkel, and Edelman (1987, and Chapter 7). We will discuss these models in some depth, with the intents both of clarifying the similarities and differences between the models, and of illustrating how our methods can yield insight into the behavior and meaning of these models.

Linsker's Model of Development of Orientation Selectivity Linsker (1986a-c, and Chapter 10) developed a model of plasticity very much like ours. He used it to study the development of orientation selectivity in visual cortex. He initially studied the development of a single, isolated postsynaptic cell, and considered inputs from only a single eye. He assumed a gaussian arbor function, but one can formulate his model in terms of a general arbor function. Let α represent the location of the presynaptic cell, relative to a postsynaptic cell at x=0. Then Linsker's equation, after explicit inclusion of the arbor function, can be written

$$\frac{dS(\alpha,t)}{dt} = A(\alpha)\left\{ \sum_{\beta} \frac{C(\alpha-\beta)}{N} S(\beta,t) - k_2[\bar{S}(t) - S_{TARGET}] \right\}$$

where $N = \sum_{\alpha} A(\alpha)$, $\bar{S}(t) = \frac{\sum_{\alpha} S(\alpha,t)}{N}$, $S_{TARGET} = k_1/k_2$ in Linsker's notation, and k_1 and k_2 are constants that Linsker defines[23]. $S(\alpha,t)$ is

22. We will be concerned here only with Linsker's work on the development of orientation selectivity in visual cortex (Linsker, 1986a-c), and not with the later work on information theory, which comprises much of his chapter in this volume.

23. We have changed the sign of k_2 compared to Linsker's equation so that the negative sign in front of k_2 can be seen explicitly. Linsker uses a positive sign, but a k_2 that is negative.

allowed to range between $-0.5A(\alpha)$ and $0.5A(\alpha)$.

Linsker's model differs from ours in several important respects. First, in his model modifiable synapses S may have either positive or negative strength, whereas we require input synapses to have positive strengths. Note that in our model, the variable S^D, the ocular dominance or difference between the two eye's strength, may become either positive or negative. Hence, we can mathematically use our analysis of S^D to understand Linsker's S.

Linsker offers two possible interpretations of the positive and negative synapses. In one interpretation, each individual synapse may become positive or negative. In contrast, biological synapses are either exclusively excitatory or exclusively inhibitory. In the alternative interpretation, there are equal numbers of excitatory and inhibitory synapses, with identical spatial distributions and connectivities, and with indistinguishable statistics of activation so that an excitatory input is equally correlated to either an excitatory or an inhibitory input at a fixed distance. Furthermore the excitatory and inhibitory populations obey mirror-image plasticity rules, so that excitatory synapses are strengthened by correlation between postsynaptic and presynaptic activity and weakened by anticorrelation, while negative synapses are equally weakened (made less negative) by correlation and strengthened by anticorrelation. These conditions together mean that the "excitatory" and "inhibitory" populations are just positive and negative halves of a single statistical population. The geniculocortical projection is exclusively excitatory, so this second interpretation is also ruled out for that projection. The projection in monkeys from the unoriented cells of layer 4C to the oriented cells of the upper cortical layers includes both excitatory and inhibitory inputs, but it is likely that the two populations are distinct in crucial ways. For example, inhibitory cells are often interneurons which, when active, inhibit nearby excitatory cells, so that the inhibitory and excitatory populations would be expected to have statistically distinct firing patterns. Similarly, while there is extensive evidence that excitatory synapses can be modified in a Hebbian manner (Nicoll, Kauer, and Malenka, 1988), current evidence suggests that there may be little modification of inhibitory synapses under the same stimulus paradigms (Abraham, Gustafsson, and Wigstrom, 1987; Griffith, Brown and Johnston, 1986).

We have offered a third alternative intepretation of the negative synapses (Miller, 1989b). ON- and OFF-cells are known, in adults, to be anticorrelated with one another over distances similar to those over which each individual cell type is correlated (Mastronarde, 1983a,b). Consider populations of ON-center and OFF-center cells making strictly positive synapses onto cortex and having identical spatial distributions (which is at least approximately biologically correct). Suppose they are exactly anticorrelated with one another, so that $C^{ON:ON}(\alpha)=C^{OFF:OFF}(\alpha)=-C^{OFF:ON}(\alpha)$. Then the mathematics of Linsker's model of an "excitatory" and "inhibitory" population is identical to the mathematics of these ON- and OFF-cells, except for the important distinction that different constraints are needed[24]. One can relax the requirement of anticorrelations being exactly equal and opposite to the correlations without substantially altering the results.

Second, Linsker's model differs from ours in using constraints that ultimately fix both the summed excitatory and the summed inhibitory input to a cortical cell. In our model, the summed excitatory input is fixed, and the inhibitory input is fixed at zero. However S^D, the variable in our model that can be either positive or negative and hence is mathematically analagous to Linsker's, is unconstrained in its total onto a cortical cell; instead we consider the effect of constraints on the total that comes from each geniculate location. Linsker's constraints emerge as follows: the effect of the second term in $\frac{dS}{dt}$, for sufficiently large k_2, is to constrain S so that $\bar{S}(t) \approx S_{TARGET}$. If one begins in a condition in which the constraint is not satisfied, and if k_2 is sufficiently large (as Linsker's k_2 is), the second term in $\frac{dS}{dt}$ dominates the term involving correlations. Then

---

24. See description of Linsker's constraints in next paragraph in text. ON- and OFF-center geniculate cells appear to be entirely identical histologically, and are distinguished only by their inputs and their patterns of neural activity. Therefore, the postsynaptic cell could not separately limit the synaptic strength of each type of input. Thus, it is not appropriate to fix the total ON-synaptic strength and the total OFF-synaptic strength onto each cortical cell, which would be equivalent to Linsker's constraints. If one assumes that each geniculate afferent maintains a fixed amount of synaptic strength, one would then constrain the total ON-synaptic strength and the total OFF-synaptic strength coming from each geniculate location. This is the type of constraint we used for S^D.

either all synapses increase in strength, or all synapses decrease in strength, until the constraint is satisfied. Furthermore Linsker shows that development will continue on a single cell until all, or all but one, synapses are at their maximum or minimum allowed value, $\pm 0.5A(\alpha)$. Hence in the final state $\sum_\alpha |S(\alpha)| \approx 0.5N$, while $\sum_\alpha S(\alpha) \approx NS_{TARGET}$. From this, one can conclude that in the final state $\sum_{\alpha:S\ \mathrm{excitatory}} S(\alpha) \approx \frac{N}{2}(0.5+S_{TARGET})$, $\sum_{\alpha:S\ \mathrm{inhibitory}} S(\alpha) \approx -\frac{N}{2}(0.5-S_{TARGET})$. Linsker's model thus separately fixes the final total excitatory and total inhibitory synaptic strengths.

Our eigenfunction analysis of S^D can provide insight into his results. Linsker starts with a layer of cells whose activities are uncorrelated. He sets $S_{TARGET} > 0.5$, which causes each synapse to grow independently until it reaches its maximum positive strength. The resulting postsynaptic cell is thus an "all-excitatory" cell. In a layer of such cells whose inputs have overlapping gaussian arbors, neighboring cells will receive common excitatory inputs and thus be correlated in their activity. The degree of correlation between two cells will be proportional to the degree of overlap of their inputs and, hence, will decrease with the distance between the two cells in a gaussian manner. Next, a layer of such cells sends inputs to a subsequent layer, and S_{TARGET} is set somewhat less than 0.5 so that a significant number of synapses are forced to become negative. In this case, "center-surround" cells develop. This can be understood simply from the fact that the fastest-growing eigenfunctions for S^D, for a gaussian correlation function (i.e., "same-eye correlations"), have receptive fields that concentrate their strength centrally. When combined with constraints that, for example, force 35 percent of final synaptic strength to be negative ($S_{TARGET}=0.15$), this can lead to a center of positive synapses with a surround of negative synapses[25]. Hence, formation of these "center-surround" cells depends critically on the negative synapses and the constraints. The robust feature is that central synapses grow faster than peripheral ones.

---

25. A thorough analysis of this process will be found in D.J.M. MacKay and K.D. Miller, in preparation.

A layer of such center-surround cells, if all cells have identical phase (positive centers, negative surrounds), has a "Mexican hat" correlation function: nearby cells have correlated patterns of activity, while more distantly spaced cells in the layer are anticorrelated with one another. Nearby cells are correlated because they receive common excitatory inputs. More distantly spaced cells have overlap between one's excitatory center and the other's inhibitory surround, and thus anticorrelated firing. The development of anticorrelations in the afferent correlation function thus depends upon the development of "center-surround" cells, which in turn depends upon both the negative synapses and the constraints.

Linsker uses repeated layers with similar values of S_{TARGET} to obtain repeated layers of cells that are "center-surround" in terms of the previous layers, and that have identical phase. This strengthens the anticorrelations between more distant cells in each subsequent layer, and introduces further oscillations in sign of the correlation function with increasing distance between cells. Retinal ganglion cells and cells in the LGN and, in monkeys, layer 4C of cortex have center-surround receptive fields in terms of the visual field. However, Linsker's mechanism for strengthening of anticorrelations requires that geniculate cells have center-surround receptive fields in terms of retinal ganglion cells, so that the center responses of geniculate cells originate from the center responses of one set of retinal ganglion cells, and the surround responses of geniculate cells originate from the center responses of surrounding retinal ganglion cells; or similarly, that monkey 4C cells have center-surround receptive fields in terms of geniculate cells. There is as yet little evidence on whether either of these is the case.

Given an afferent correlation function with strong anticorrelations, and a sufficiently broad arbor function so that inputs in the center of the receptive field "see" a substantial number of anticorrelated inputs, oriented cells develop. Linsker found that multibanded oriented cells will develop, that is cells with a central band of synapses flanked by one or more side bands of alternating sign. Development of such multilobed structures appears to depend upon both multiple oscillations of sign in the correlation function, and on very fine tuning of the constraints (unpublished observations). However, we have seen that if a Mexican hat correlation function is sufficiently narrow with respect to the arbor function, the fastest-growing receptive field pattern has a positive stripe adjacent to a negative stripe. Hence, development of an oriented cell consisting of two bands, one excitatory and one inhibitory, can occur simply with anticorrelations and in the absence of constraints. When multiple cells are connected into a cortical layer, an alternation of two-banded and three-banded structures can result, due to the oscillation in phase across cortex of the eigenfunctions of S^D (Miller, 1989b).

There are again several difficulties with interpreting these results biologically. First, it is not clear that orientated cells can be achieved either without negative synapses or without unrealistic assumptions about the statistics of activity of the positive and negative synapses[26]. If synapses are solely positive and the arbor function does not decrease with distance sufficiently quickly, central synapses grow more *slowly* than peripheral synapses in the presence of a Mexican hat correlation function, leading to bizarre receptive fields with synaptic strength confined to peripheral clumps. This occurs because central synapses then have more anticorrelated neighbors than peripheral ones. If the arbor function tapers with distance sufficiently quickly, central synapses will grow more quickly than peripheral ones, but the synapses tend to remain confined within the correlated central region in a circularly symmetric manner (Miller, 1989a,c). However, in some conditions it appears possible to obtain oriented cells by this mechanism with strictly positive synapses (Linsker, personal communication; Miller, 1989a,c)[27]. Second, when two eyes are present, anticorrelations among afferents within each eye will destroy monocularity (figure 26). A combination of same-eye and opposite-eye anticorrelations can eliminate anticorrelations from C^D, while

---

26. If our interpretation in terms of ON- and OFF-cells is utilized, this problem is avoided. However, it is replaced by another problem. Linsker's result requires ON-ON and OFF-OFF correlations to be positive among near neighbors and negative among more distant neighbors, while OFF-ON correlations are negative among near neighbors and positive among more distant neighbors. The changes of sign with distance have thus far not been observed.

27. If total synaptic strength is conserved by multiplication of each synapse by a constant, rather than subtraction from each synapse of a constant, after each iteration, at least weakly orientated cells can emerge (Miller, 1989a,c).

causing deep anticorrelations in the correlation function that drives development of the sum of the synaptic strengths of the two eyes, $C^S \equiv C^{LL} + C^{LR}$. Again, at least under special circumstances this can yield cells that are monocular and at least weakly oriented (Miller, 1989a,c).

Finally, returning to synapses that can be either negative or positive, suppose that a layer of such oriented cells develops in the presence of weak but long-distance excitatory connections between cells within the layer. Then Linsker has proposed an analysis showing that oriented cells will organize into a global arrangement of orientations that in certain respects closely resembles biological results found subsequently (Blasdel and Salama, 1986)[28]. He assumes that the excitatory connections are weak enough not to disturb the receptive field structure that develops in isolated cells, but strong enough to nonetheless organize these cells with respect to one another during development. The eigenfunction analysis presented here for S^D shows that, in the absence of constraints, even weak cortical interactions cause patterns to develop that oscillate in phase across cortex. That is, if one part of cortex has oriented receptive fields with an excitatory central band and inhibitory side bands, nearby parts of cortex would have receptive fields with an inhibitory central band and excitatory side bands. For this reason, Linsker's assumption that receptive field structure is not disturbed is not valid in the absence of his constraints. His constraints cause this oscillation to be suppressed by causing a pattern with infinite wavelength – no oscillation – to grow faster than other patterns. Hence, the cortical organization he finds also depends critically on his constraints (to be discussed in Miller and MacKay, in preparation). Such constraints are not present when the model is interpreted in terms of ON and OFF cells, so such an interpretation leads to different cortical organization of orientation than that proposed by Linsker (Miller, 1989b).

---

28. One point of discrepency is that Linsker's mechanisms leads groups of vertically oriented cells to be extended across cortex in the retinotopically vertical direction, and similarly for all other orientations. No such arrangement has been seen in several studies (Blasdel and Salama, 1986; Swindale, Matsubara, and Cynader, 1987).

In sum, Linsker's work has been of great importance in pointing out in general how significant organization of individual receptive fields, as well as of cortex as a whole, can arise from simple Hebbian mechanisms. In particular, Linsker pointed out that symmetry-breaking could occur, so that oriented receptive fields could arise from circularly symmetric arbors and correlations. It is important to understand the elements of the model that are responsible for each element of the results. This allows one to understand the specific results that might be biologically realistic. The development of center-surround organization and of anticorrelations in Linsker's model seems nonbiological, because of their dependence upon the negative synapses and the constraints. However, anticorrelations may arise by other means, and in particular, an interpretation in terms of ON and OFF-cells may provide a biological interpretation for some of the results on the development of oriented cells.

Pearson, Finkel, and Edelman's Model of Somatosensory Development Pearson, Finkel, and Edelman (1987; Chapter 7) developed a similar, but more complex, model to study somatosensory development and plasticity. We will discuss the model as presented in the 1987 paper. It is a "real-time" model, in that postsynaptic voltages develop and decay over time. Thus individual synaptic changes are followed, whereas our model only follows synaptic changes averaged over the ensemble of input patterns. The plasticity rule used by Pearson and colleages assumes that synapses are modifiable only during postsynaptic depolarization. Modifiable synapses tend to be converted from a weak to a strong form if sufficient "modifying substance," released by presynaptic activity, has accumulated; otherwise, they tend to be converted from the strong form to the weak form. Thus, synaptic plasticity occurs only with sufficient postsynaptic depolarization, and the sign of plasticity is then determined by the amount of presynaptic activity. The essence of the plasticity model can thus be expressed

$$\frac{dS(x,\alpha,t)}{dt} \propto f_1[POST(x)]f_2[PRE(\alpha)]$$

where POST is postsynaptic depolarization, PRE is presynaptic activity (both may have to be integrated over some relevant time), f_1 is a function that is zero at or below resting potential and becomes positive with depolarization, and f_2 is a function that is negative for low presynaptic activities and positive for large presynaptic activities. In this form, it is easy to see that the model of plasticity is a particular form of a Hebbian rule. The most significant variation in this version is that the function of postsynaptic activity can never be negative. Such a variation was proposed by Stent (1973) but would seem to be contradicted in visual cortex by experiments that infuse muscimol to hyperpolarize postsynaptic cells, leading to an ocular dominance shift in favor of the *less* active eye (Reiter and Stryker, 1988), as was discussed previously.

The model has nearly 30 parameters, complicating analysis and rendering difficult the determination of the parameter regimes of which simulation results are representative. By considering the averaged changes to be expected in terms of the effective arbor function, correlation function and cortical interaction function, the present analysis can shed light on their results. The arbor functions used are identical: input cells connect to a 7×7 square arbor of excitatory cortical cells, with uniform connectivity within the arbor. The model of Pearson and colleagues uses two identical input grids, one representing glabrous and one representing dorsal input. These can equally be considered "right eye" and "left eye". The inputs are activated with 3×3 square stimuli, from which the average correlation in input can be easily calculated: within glabrous or dorsal, it is a function that is strictly positive and is nonzero over a square of ±2 gridpoints; between glabrous and dorsal, it is zero[29].

29. Glabrous and dorsal grids are actually connected on one side, so along this border there is some glabrous/dorsal correlation. Otherwise, with the sequential presentation of stimuli used in the model, the description given here of the correlation function is only strictly true in the limit of a large number of cells. Let correlations be defined as

$$C(\alpha,\beta) = \frac{1}{\#\text{stimuli}} \sum_{\text{stimuli}} [a(\alpha) - \bar{a}][a(\beta) - \bar{a}] = \frac{\sum_{\text{stimuli}} [a(\alpha)a(\beta)]}{\#\text{stimuli}} - \bar{a}\,\bar{a}$$

where $a(\alpha)$ is the activity of the input from α, and $\bar{a}$ is the mean activity (averaged over all stimuli) of each input. Let $a(\alpha)$ be 1 for stimuli in which the input at α is stimulated and zero otherwise. If two cells are simultaneously activated by stimuli centered at n cells, and individually activated by stimuli centered at 9 cells, out of N total cells stimulated, then this expression is

The cortical interaction function must be estimated more roughly as there are many parameters underlying it[30]. Cortical cells are either excitatory (e) or inhibitory (i) cells. There are two excitatory and one inhibitory cortical cells at each point in the cortical grid. Input cells contact only e cells, and only the responses of e cells are studied. There are assumed to be e→e, i→e, and e→i intracortical connections. The e→e connections are themselves assumed to be modifiable by the same plasticity rule as the input synapses. Connectivity is established in terms of "square rings" of grid points around each point. At distances of 0–1 square rings, intracortical connections are strongly excitatory (that is, e→e connections are strongly dominant over [e→i→e] connections); at a distance of 2 square rings, connections are more weakly excitatory; and at greater distances (3 to 6 square rings), connections are exclusively inhibitory and extremely strong.

$C(\alpha, \beta) = n/N - (9/N)^2$. When the number of cells N is large compared to 9, so that each cell's mean activity $9/N$ approaches zero, the description given in the text becomes accurate; since there were 1024 input cells, this is a very good approximation. The description of the correlation function in the text would also be true if stimuli were presented in parallel, with random times of activation at each site based on the mean activity of each.

In the model of Pearson and colleagues, significant synaptic changes occur over times short compared to the times in which all input patterns are activated. That is, a stimulus consisted of a prolonged set of activations (3 cycles of 6 steps on, 4 steps off) of a single 3-grid by 3-grid square. In one "pass" over the hand, a complete set of non-overlapping squares was sequentially given one such stimulus each. There were 4 passes. The 3×3 stimulus squares overlapped one another between passes, but not within passes. The result is that a local pattern could be formed during the prolonged stimulation of a single square, and during a pass in which only nonoverlapping squares are stimulated, rather than through an average over all stimuli. Because the size of such a local pattern is determined by the number of cortical cells activated by stimulation of a square, and this in turn should be determined by the intracortical interactions with the stimuli used, we have ignored this fact in our discussion.

30. i cells are set to have much lower and much sharper thresholds then e cells. i cells produce shunting inhibition, so that a sufficient level of i-cell activation onto a cortical cell will prevent it from firing. e→e synapses are made 4.5 times more effective than e→i synapses or input synapses, and there are twice as many e cells as i cells.

The correlation function is essentially a "same-eye correlation" function nonzero over slightly less than an arbor radius. Therefore, one expects a largely or fully "monocular" cortex. That is, most cells should be largely or exclusively driven by either glabrous or dorsal inputs. This is found: 94 percent of nonboundary cells in the simulation presented are "monocular" (boundary cells form a special case in their model). From the intracortical interaction and the fact that there are no constraints on arbors, one predicts that glabrous and dorsal bands of approximately 5 grid intervals each (±2 grid intervals about a central point) will become established[31]. This appears to be approximately consistent with the results seen[32]. In particular, because $e \rightarrow e$ connections are modifiable, squarish clumps of cells develop that are densely connected to one another by $e \rightarrow e$ synapses, but only weakly connected to other cortical cells by $e \rightarrow e$ synapses. Each clump is "monocular," responding to glabrous or to dorsal but not to both. Approximately 3 x 5 clumps form on the 12 x 28 nonboundary region of cortex, suggesting a linear clump size of between 4 and 5.5 grid intervals.

These clumps are termed "groups" by the authors, in part because each cell in a clump has an essentially identical receptive field. The authors argue that the formation of groups is the cause of the cortical organization observed in the model. We would argue, instead, that cortical organization develops in the model due to correlated inputs, intracortical interactions and a Hebbian type of learning rule. The groups are a possible *result* of such cortical organization that may be induced for particular choices of model parameters. One factor contributing to formation of groups may be that $e \rightarrow e$ synapses are assumed to be strongly dominant over thalamic inputs: $e \rightarrow e$ synapses are assigned 4.5 times the effectiveness of thalamic synapses for an equal synaptic strength. This may cause intracortical connections to

31. Note that the large periodicity, compared to the arbor size, suggests that some glabrous and dorsal sites may lose much of their input to cortex.

32. In examining pictures of the simulation, it is important to note that there are two excitatory cells at each grid position, but each is presented as though at a separate grid position. Thus pictures are presented with half the scale in one dimension as in the other dimension.

strongly dominate thalamic inputs in determining a cell's responses and cause a "clump" of strongly connected cells, developed by the Hebbian interactions, to become a "group" of identically responding cells.

Given an understanding of the sources of the model's behavior, a comparison to biology can be made. Two scales emerge from the model that can be compared with biology. One is the linear extent of the groups, which represents the scale of topographic disorder. On this scale, receptive fields appear to remain invariant across cortex, then to jump suddenly to a new location, so that the topographic map appears disorderly and discontinuous (see figure 8 of Pearson, Finkel and Edelman, 1987). The second scale is the linear extent of glabrous or dorsal patches. Our analysis demonstrates that these two scales must be roughly the same in the model, since both are determined identically by the intracortical interactions. This is true independently of the many details of the model and its implementation.

Somatosensory receptive fields are known to vary continuously and smoothly with distance across cortex over all distances larger than 50–100 microns (compare figure 3 of Stryker, Jenkins, and Merzenich, 1987 with figure 8 of Pearson, Finkel, and Edelman, 1987; also see Merzenich et al, 1987). Hence, the scale of topographic disorder represented by groups must be smaller than 100 microns. Yet dorsal patches typically are several or many 100's of microns in extent (Merzenich et al., 1987; Stryker, Jenkins, and Merzenich, 1987). The prediction, inherent in the process of group formation in the model, that topography should be disordered on the scale of dorsal patches as well as on the scale of the intracortical interactions[33], appears inconsistent with the biology.

---

33. A periodic scale determined by the intracortical interactions should be at least twice the horizontal extent over which cortical interactions are predominantly excitatory in the developing animal (the size will be larger than this if surrounding inhibitory interactions are relatively weak). The biological size of dorsal patches is likely to be consistent with this. In contrast, a scale of less than 50-100 microns, attributed by the model to the groups by identification of that scale with the biological scale of topographic disorder, seems likely not to be. Measurements of the intracortical interactions in the developing animal can provide an important independent check to the arguments made here.

The model is valuable in showing how the types of organization expected from a Hebbian mechanism can develop with a rather complex and detailed implementation of the Hebbian rule, and in applying these results to the somatosensory cortex. However, parameters that force groups to form may be precisely the wrong parameters, biologically. Rather, it seems likely that the cortical organization resulting from Hebbian interactions should develop in a way that allows smooth, continuous topographic variation of receptive fields across cortex.

More generally, we have shown that by simplifying a model, so that it incorporates only the simplest and best-established features of a system, and so that the essential factors responsible for its behavior can be understood, biological understanding may be deepened. While it is important to grapple with the realities of biological and biophysical complexity, a more complex model is not necessarily a more realistic one.

Conclusion

Ocular dominance segregation has provided perhaps the simplest and best characterized model of activity-dependent competition in neural development. We have shown that a minimal model of this process, incorporating only a competitive rule, correlations among inputs serving each eye, interactions across cortex, and limits to geniculocortical connectivity, is sufficient to account in surprising detail for much of the experimentally observed phenomena, as well as to predict the results of experiments yet undone. In particular, such a simple model is sufficient to predict the width of ocular dominance stripes to be expected from measurable biological parameters.

More detailed models will no doubt arise in the future that address the many features omitted from the current model. One would like to understand and incorporate more biophysical and temporal detail in modeling plasticity rules and geniculate and cortical activations. There are many classes of geniculate afferents, and the correlation between the activities of two afferents depends on the class to which each belongs (Mastronarde, 1989). Similarly, there are many classes of visual cortical cells (Lund, 1988; Martin, 1988). Initial geniculocortical and corticocortical connectivities are both more specific and more stochastic than the simple distance-dependent rules we have used here (Lund, 1988; Martin, 1988). Cortex has three dimensions, not two. The retinotopic mapping to cortex is not linear, and the grid of retinal inputs is more nearly hexagonal than square, each of which could have important impacts on receptive field structure (Mallot, 1985; Soodak, 1987). The development of ocular dominance must be understood in the presence of orientation selectivity and other cortical properties. Sprouting and retraction of synapses may play a fundamental role, rather than a secondary role following physiological changes as assumed here. Better understandings of limitations on individual and total synaptic weights may be achieved in the future. Thus, the present model is only a beginning.

We believe nonetheless that this model can provide a framework for future endeavors. As one or another complexity is studied, one must ask whether it fundamentally alters or adds to the simple dynamics presented here, and if so how and why? The fact that so much of visual cortical development appears implicit in the simplest competitive dynamics makes plausible the idea that such simple dynamics may underlie the segregation that occurs amidst the complexity of the visual cortex.

Appendix

Derivation of the Model Equations From Nonlinear Rules

Consider a nonlinear version of the Hebb rule,

$$\frac{dS^L(x,\alpha,t)}{dt} = \lambda A(x-\alpha) f_c[c(x,t)] f_a[a^L(\alpha,t)] \tag{A1}$$

(and similarly for S^R), with a nonlinear activation rule,

$$c(x,t) = g[\theta_x(t) + \textstyle\sum_y B(x,y)\, c(y,t)] \tag{A2}$$

where f_a, f_c and g are nonlinear functions and $\theta_x(t) = \sum_\beta [S^L(x,\beta,t)a^L(\beta,t) + S^R(x,\beta,t)a^R(\beta,t)]$. c(x,t) is a function of all of the input strengths $\theta_y(t)$, although this is not indicated explicitly. We will see that the linearization of these equations can be taken to be of the form of eq. 5 for S^D.

The Hebb rule for the difference, $S^D(x,\alpha,t) \equiv S^L(x,\alpha,t) - S^R(x,\alpha,t)$ is now

$$\frac{dS^D(x,\alpha,t)}{dt} = \lambda A(x-\alpha) f_c[c(x,t)] f_a^D(\alpha,t) \tag{A3}$$

where $f_a^D(\alpha,t) \equiv f_a[a^L(\alpha,t)] - f_a[a^R(\alpha,t)]$. Let

$$S^S(x,\alpha,t) \equiv S^L(x,\alpha,t) + S^R(x,\alpha,t),$$

$$a^S(x,\alpha,t) \equiv a^L(x,\alpha,t) + a^R(x,\alpha,t),$$

$$a^D(x,\alpha,t) \equiv a^L(x,\alpha,t) - a^R(x,\alpha,t),$$

$$\theta_x^S(t) \equiv \tfrac{1}{2}\sum_\beta S^S(x,\beta,t)a^S(\beta,t),$$

$$\theta_x^D(t) \equiv \tfrac{1}{2}\sum_\beta S^D(x,\beta,t)a^D(\beta,t).$$

Then $\theta_x(t) = \theta_x^S(t) + \theta_x^D(t)$. Equation A3 will be expanded about $\theta_x^D(t) = 0$, to first order in $\theta_x^D(t)$.

Let $c^S(x,t)$ be the solution of

$$c^S(x,t) = g[\theta_x^S(t) + \sum_y B(x,y)\, c^S(y,t)]. \tag{A4}$$

Then

$$\frac{dS^D(x,\alpha,t)}{dt} \equiv \tag{A5}$$
$$\lambda A(x-\alpha)f_a^D(\alpha,t)\left\{f_c[c^S(x,t)] + f'_c[c^S(x,t)]\sum_y \left.\frac{\partial c(x,t)}{\partial \theta_y(t)}\right|_{c\equiv c^S} \theta_y^D(t)\right\} + O[(\theta^D)^2].$$

Here f'_c is the derivative of the function f_c.

The partial derivative can be evaluated; letting $g'^S_x(t) \equiv g'[\theta_x^S(t) + \sum_y B(x,y)\, c^S(y,t)]$, where $'$ again indicates derivative, it is

$$\left.\frac{\partial c(x,t)}{\partial \theta_y(t)}\right|_{c\equiv c^S} = g'^S_x(t)\,[\mathbf{1} + \tilde{B} + (\tilde{B})^2 + \cdots]_{xy} \tag{A6}$$

where $\mathbf{1}$ is the identity matrix, $\tilde{B}$ is the matrix with elements $\tilde{B}_{xy} = B(x,y)g'^S_y(t)$, and $[...]_{xy}$ means the xy element of the matrix in brackets. Let $I(x,y,t) = [\mathbf{1} + \tilde{B} + (\tilde{B})^2 + \cdots]_{xy}$, $C(\alpha,\beta,t) = \frac{1}{2} f_a^D(\alpha,t)a^D(\beta,t)$, $D(x,t) = f'_c[c^S(x,t)]g'^S_x(t)$. Then eq. A5 becomes, to first order in θ^D,

$$\frac{\partial S^D(x,\alpha,t)}{\partial t} = \lambda A(x-\alpha)f_c[c^S(x,t)]f_a^D(\alpha,t) + \tag{A7}$$
$$\lambda A(x-\alpha)D(x,t)\sum_{y,\beta} I(x,y,t)C(\alpha,\beta,t)S^D(y,\beta,t).$$

After averaging, by equality of the two eyes the first term must be zero. The lowest order term resulting from averaging of the second term is

$$\lambda A(x-\alpha)\sum_{y,\beta}\langle D(x,t)I(x,y,t)C(\alpha,\beta,t)\rangle\langle S^D(y,\beta,t)\rangle.$$

Suppose that $\langle D(x)I(x,y)C(\alpha,\beta)\rangle$ can be approximated by $\langle D(x)I(x,y)\rangle\langle C(\alpha,\beta)\rangle$. This will be true if the sum of the two eyes' inputs is statistically independent of the difference between the two eyes' inputs. By the assumption of homogeneity of cortex, $\langle D(x)I(x,y)\rangle$ can only depend on x–y, while $\langle C(\alpha,\beta)\rangle$ can only depend on $\alpha-\beta$. With these assumptions, then, the linearized version of this nonlinear model is given by eq. 5 in the text for S^D.

References

Abraham, W.C., Gustafsson, B., and Wigstrom, H. (1987). Long-term potentiation involves enhanced synaptic excitation relative to synaptic inhibition in guinea-pig hippocampus. Journal of Physiology (London), 394:367-380.

Altmann, L., Luhmann, H.J., Greuel, J.M., and Singer, W. (1987). Functional and neuronal binocularity in kittens raised with rapidly alternating monocular occlusion. Journal of Neurophysiology, 58:965-980.

Anderson, P.A., Olavarria, J., and Van Sluyters, R.C. (1988). The overall pattern of ocular dominance bands in cat visual cortex. Journal of Neuroscience, 8:2183-2200.

Bienenstock, E.L., Cooper, L.N., and Munro, P.W. (1982). Theory for the development of neuron selectivity: Orientation specificity and binocular interaction in visual cortex. Journal of Neuroscience, 2:32-48.

Bishop, P.O., Henry, G.H., and Smith, C.J. (1971). Binocular interaction fields of single units in the cat striate cortex. Journal of Physiology (London), 216:39-68.

Blasdel, G.G. and Pettigrew, J.D. (1979). Degree of interocular synchrony required for maintenance of binocularity in kitten's visual cortex. Journal of Neurophysiology, 42:1692-1710.

Blasdel, G.G. and Salama, G. (1986). Voltage-sensitive dyes reveal a modular organization in monkey striate cortex. Nature, 321:579-585.

Brown, M.C., Jansen, J.K.S., and Van Essen, D. (1976). Polyneuronal innervation of skeletal muscle in new-born rats and its elimination during maturation. Journal of Physiology, 261:387-422.

Changeux, J.P. and Danchin, A. (1976). Selective stabilization of developing synapses as a mechanisms for the specification of neuronal networks. Nature, 264:705-712.

Chapman, B., Jacobson, M.D., Reiter, H., and Stryker, M.P. (1986). Ocular dominance shift in kitten visual cortex caused by imbalance in retinal electrical activity. Nature, 324:154-156.

Des Rosiers, M.H., Sakurada, O., Jehle, J., Shinohara, M., Kennedy, C., and Sokoloff, L. (1978). Demonstration of functional plasticity in the immature striate cortex of the monkey by means of the ^{14}C-deoxyglucose method. Science, 200:447-449.

Ferster, D. and LeVay, S. (1978). The axonal arborizations of lateral geniculate neurons in the striate cortex of the cat. Journal of Comparative Neurology, 182:923-944.

Fladby, T. and Jansen, J.K.S. (1987). Postnatal loss of synaptic terminals in the partially denervated mouse soleus muscle. Acta Physiologica Scandinavia, 129:239-246.

Fraser, S.E. (1980). Differential adhesion approach to the patterning of nerve connections. Developmental Biology, 79:453-464.

Gilbert, C.D. and Wiesel, T.N. (1983). Clustered intrinsic connections in cat visual cortex. Journal of Neuroscience, 3:1116-1133.

Griffith, W.H., Brown, T.H., and Johnston, D. (1986). Voltage-clamp analysis of synaptic inhibition during long-term potentiation in hippocampus. Journal of Neurophysiology, 55:767-775.

Guillery, R.W. (1972). Binocular competition in the control of geniculate cell growth. Journal of Comparative Neurology, 144:117-130.

Gustafsson, B., Wigstrom, H., Abraham, W.C., and Huang, Y.-Y. (1987). Long-term potentiation in the hippocampus using depolarizing current pulses as the conditioning stimulus to single volley synaptic potentials. Journal of Neuroscience, 7:774-780.

Hata, Y., Tsumoto, T., Sato, H., Hagihara, K., and Tamura, H. (1988). Inhibition contributes to orientation selectivity in visual cortex of cat. Nature, 335:815-817.

Hayes, W.P. and Meyer, R.L. (1988a). Retinotopically inappropriate synapses of subnormal density formed by misdirected optic fibers in goldfish tectum. Developmental Brain Research, 38:304-312.

Hayes, W.P. and Meyer, R.L. (1988b). Optic synapse number but not density is constrained during regeneration onto surgically halved tectum in goldfish: HRP-EM evidence that optic fibers compete for fixed numbers of postsynaptic sites on the tectum. Journal of Comparative Neurology, 274:539-559.

Hayes, W.P. and Meyer, R.L. (1989a). Normal numbers of retinotectal synapses during the activity-sensitive period of optic regeneration in goldfish: HRP-EM evidence implicating synapse rearrangement and collaterral elimination during map refinement. Journal of Neuroscience, 9:1400-1413.

Hayes, W.P. and Meyer, R.L. (1989b). Impulse blockade by intraocular tetrodoxin during optic regeneration in goldfish: HRP-EM evidence that the formation of normal numbers of optic synapses and the elimination of exuberant optic fibers is activity independent. Journal of Neuroscience, 9:1414-1423.

Hebb, D.O. (1949). The Organization of Behavior. John Wiley and Sons, Inc., New York.

Hess, R., Negishi, K., and Creutzfeldt, O. (1975). The horizontal spread of intracortical inhibition in the visual cortex. Experimental Brain Research, 22:415-419.

Hubel, D.H. and Wiesel, T.N. (1962). Receptive fields, binocular interaction and functional architecture in the cat's visual cortex. Journal of Physiology, 160:106-154.

Hubel, D.H. and Wiesel, T.N. (1963). Receptive fields of cells in striate cortex of very young, visually inexperienced kittens. Journal of Neurophysiology, 26:994-1002.

Hubel, D.H. and Wiesel, T.N. (1965). Binocular interaction in striate cortex of kittens reared with artificial squint. Journal of Neurophysiology, 28:1041-1059.

Hubel, D.H. and Wiesel, T.N. (1970). The period of susceptibility to the physiological effects of unilateral eye closure in kittens. Journal of Physiology, 206:419-436.

Hubel, D.H. and Wiesel, T.N. (1972). Laminar and columnar distribution of geniculocortical fibers in the macaque monkey. Journal of Comparative Neurology, 146:421-450.

Hubel, D.H., Wiesel, T.N., and LeVay, S. (1977). Plasticity of ocular dominance columns in monkey striate cortex. Philosophical Transactions of the Royal Society of London Series B, 278:377-409.

Humphrey, A.L., Sur, M., Uhlrich, D.J., and Sherman, S.M. (1985a). Projection patterns of individual X- and Y-cell axons from the lateral geniculate nucleus to cortical area 17 in the cat. Journal of Comparative Neurology, 233:159-189.

Humphrey, A.L., Sur, M., Uhlrich, D.J., and Sherman, S.M. (1985b). Termination patterns of individual X- and Y-cell axons in the visual cortex of the cat: Projections to area 18, to the 17/18 border region, and to both areas 17 and 18. Journal of Comparative Neurology, 233:190-212.

Kato, H., Bishop, P.O., and Orban, G.A. (1981). Binocular interaction on monocularly discharged lateral geniculate and striate neurons in the cat. Journal of Neurophysiology, 46:932-951.

Katz, L.C. & Constantine-Paton, M. (1988). Relationships between segregated afferents and postsynaptic neurons in the optic tectum of three-eyed frogs. Journal of Neuroscience, 8:3160-3180.

Keller, J.B. (1977). Effective behavior of heterogeneous media. In Landman, U. (Ed.), Statistical Mechanics and Statistical Methods in Theory and Application, pp. 631-644, Plenum.

Larson, J. and Lynch, G. (1986). Induction of synaptic potentiation in hippocampus by patterned stimulation involves two events. Science, 232:985-988.

Legendy, C.R. (1978). Cortical columns and the tendency of neighboring neurons to act similarly. Brain Research, 158:89-105.

LeVay, S. and Stryker, M.P. (1979). The development of ocular dominance columns in the cat. In Ferrendelli, J.A. (Ed.), Aspects of Developmental Neurobiology, pp. 83-98, Society for Neuroscience, Bethesda.

LeVay, S., Stryker, M.P., and Shatz, C.J. (1978). Ocular dominance columns and their development in layer IV of the cat's visual cortex: A quantitative study. Journal of Comparative Neurology, 179:223-244.

LeVay, S., Wiesel, T.N., and Hubel, D.H. (1980). The development of ocular dominance columns in normal and visually deprived monkeys. Journal of Comparative Neurology, 191:1-51.

Levick, W.R. and Williams, W.O. (1964). Maintained activity of lateral geniculate neurones in darkness. Journal of Physiology, 170:582-597.

Linsker, R. (1986a). From basic network principles to neural architecture: Emergence of spatial opponent cells. Proceedings of the National Academy of Sciences, 83:7508-7512.

Linsker, R. (1986b). From basic network principles to neural architecture: Emergence of orientation-selective cells. Proceedings of the National Academy of Sciences, 83:8390-8394.

Linsker, R. (1986c). From basic network principles to neural architecture: Emergence of orientation columns. Proceedings of the National Academy of Sciences, 83:8779-8783.

Lowel, S. and Singer, W. (1987). The pattern of ocular dominance columns in flat-mounts of visual cortex. Experimental Brain Research, 68:661-666.

Luhmann, H.J., Martinez Millan, L., and Singer, W. (1986). Development of horizontal intrinsic connections in cat striate cortex. Experimental Brain Research, 63:443-448.

Lund, J.S. (1988) Anatomical organization of macaque monkey striate visual cortex. Annual Review of Neuroscience, 11:253-288.

Mallot, H.A. (1985). An overall description of retinotopic mapping in the cat's visual cortex areas 17, 18, and 19. Biological Cybernetics, 52:45-51.

Malsburg, C. von der (1973). Self-organization of orientation sensitive cells in the striate cortex. Kybernetik, 14:85-100.

Malsburg, C. von der (1979). Development of ocularity domains and growth behavior of axon terminals. Biological Cybernetics, 32:49-62.

Malsburg, C. von der and Willshaw, D.J. (1976). A mechanism for producing continuous neural mappings: Ocularity dominance stripes and ordered retino-tectal projections. Experimental Brain Research, Supplement 1:463-469.

Malsburg, C. von der and Willshaw, D.J. (1977). How to label nerve cells so that they can interconnect in an ordered fashion. Proceedings of the National Academy of Sciences USA, 74:5176-5178.

Martin KA (1988). The Wellcome Prize lecture: From single cells to simple circuits in the cerebral cortex. Quarterly Journal of Experimental Physiology, 73:637-702.

Mastronarde, D.N. (1983a). Correlated firing of cat retinal ganglion cells. I. Spontaneously active inputs to X and Y cells. Journal of Neurophysiology, 49:303-324.

Mastronarde, D.N. (1983b). Correlated firing of cat retinal ganglion cells. II. Responses of X- and Y-cells to single quantal events.

Journal of Neurophysiology, 49:325-349.
Mastronarde, D.N. (1989). Correlated firing of retinal ganglion cells. Trends in Neurosciences, 12:75-80.
Merzenich, M.M., Nelson, R.J., Kaas, J.H., Stryker, M.P., Jenkins, W.M., Zook, J.M., Cynader, M.S., and Schoppmann, A. (1987). Variability in hand surface representations in areas 3b and 1 in adult owl and squirrel monkeys. Journal of Comparative Neurology, 258:281-296.
Miller, K.D. (1989a). Correlation-based models of neural development. In Gluck, M.A. and Rumelhart, D.E. (Eds.), Neuroscience and Connectionist Theory, Lawrence Erlbaum, Hillsboro, NJ.
Miller, K.D. (1989b). Orientation-selective cells can emerge from a Hebbian mechanism through interactions between ON- and OFF-center inputs. Society for Neuroscience Abstracts, 15:794.
Miller, K.D. (1989c). Ph. D. Thesis, Stanford University Medical School. University Microfilms, Ann Arbor.
Miller, K.D., Keller, J.B., and Stryker, M.P. (1986). Models for the formation of ocular dominance columns solved by linear stability analysis. Society for Neuroscience Abstracts, 12:1373.
Miller, K.D., Keller, J.B., and Stryker, M.P. (1989). Ocular dominance column development: Analysis and simulation. Science, 245:605-615.
Murray, M., Sharma, S., and Edwards, M.A. (1982). Target regulation of synaptic number in the compressed retinotectal projection of goldfish. Journal of Comparative Neurology, 209:374-385.
Nicoll, R.A., Kauer, J.A., and Malenka R.C. (1988). The current excitement in long-term potentiation. Neuron, 1, 97-103.
Ohzawa, I. and Freeman, R.D. (1986). The binocular organization of simple cells in the cat's visual cortex. Journal of Neurophysiology, 56:221-242.
Oja, E. (1982). A simplified neuron model as a principal component analyzer. Journal of Mathematical Biology, 15:267-273.
Pearson, J.C., Finkel, L.H., and Edelman, G.M. (1987). Plasticity in the organization of adult cerebral cortical maps: A computer simulation based on neuronal group selection. Journal of Neuroscience, 7:4209-4223.
Rakic, P. (1977). Prenatal development of the visual system in the rhesus monkey. Philosophical Transactions of the Royal

Society of London Series B, 278:245-260.
Reiter, H.O. and Stryker, M.P. (1988). Neural plasticity without postsynaptic action potentials: Less-active inputs become dominant when kitten visual cortical cells are pharmacologically inhibited. Proceedings of the National Academy of Sciences USA, 85:3623-3627.
Reiter, H.O., Waitzman, D.M., and Stryker, M.P. (1986). Cortical activity blockade prevents ocular dominance plasticity in the kitten visual cortex. Experimental Brain Research, 65:182-188.
Rodieck, R.W. and Smith, P.S. (1966). Slow dark discharge rhythms of cat retinal ganglion cells. Journal of Neurophysiology, 29:942-953.
Rose, G.M. and Dunwiddie, T.V. (1986). Induction of hippocampal long-term potentiation using physiologically patterned stimulation. Neuroscience Letters, 69:244-248.
Schneider, G.E. (1973). Early lesions of superior colliculus: Factors affecting the formation of abnormal retinal projections. Brain, Behavior and Evolution, 8:73-109.
Sejnowski, T.J. (1977). Storing covariance with nonlinearly interacting neurons. Journal of Mathematical Biology, 4:303-321.
Senden, Marius von (1960). Space and Sight; the Perception of Space and Shape in the Congenitally Blind Before and After the Operation. Translated by P. Heath. Free Press, Glencoe, Ill.
Shatz, C.J., Lindstrom, S., and Wiesel, T.N. (1977). The distribution of afferents representing the right and left eyes in the cat's visual cortex. Brain Research, 131:103-116.
Shatz, C.J. and Stryker, M.P. (1978). Ocular dominance in layer IV of the cat's visual cortex and the effects of monocular deprivation. Journal of Physiology, 281:267-283.
Sherman, S.M., Guillery, R.W., Kaas, J.H., and Sanderson, K.J. (1974). Behavioral, electrophysiological, and morphological studies of binocular competition in the development of the geniculocortical pathways of cats. Journal of Comparative Neurology, 158:1-18.
Sherman, S.M. and Spear, P.D. (1982). Organisation of visual pathways in normal and visually deprived cats. Physiological Review, 62:738-855.
Singer, W. (1985). Central control of developmental plasticity in the

mammalian visual cortex. Vision Research, 25:389-396.

Soodak, R.E. (1987). The retinal ganglion cell mosaic defines orientation columns in striate cortex. Proceedings of the National Academy of Sciences USA}, 84:3936-3940.

Stent, G.S. (1973). A physiological mechanism of Hebb's postulate of learning. Proceedings of the National Academy of Sciences USA, 70:997-1001.

Stryker, M.P. (1986). The role of neural activity in rearranging connections in the central visual system. In Ruben, R.J., Water, T.R. Van De, and Rubel, E.W. (Eds.), The Biology of Change in Otolaryngology, pp. 211-224, Elsevier Science B.V., Amsterdam.

Stryker, M.P. and Harris, W. (1986). Binocular impulse blockade prevents the formation of ocular dominance columns in cat visual cortex. Journal of Neuroscience, 6:2117-2133.

Stryker, M.P., Jenkins, W.M., and Merzenich, M.M. (1987). Anesthetic state does not affect the map of the hand representation within area 3b somatosensory cortex in owl monkey. Journal of Comparative Neurology, 258:297-303.

Swindale, N.V. (1980). A model for the formation of ocular dominance stripes. Proceedings of the Royal Society of London Series B., 208:243-264.

Swindale, N.V. (1988). Role of visual experience in promoting segregation of eye dominance patches in the visual cortex of the cat. Journal of Comparative Neurology, 267:472-488.

Swindale, N.V., Matsubara, J.A., and Cynader, M.S. (1987). Surface organization of orientation and direction selectivity in cat area 18. Journal of Neuroscience, 7:1414-1427.

Toyama, K., Kimura, M., and Tanaka, K. (1981a). Cross-correlation analysis of interneuronal connectivity in cat visual cortex. Journal of Neurophysiology, 46:191-201.

Toyama, K., Kimura, M., and Tanaka, K. (1981b). Organization of cat visual cortex as investigated by cross-correlation technique. Journal of Neurophysiology, 46:202-214.

Tso, D.Y., Gilbert, C.D., and Wiesel, T.N. (1986). Relationships between horizontal interactions and functional architecture in cat striate cortex as revealed by cross correlation analysis. Journal of Neuroscience, 6:1160-1170.

Tusa, R.J., Palmer, L.A., and Rosenquist, A.C. (1978). The

retinotopic organization of area 17 (striate cortex) in the cat. Journal of Comparative Neurology, 177:213-235.

Van Sluyters, R.C. and Levitt, F.B. (1980). Experimental strabismus in the kitten. Journal of Neurophysiology, 43:686-699.

Whitelaw, V.A. and Cowan, J.D. (1981). Specificity and plasticity of retinotectal connections: A computational model. Journal of Neuroscience, 1:1369-1387.

Wiesel, T.N. and Hubel, D.H. (1963a). Single-cell responses in striate cortex of kittens deprived of vision in one eye. Journal of Neurophysiology, 26:1003-1017.

Wiesel, T.N. and Hubel, D.H. (1963b). Effects of visual deprivation on morphology and physiology of cells in the cat's lateral geniculate body. Journal of Neurophysiology, 26:978-993.

Wiesel, T.N. and Hubel, D.H. (1965). Comparison of the effects of unilateral and bilateral eye closure on cortical unit responses in kittens. Journal of Neurophysiology, 28:1029-1040.

Wiesel, T.N., Hubel, D.H., and Lam, D.M.K. (1974). Autoradiographic demonstration of ocular dominance columns in the monkey striate cortex by means of transneuronal transport. Brain Research, 79:273-279.

Willshaw, D.J. and Malsburg, C. von der (1976). How patterned neural connections can be set up by self-organization. Proceedings of the Royal Society of London Series B, 194:431-445.

Willshaw, D.J. and Malsburg, C. von der (1979). A marker induction mechanism for the establishment of ordered neural mappings: Its application to the retinotectal problem. Philosophical Transactions of the Royal Society of London Series B, 287:203-243.

Worgotter, F. and Eysel, U. Th. (1989). On the influence and topography of excitation and inhibition at orientation-specific visual cortical neurons in the cat. Submitted for publication.

CHAPTER 10

Self-Organization in a Perceptual System:
How Network Models and Information Theory May Shed Light on Neural Organization

R. Linsker

Introduction

A young animal or child perceives and identifies features in its environment in an apparently effortless way. No presently known algorithms even approach this type of flexible and general-purpose perceptual capability. Discovering the principles that may underlie perceptual processing is important both for neuroscience and for the development of synthetic perceptual systems.

Two important aspects of the mystery of perception are:

- What processing functions does the neural "machinery" perform on perceptual input, and what is the circuitry that implements these functions?
- How does this "machinery" come to be?

Unlike conventional computer hardware designs, neural circuitry is not "hard-wired" or specified as an explicit set of point-to-point connections. Instead, it develops under the influence of a genetic specification (which codes for the production of various enzymes, for example) and epigenetic factors (such as electrical activity) both before and after birth. How this happens is in large part unknown.

Biological development processes are far too complex to hope that a relatively complete understanding of perceptual system development and function will emerge any time soon. But we are familiar with complex synthetic systems (such as computers) whose principles of organization can be understood without one's knowing in detail how the components work. Furthermore, the same principles can be used to build computers in any of several different technologies. Might there be organizing principles: (1) that explain essential aspects of how a perceptual system develops and functions; (2) that we can attempt to infer without waiting for far more detailed experimental information; and (3) that can lead to profitable experimental programs, testable predictions, and applications to synthetic perception as well as neuroscientific understanding? I believe the answer is yes, and that the use of theoretical neural networks embodying biologically-motivated rules and constraints is a powerful tool in this study.

This optimism is encouraged by recent work (Linsker, 1986) in which I have found that a multilayered network, developing according to simple yet biologically plausible rules (including a "Hebb-type" connection strength modification rule (Hebb, 1949) to be discussed), self-organizes to produce feature-analyzing "cells." These "cells" have response properties that are qualitatively similar to those of cells of the first few processing stages of the mammalian visual system (Hubel and Wiesel, 1977; 1979). These properties include sensitivity to light-dark contrast and sensitivity to the orientation of an edge or bar. These properties develop before birth (hence before structured visual experience) in certain animals, and in the theoretical network the corresponding properties develop even in the absence of structured input (with only random signaling activity in the input layer of the network).

Why does feature-analyzing function emerge from these rules? Is it a mere accident or curiosity? Or are the development rules perhaps acting to optimize some quantity that is important to the information processing function of a perceptual system?

In this chapter, I first summarize the network ideas introduced previously (Linsker, 1986) and some of the main results. This sets the stage for exploring why feature-analyzing function emerges. I then show that even a single cell of such a layered network, developing according to a Hebb-type rule, exhibits a remarkable set of optimization properties. These properties are closely related to issues in statistics, theoretical physics, adaptive signal processing, the formation of knowledge representations in artificial intelligence, and information theory.

Next, I use these results to infer an information-theoretic

principle that can be applied to the network as a whole, rather than a single cell. In the type of network we mainly consider, there are a number of cell layers. The layers may be interconnected in various ways, but there is no feedback to previous layers. The signal activities in any given layer and at any particular time comprise an array of real numbers. This array of activity values is transformed by the connections from each layer to the next, and by connections within each layer, to produce an array of activity values in the following layer(s). Each such array of activity values conveys a certain amount of information (about the external environment and the internal state and structure of the network). This point will be explained later in more detail.

The organizing principle I propose is that these connections develop in such a way as to *maximize the amount of information that is preserved* in the passage from each layer to the next, subject to certain constraints that reflect, for example, the "costs" of making connections.

I illustrate how this principle works for some simple cases. More work is needed to apply this principle to practical computations of biologically important cases, but the approach appears very promising. I conclude with some speculative comments on why this principle, or some variant of it, may be important for the emergence of perceptual function in biological and synthetic systems.

While this chapter introduces new concepts and results, it is tutorial in nature. Therefore, where connections are made to work in other fields (biology, information theory, statistics), the necessary background material from those fields is explained as needed.

A Layered Self-Adaptive Network

Biological Background The visual system is the most extensively studied perceptual system in mammals, and we shall focus on it for this reason. Visual information is processed in stages. Relatively simple aspects of form, such as contrast and edge orientation, are analyzed in the earlier stages, and more complex aspects later. Both the retina and cortex are organized into layers of cells with interconnections within and between layers. Within an anatomical layer, at least for the early processing stages, there is a population of cells each of which performs approximately the same processing function on its inputs. One can think of such a population as an array of filters, each of which "sees" a limited region of visual space (called the "receptive field" of that cell). More than one population of cells can share an anatomical layer. Other aspects of visual processing, such as color and motion analysis, proceed in parallel with the analysis of form.

Many cells respond to input activity by firing an electrical pulse, or "action potential," that travels down the output fiber, or axon. These pulses cause a chemical neurotransmitter substance to be released at synapses, which are regions of near-contact with other cells. The latter cells receive and process these chemical input signals. Some cells, for example in the retina, do not produce action potentials, but rather exhibit more graded electrochemical phenomena that can be used for signaling.

Although a cell's response function is in general nonlinear, visual neurophysiologists have found that for many cells it is appropriate to make a "linear summation" approximation, in which the cell's output response varies monotonically with some linear combination of the cell's input signal activities. An example of such an output response measure is the "firing rate": the number of action potential pulses (or "spikes") generated by a cell within some time window (in the case of cells that produce spikes).

Specification of the Network Will a simple self-adaptive network develop feature-analyzing cells without our specifying which features are to be analyzed? If it does, are these cell types related to those observed in biological systems? To address these questions, I have studied a network that captures some of the important biological properties described above, while intentionally avoiding many complicating factors. Such simplification is useful both because many of the details are unknown, and because our goal is to understand what principles are most important for the development of perceptual function. If, for example, one wants to know in what way nonlinearity of response may be important for development, it is valuable to start by studying a linear response system, and seeing whether or not the network exhibits the main feature-analyzing properties observed in the biological system.

A similar comment applies to feedback connections (from a later to an earlier layer of cells in the processing sequence). Such feedback is known to exist, but it is unclear how important (or not) it may be for the development of feature-analyzing function. (There are many other functions that feedback may serve, such as control of dynamic range, attentional mechanisms, etc.) We therefore analyze networks without feedback, to understand their developmental properties first.

The interconnections within the retina are known to be more complicated than the simple feedforward arrangement we shall discuss. Furthermore, mechanisms that are not dependent on neural activity appear to be involved in the development of some feature-analyzing properties in biological systems. The main purpose of the simulations we will describe is to explore what types of simple yet biologically plausible development rules *suffice* to generate feature-analyzing cell assemblies, rather than to rule out other ways of generating these properties. The results of the simple model we will describe will then lead us to infer a potential organizing principle that can encompass several of these broader issues of mechanism, including nonlinearity, more complex connectivity, and both activity-dependent and activity-independent ways of forming and modifying connections.

The network we consider is shown in figure 1. The cells are organized into two-dimensional layers A, B, C,..., with feedforward connections to each cell of layers B, C,... from an overlying neighborhood of cells of the previous layer. Layer A receives input from the visual world (if there is any such input). I have focused especially on the case in which there is no such input, and in which there is only random activity of the cells of layer A, with no correlation of activity from one cell to the next (random noise akin to "snow" on a TV screen). I have done this in order to understand how certain feature-analyzing cells may emerge even before birth, as has been observed in certain primates.

The positions of the connections to each cell need not be regular as in figure 1, but can be chosen randomly according to a density distribution (e. g., a Gaussian) that favors connections from nearby cells of the previous layer.

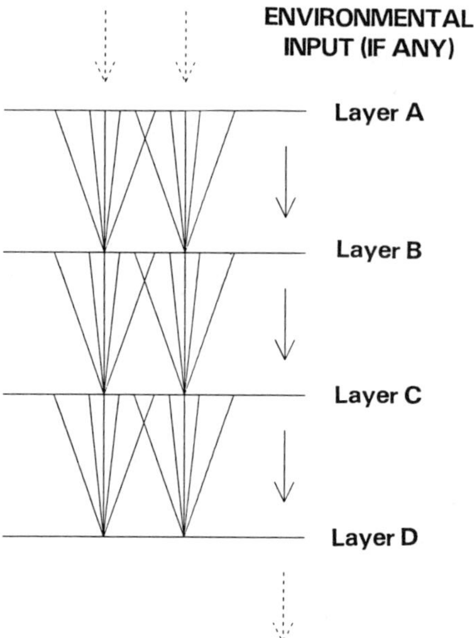

Figure 1: A layered self-adaptive network with local feedforward connections. Each two-dimensional layer contains many cells. Five input connections to each of two cells in layers B, C, and D are shown. Several hundred inputs to each cell are used in simulations. Each cell also provides input to many cells of the following layer. Lateral connections (within a layer), discussed in the text, are not indicated here.

These positions are fixed (for simplicity) for the duration of the development process. Each cell, at each time, has some signaling activity which we denote by a real number. Each cell exhibits a simple linear response: that is, the output is a linear combination of the inputs, each input being weighted by a "connection strength" which will develop in a certain way. Each model cell thus acts as a linear filter.

Two points should be noted. First, defining the output response instead as a nonlinear (e. g., sigmoid) function of the weighted sum of the inputs would allow one to more closely approximate firing rates in biological neurons (which are always nonnegative, and saturate at some maximum rate). However, we will see that even a linear response rule can lead to the formation of feature-analyzing cells, and we will explore what properties of linear adaptive filters are responsible for this result. Some of the insights gained will be applicable to the nonlinear response case as well. Second, any transformation implemented by a feedforward sequence of layers of linear filters is a linear transformation, and hence could be implemented by a single layer of connections with properly chosen connection strengths. However, our purpose is not to implement particular desired transformations, but rather to study what transformations are *learned* by a network *without supervision.* This multistage learning process depends upon the presence of multiple network layers.

A Hebb Rule For the development process, we use a version of an idea proposed by the neuropsychologist Donald Hebb in 1949. This idea has been central to much work on synthetic neural networks over the years, as well as to neuroscientists' thinking about how the development of synaptic connections may relate to memory and learning phenomena. Hebb's idea was that if cell 1 is one of the cells providing input to cell 2, and if cell 1's activity tends to be "high" whenever cell 2's activity is "high," then the future contribution that the firing of cell 1 makes to the firing of cell 2 should be increased. In the language of neural networks, the connection strength is increased (made more positive). A mathematical formulation needs to be more precise than this, and to state under what conditions the strength may decrease. We use a form in which the change in strength contains a term proportional to the product of input and output activities at that connection.

The Hebbian idea of modifying connection strengths according to the degree of correlated activity between input and output is central to what follows. It is therefore useful to give an analogy that may help to understand it intuitively. Consider a group of people whose collective opinion on a question is by definition the weighted average of its members' opinions. If, over time, a member's opinion tends to agree with the group's opinion, then the analog of the Hebb rule states that that member's vote on future issues is to be weighted more strongly. The member's vote is given less weight, or even negative weight, if he consistently disagrees with the group's opinion. This type of positive-feedback control of the weighting factors tends to lead to consensus within the group. As we shall see, it has other, surprising consequences for the properties of the "group," or output-cell, response.

Mathematical Formulation This subsection and the next summarize simulations that are described in detail in (Linsker, 1986).

Consider a cell M and the cells $L_1, L_2, ..., L_N$ that provide input to M. For simplicity, we avoid treating effects that depend upon the time sequence of signal activity values. We instead think of the activity history of a layer as a set of "snapshots," with the ordering of the "snapshots" playing no role. That is, we suppose that a set of activity values, denoted by $(L_1^\pi, L_2^\pi, ..., L_N^\pi)$, is "presented" as input to the M cell, the M cell generates an output activity value M^π, and a new set of input activities is then presented. The superscript π indexes the "presentation" of inputs (that is, the particular "snapshot") and the corresponding output. Then the linear response rule is:

$$M^\pi = a_1 + \Sigma_j L_j^\pi c_j \qquad (1)$$

where c_j is the strength of the j'th input connection to the M cell. Our Hebb-type rule is

$$(\Delta c_i)^\pi = a_2 L_i^\pi M^\pi + a_3 L_i^\pi + a_4 M^\pi + a_5, \qquad (2)$$

where the a's are arbitrary constants ($a_2 > 0$). We assume that the c values change slowly from one presentation to the next. Then we can average equation 2 over an ensemble of many presentations, and use equation 1 to express M^π in terms of the $\{L_j^\pi\}$, to obtain the rate of change of each c value. Some algebraic manipulation (Linsker, 1986) gives

$$\dot{c}_i = \Sigma_j Q_{ij} c_j + [k_1 + (k_2/N)\Sigma c_j] \qquad (3)$$

where $k_{1,2}$ are particular combinations of the constants a_{1-5}. Apart from determining the values of $k_{1,2}$, the values of a_{1-5} play no further role in what follows. Here

$$Q_{ij} \equiv <(L_i^\pi - \bar{L}) \times (L_j^\pi - \bar{L})> \qquad (4)$$

is the covariance of the activities of input cells i and j, where <...> and the overbar both denote the ensemble average. (For present purposes, $\bar{L}$, the ensemble average of the input activity at a synapse, can be taken to be the same for all synapses i,j.) The appearance of the input covariance matrix Q does not mean that there is any direct interaction between synapses i and j. Q appears simply because the Hebb rule causes $\dot{c}_i$ to depend upon the product $<L_i^\pi M^\pi>$, and M^π in turn depends upon all the $\{L_j^\pi\}$ values (via equation 1). The Q matrix will play an important role in what follows.

To prevent c values from becoming infinite during the development process, a saturation constraint is imposed. Each c value is constrained to lie between two values c_- and c_+. In a more biologically realistic case, there are excitatory synapses that have $0 \leq c \leq c_+$ and inhibitory synapses that have $c_- < c < 0$. The analysis of this case gives the same results.

First the connections from layer A to B "mature," or develop to their final values. That is, the initial c values are chosen at random, and the set of differential equations given by equation 3 (for $i = 1,2,...,N$) is solved, using the Q_{ij} function that applies to layer A activity. (For random "snow" activity in layer A, Q_{ij} is 1 when i and j are the same A cell, and 0 otherwise.) Knowing the mature c values for the A-to-B connections, as well as the Q_{ij} function for layer A, then allows us to compute the Q_{ij} function for the mature layer B. Then the development of the B-to-C connections is computed, using the Q_{ij} function appropriate to layer B. By repeating the process, we compute in turn the connection strengths for successive layers of connections.

Simulation Results A few parameters for each layer of cells determine the mature c values of the cells in that layer. These parameters include k_1 and k_2 and the breadth of the region in the previous layer that provides input to a cell of the developing layer (figure 1). As we shall see, the choice of $k_{1,2}$ values determines the mature value of the total connection strength Σc_j of the inputs to the M cell.

When we explore the parameter space, we find that there are only a few different ways in which each layer can develop. Briefly, we find that a sequence of feature-analyzing cell types emerges as one layer after another matures.

The first such type emerges in layer B. There is a parameter regime in which each c value reaches its excitatory limit c_+. In this case each B cell, once it has matured, computes the local average of the activity in the overlying region of layer A from which it receives input.

Once the B cells have matured in this way, nearby B cells have correlated activity. Each activity pattern in layer B is a blurred image of random "snow." If one B cell's activity happens to be "high" at a given time, its neighbors' activities are likely to be "high" also.

As a result of this activity correlation, a new cell type emerges in layer C. This "center-surround" cell type (Linsker, 1986) acts as a contrast-sensitive filter - it responds maximally to a bright circular spot centered on the cell's receptive field, against a dark background (or the reverse).

The Q function for pairs of center-surround cells in layer C determines the developmental possibilities for the C-to-D connections, and so on. We find that the next new type of feature-analyzing cell to emerge as we pass to succeeding layers is an "orientation-selective" cell. This cell responds maximally to a bright edge or bar against a dark background (or the reverse), when the edge or bar has a particular orientation. The receptive field map for such a computed cell is shown

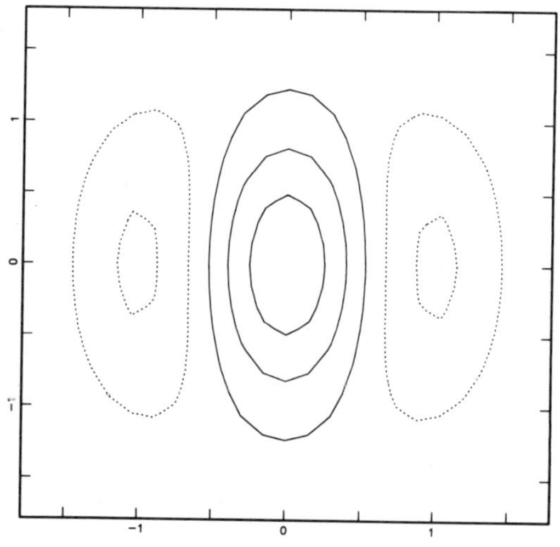

Figure 2: Receptive field map of a computed orientation-selective cell. A point of illumination at any position in the plane evokes an output response from the model cell that is proportional to the contour value at that position. Positive contour values (solid curves) denote an excitatory output response; negative values (dotted curves) an inhibitory response. Contour values range from -0.45 to +0.75 in steps of 0.30. The peak response (at the receptive field center) is normalized to unity. The parameter values that generated this particular orientation-selective cell, and the units ("r_G") of distance along the axes, are given in (Linsker, 1986; See figure 1a at p. 8780.) Axes denote distance of illumination point from receptive field center.

in figure 2. This map is a contour plot showing the response of the cell to point illumination, as a function of the position of the illumination in visual space.

Each orientation-selective cell will develop to favor an arbitrary orientation if the network contains only feedforward connections as in figure 1. However, if lateral connections between nearby cells of the orientation-selective cell layer are included in the simulation, then the orientation preferences of the cells in the layer can become organized in certain arrangements. Cells having similar orientation preferences develop to occupy irregular band-shaped regions (see figure 3 and, for an experimental comparison, Blasdel and Salama, 1986, figure 5a).

Discussion of the Simulations Center-surround cells are a prominent feature of mammalian retina. Orientation-selective cells emerge in cat and monkey visual cortex (Hubel & Wiesel, 1977; 1979). Irregular band-shaped regions of cells of similar orientation – called "orientation columns" – are a prominent feature in the orientation-

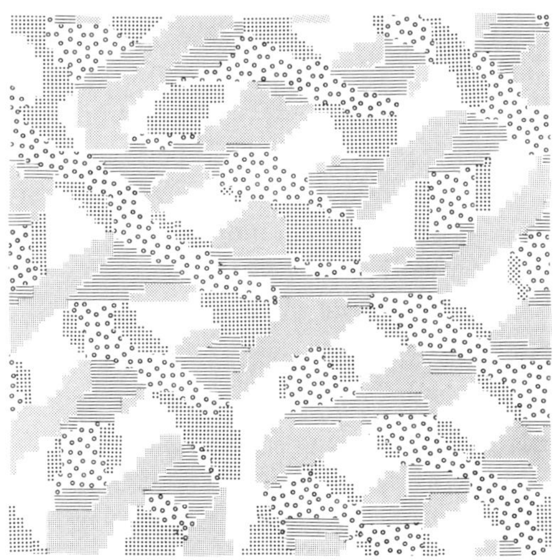

Figure 3: Example arrangement of orientation columns computed using the self-adaptive model (see text) with feedforward and lateral connections. Each model cell in a 72 x 72 array develops an orientation preference. Regions of cells having similar preference are labeled as follows (orientation angle is measured counterclockwise from the vertical): 9–45° (larger dots); 45–81° (open circles); 81–117° (horizontal lines); 117–153° (smaller dots); 153–189° (blank). (Adapted with permission from: R. Linsker, Proc. Natl. Acad. Sci. USA 83:8779-83; see original paper for details of computation.)

selective cell layers. The role that lateral connections in cortex play in the formation of orientation selectivity is at present experimentally unsettled. It is known that in certain primates, orientation selectivity is well-formed at birth – hence in the absence of any structured visual experience.

Our point is not to suggest that these feature-analyzing cells – particularly the center-surround cells – arise in animals in the same way as they do in this synthetic network. As noted above, the anatomy of interlayer connections in retina is known to be more complex than a simple feedforward arrangement. Furthermore, center-surround cells can be constructed by a simple non-adaptive model in which excitatory inputs from some narrow region, and inhibitory inputs from a broader region, both converge on a cell. (In our simulations we assumed that the breadth of the input region to a cell was the same for excitatory and inhibitory synapses, in order to avoid biasing the solution toward the formation of a center-surround cell type.)

Our point is rather that a set of progressively more complex feature-analyzing cell types develops in the layered network, and that these cell types, and their organization, qualitatively exhibit some of the most salient features found in the first few stages of mammalian visual processing. The results suggest that some properties whose origin has been mysterious – such as orientation selectivity – may have a natural explanation in terms of the functioning of a Hebb-type development process in a layered network.

Two simple examples of how *structured* input to layer A would affect the simulation results are worth noting:

- If nearby pixels have correlated intensity values, and this is the only important input correlation present, then Q in layer A would resemble the Gaussian Q that we found in layer B. The subsequent development of the model would proceed in a way similar to that which we described, except that the appearance of each feature-analyzing cell type could be advanced one layer.
- If layer A is shown an ensemble of patterns, each consisting of sinusoidal stripes with arbitrary phase and orientation, then orientation selectivity can develop as early as layer B (Linsker, 1986).

We have assumed for simplicity that the statistical properties of the ensemble of presentations, that is, the covariances Q_{ij}, are unchanged or "stationary" during development. If the ensemble statistics change, cells that had reached their apparently final mature c values may change these c values in accordance with the new ensemble characteristics. Thus, although we always speak of cell *development*, the present approach is equally applicable to studying questions of cell *plasticity* during the life of the animal.

Hebb Rules and Optimization Properties

We have seen that even a simple layered network with local feedforward connections obeying a Hebb-type rule develops a sequence of progressively more sophisticated feature-analyzing properties as we pass from one layer to the next. We will now see that a Hebb-type rule, acting on the connections to a single cell of a layered network, causes the cell to develop some remarkable optimization properties that are closely related to its information processing function. Several of these properties will be discussed in turn, and related to one another. The question of what optimization criterion may govern the development of connections from one layer of cells to another *layer*, rather than just from one set of cells to a single output cell, will be considered later, in the section entitled "Information Theory and the Principle of Maximum Information Preservation."

Maximization of Output Activity Variance Consider a cell M that receives input from cells $L_1, L_2, ..., L_N$. (Here and later, "input" means local input to cell M, not the environmental input to the network as a whole. Similarly, "output" refers to the M cell's activity value, not to the output from the network as a whole.) Let the M cell's development be described by equations 1-4 with a saturation constraint on the range of each c value, as described above. We assume that the ensemble statistical properties of the L-cell activities, that is, the Q_{ij} values for the L cells (as in equation 4), are unaffected by the choice of c values. This is true if there is no feedback from M, or the cells it influences, to the L cells. It should be a satisfactory approximation if feedback is present but is sufficiently weak, although this has not been studied quantitatively.

Define the function

$$E \equiv E_Q + E_k \tag{5}$$

where

$$E_Q \equiv -(½)<(M^\pi - \overline{M})^2> = -(½)\Sigma_i \Sigma_j Q_{ij} c_i c_j, \tag{6}$$

and

$$E_k \equiv -k_1 \Sigma c_j - (k_2/2N)(\Sigma c_j)^2. \tag{7}$$

I have constructed the function E to have the property that $-\partial E/\partial c_i = \dot{c}_i$ for each i. This means that, as the Hebb rule causes each of the c values to change with time, the value of E, as a function of the c's, decreases along a path of locally steepest, or gradient, descent. (If $\dot{c}_i > 0$, then $\partial E/\partial c_i < 0$, so c_i increases and E decreases with time. If $\dot{c}_i < 0$, then $\partial E/\partial c_i > 0$, so c_i decreases and E again decreases with time.)

The value of E thus achieves a local minimum at cell maturity. Moreover, for the cases of interest here - including those that lead to the center-surround and orientation-selective cell types described above – this minimum is a *global* near-minimum as well (Linsker, 1986). We therefore will focus on the case in which the development process does not get "stuck" in high-lying local minima. This appears to be the typical case for a perceptual network exposed to a large ensemble of presentations, although it is an empirical finding and I have not established the limits of its validity.

What is the meaning of E achieving a global, or absolute, minimum value? For any given value of total connection strength Σc_j, E is minimized when $<(M^\pi - \overline{M})^2>$ - the statistical variance of M - is maximized. Changing the values of the parameters $k_{1,2}$ adjusts, or tunes, what the mature value of Σc_j will be. (The E_k term, which is a function of Σc_j and $k_{1,2}$ only, plays a role similar to a "Lagrange multiplier" term, although E_k is parabolic rather than linear in Σc_j.)

Therefore, the development rule of equation 3 causes a cell to develop so as to *maximize the variance of its output activity,* subject to the constraint that the total connection strength have a given (parameter-determined) value and subject to the saturation bounds for each c value.

Let us see intuitively what variance maximization means to a perceptual system.

Consider first a hypothetical M cell whose c values are such that the cell's output variance is zero. That is, regardless of the input values ($L_1, L_2, \ldots, L_N$) chosen from the ensemble of presentations, the output is always the same. This cell would be useless for conveying any information about the environment to later parts of the perceptual system.

On the other hand, if the c values are chosen in a different and special way, then the M cell's output value exhibits the largest possible spread or variance (consistent with the constraints on the c's) as the set of input values ranges over its ensemble. We have shown that a Hebb-type rule tends to generate c values satisfying this special condition. In an informal sense, provided certain conditions are met, the Hebb rule acting on our described M cell tends to produce an M cell whose output activity optimally preserves the information contained in the set of input activities. Later, we will make this statement more precise, by applying some concepts from information theory, and we will modify it to accommodate the situation in which multiple M cells interact with one another.

Optimization in Another Type of Neural Network Hopfield (1982) emphasized that the dynamics of a neural network can be described in some cases by the local minimization of a function. An interesting mathematical relationship exists between the E function defined in equations 5-7 and Hopfield's "energy" function – although the network structure and behavior that each describes are very different. A great deal of physical intuition results from the formulation of the E-minimization principle in both cases, so it is very useful to show what this relationship is.

Once again, our E function is $E = E_Q + E_k$ where E_Q is shown in equation 6. The development rule causes E_Q to be minimized subject to the constraint that Σc_j have a specified value and subject to the saturation constraints on each c value. The arrangement described by equation 6 consists of one M cell with N inputs from cells $L_1, L_2, \ldots, L_N$ and is shown in figure 4a. The $N \times N$ matrix of elements Q_{ij} is the covariance matrix of the input cell activities. The c's are the connection strengths from each input cell to the output cell. The minimization of E describes the development of the c's under the influence of the ensemble of inputs characterized by the covariance matrix Q.

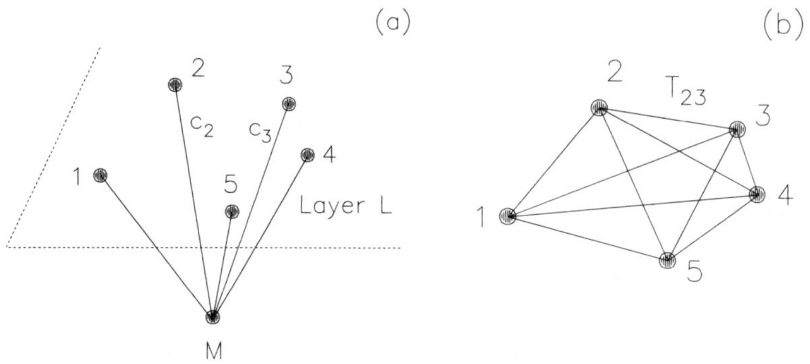

Figure 4: Relationship between (a) a single M cell (with its N inputs) of a layered self-adaptive network, and (b) a Hopfield network with N cells and N(N−1)/2 connections; for N = 5. See text for discussion.

In Hopfield's case (1982), as illustrated in figure 4b, there are N cells and the activity state of the i'th cell is called V_i. Each pair of cells is connected with fixed connection strength T_{ij}, so the number of connections is of order $N^2/2$, and the "energy" function is

$$E' \equiv -(½)\Sigma_i \Sigma_j T_{ij} V_i V_j. \qquad (8)$$

The activities V_i change with time according to a linear summation rule with a threshold: V_i increases (unless it is already at its upper limit) if $\Sigma T_{ij} V_j > 0$, and decreases if $\Sigma T_{ij} V_j < 0$. If the T_{ij} matrix is symmetric, then the V_i's change so as to decrease the value of E' to a local minimum. Connection strengths are fixed; there is no learning or network development. The dynamical process described by equation 8 is the change in the activities $\{V_i\}$ from some initial state to a final state of locally minimum E'. If one wants to use the network for memory retrieval, a suitable choice of T_{ij} is given by an expression that is essentially the covariance of V_i^k and V_j^k over the ensemble of memories (indexed by k) to be stored.

Note that E' has the identical structure as our E_Q if we identify V_i with c_i and T_{ij} with Q_{ij}. What is the significance of this?

When T is a covariance matrix, Hopfield's network computes a local minimum of E' using N cells and order $N^2/2$ connections, explicitly embodying the T values. The state for which E' is minimal is the set of final activities $(V_1, V_2, ..., V_N)$.

One cell of our network computes a local minimum of the same function, our E_Q, using N connections. The Q function, which corresponds to T, is nowhere explicitly represented in the network. The Hebb rule implicitly responds to the covariance matrix, Q, as the ensemble of input patterns is presented to the M cell. The state for which E_Q is minimal is not a set of activities, but a set of mature connection strengths $(c_1, c_2, ..., c_N)$.

To summarize: For T matrices that are covariance matrices, one cell of our network can locally optimize the same function as a fully connected Hopfield-type network. In our network, this optimization process consists of developing a final set of c-values starting with some initial set of values, under the influence of a statistically stationary ensemble of input patterns having covariance matrix T. In the Hopfield-type network case, the process consists of seeking a final set of cell activity values starting with some initial set of values, in a network whose connection strengths are fixed and prespecified to be the T values themselves.

These considerations lead to an interesting connection, only briefly outlined here, between memory retrieval and perception in a network model.

Memory Retrieval and Perception in a Network Model If there are sufficiently few memory patterns to be stored, relative to N, then E' or E_Q will tend to have minima at the $\{V_i\}$ or $\{c_i\}$ values, respectively, corresponding to those memories. Depending upon the initial choice of the V's or c's, one or another of these memory states will be activated or selected. In the case of Hopfield's network, "activated" means that the final activity state will match one of the stored memories. In the case of a cell in a layered self-adaptive network, "selected" means that the final set of c values will cause the M cell to be a *matched filter* for one of these memories. That is, the mature M cell will respond most strongly when presented with the set of input activities corresponding to that memory.

If the number of patterns in the ensemble is large, then the E_Q function will no longer capture details of any one of the patterns. The structure of the E_Q function may become simpler. The global minimum of E_Q will lie at the $(c_1, c_2, ...)$ value for which the M cell's variance is maximized. The mature M cell will function as a *feature-analyzing* cell, rather than a matched filter to a particular memory. The particular feature or pattern element to which the mature cell will optimally respond, such as an oriented edge, need not even appear in any of the presented patterns.

Principal Component Analysis There is a special case in which variance maximization corresponds to an important, and widely-used, statistical method for feature extraction. This is the case in which the output variance is maximized subject to the constraint that $\Sigma c_i^2 = 1$. Oja (1982) showed that one can achieve this maximization by using a particular form of Hebb rule, equivalent to (in our notation):

$$\dot{c}_i \propto \langle M^\pi (L_i^\pi - M^\pi c_i) \rangle. \tag{9}$$

For this expression, we put $M^\pi \equiv \Sigma L_i^\pi c_i$ and define the activities, subtracting nonzero mean values if necessary, so that $\langle L_i^\pi \rangle = 0$ for all i. The additional term in the Hebb-type rule, proportional to c_i, causes Σc_i^2 to be close to 1, and no explicit constraint needs to be imposed.

In statistics, principal component analysis (PCA) is a standard method, reviewed by Huber (1985), for identifying "interesting" but unanticipated structure, such as clustering, in high-dimensional data sets. For example, an economist confronted with 1000 dimensions of data, such as the prices of different commodities, may want to know which several features of the data, for example, which several linear combinations of the 1000 quantities, are most salient.

PCA works as follows. Consider a set of data points indexed by π, each point $\mathbf{L}^\pi$ having coordinates $(L_1, L_2, ..., L_N)$. For PCA we compute a vector **c** for which the projection of the set of data points onto the axis parallel to **c** has maximum variance. The projection of $\mathbf{L}^\pi$ onto **c**, when $\Sigma c_i^2 = 1$, is just $M^\pi = \Sigma_i L_i^\pi c_i$, and the variance of the projected distribution is identical to the variance of M^π.

An example of PCA is illustrated in figure 5. Projecting the cloud of data points onto line 1 captures the salient feature of the data – that there are two clusters. The variance, or spread, of the data points along this axis is greater than for any other projection axis. Projecting the cloud onto line 2 would obscure the cluster structure. While the cluster structure is evident in the raw data of the two-dimensional plot shown here, such structure is often totally concealed

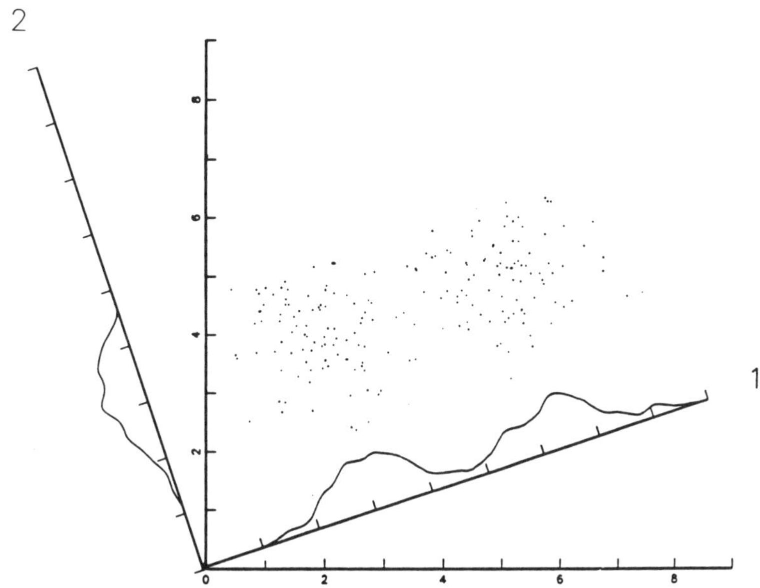

Figure 5: Illustration of principal component analysis. A cloud of data points is shown in two dimensions, and the density plots formed by projecting this cloud onto each of two axes 1 and 2 are indicated. The projection onto axis 1 has maximum variance, and clearly shows the bimodal (clustered) character of the data.

in high-dimensional data sets, until an analysis method such as PCA is applied.

Since the PCA method corresponds to choosing **c** so as to maximize the variance of M^π subject to $\Sigma c_i^2 = 1$, it follows that the mature M cell generated by Oja's version of the Hebb rule performs PCA on its set of inputs (Oja, 1982).

Optimal Inference Consider an arbitrary M cell characterized by a set of c values and having the linear response rule $M^\pi = \Sigma_i L_i^\pi c_i$ with $\langle L_i^\pi \rangle = 0$ for all i. Suppose we know the c values, and are told a particular value of the output M^π. We are asked to estimate the input activities $(L_1, L_2, \ldots, L_N)$ for that presentation. Let us "score" any such estimate by

1. computing the difference between the estimate $L_i^\pi(\text{est})$ and the true value of L_i^π,

2. squaring this difference, and
3. summing this squared error over i.

Averaging this score over an ensemble of presentations gives the mean squared error:

$$\mathrm{MSE} \equiv \Sigma_i \langle [L_i^\pi - L_i^\pi(\mathrm{est})]^2 \rangle. \tag{10}$$

What estimation rule will give the best, meaning the minimum, MSE? For a linear estimation rule of the form $L_i^\pi(\mathrm{est}) \equiv g_i M^\pi$, where we want to know what g values to use, the answer is found by minimizing MSE with respect to each of the g_i's. This is easily done by differentiating MSE, and is also a simple case of the Gauss-Markoff theorem (Liebelt, 1967), which applies more generally to the optimal estimation of a set of inputs given a set of outputs, rather than just one output. The result is:

$$L_i^\pi(\mathrm{opt\ est}) = M^\pi \times (\Sigma_j Q_{ij} c_j) / (\Sigma_i \Sigma_j c_i Q_{ij} c_j). \tag{11}$$

The MSE corresponding to this optimal estimate is then:

$$\mathrm{MSE(opt)} \equiv \Sigma_i \langle [L_i^\pi - L_i^\pi(\mathrm{opt\ est})]^2 \rangle = \Sigma_i \langle (L_i^\pi)^2 \rangle - H, \tag{12}$$

where

$$H \equiv [\Sigma_i (\Sigma_j Q_{ij} c_j)^2] / (\Sigma_i \Sigma_j c_i Q_{ij} c_j). \tag{13}$$

Expressed in matrix form, with **c** denoting the column vector $(c_1, c_2, \ldots)$ and **Q** denoting the matrix (Q_{ij}), we have $H \equiv (\mathbf{c}^T\mathbf{QQc})/(\mathbf{c}^T\mathbf{Qc})$ (superscript **T** denotes the matrix transpose).

The calculation so far involves a standard use of optimal estimation theory (Liebelt, 1967). The linear filter, represented here by the set of c values, is specified. The result of a "measurement" using the filter – that is, the output value – is given. The task is to reconstruct the input values with minimum error, using a simple mean squared error criterion.

We now go beyond this simple framework to ask (Linsker, 1988a): For what linear filter – what set of c values - is this minimum-error reconstruction the most accurate? That is, what choice of c's minimizes MSE(opt) of equation 12?

Since the first term on the right-hand side of equation 12 is independent of the c's, minimizing MSE(opt) is accomplished by maximizing H. The mathematical condition for this to occur is that the vector **c** be an eigenvector of **Q** having maximal eigenvalue. This is identical to the condition that **c** needs to satisfy in order for the M cell to perform PCA on its input values.

Therefore the PCA condition and the principle of optimal inference – namely, that MSE(opt) be minimized, or H be maximized – lead to the same set of c values. A Hebb rule of the form of equation 9 generates an M cell that satisfies both conditions. In the presence of other constraints, or additional cost terms, there is no guarantee that PCA and H-maximization are equivalent, since the PCA principle maximizes the quantity $(\mathbf{c}^T\mathbf{Q}\mathbf{c})/(\mathbf{c}^T\mathbf{c})$ which is not identical to the expression for H in equation 13.

Optimization in the Presence of Processing Noise and Constraints on Output Variance We have identified several optimization properties related to the cell's output variance. Suppose, however, that for some reason the variance, or spread of output activities, is itself constrained. For example, the output activity may be confined to lie within some operating range. This is a biologically plausible situation. In that case, what is optimized by a suitable Hebb-type rule? We will discuss this case for a particular processing model, giving only the main results and omitting the details.

Suppose the signal L_j^π on the j'th input line or connection is corrupted by noise v_j^π, where v_j^π has a mean of zero and a variance B, and is uncorrelated both with the noise on other input lines and with any of the input signals L_j^π. The cell computes the weighted sum $x \equiv \Sigma_j(L_j^\pi+v_j^\pi)c_j$. The variance of x is the sum of two terms: the variance due to the signal in the absence of noise, $\Sigma_{ij}Q_{ij}c_ic_j$; and the variance due to the noise, $B\Sigma c_j^2$. Consider a suitable synaptic modification rule in which $\dot{c}_i$ contains a term of the form $\langle L_i x \rangle$. This rule causes the model cell to develop such that the variance of x due to the signal is maximized relative to the variance due to the noise. This type of signal-to-noise optimization property can also emerge when the cell's output M is a monotonic nonlinear function of x, such as a sigmoid function, if the synaptic modification rule is of the form described.

Adaptive Signal Processing; Supervised and Unsupervised Learning Returning to the case of a linear-response model neuron, suppose we wish to train a linear cell to respond to each of a set of prescribed input vectors by generating an output that best matches a prescribed desired output. An input vector is denoted $\mathbf{L}^{\pi} \equiv (L_1, L_2, ..., L_N)$ and each desired output is a scalar number M^{π}_{des}. The actual output is $M^{\pi} = \Sigma_j L^{\pi}_j c_j$ where each $<L^{\pi}_i> = 0$ and the optimal c values are to be determined by a learning process. A mean square measure of error is used:

$$MSE' = <(M^{\pi} - M^{\pi}_{des})^2> \tag{14}$$

where <...> again indicates the ensemble average. Then MSE′ is a minimum when the c values are chosen to satisfy $<M^{\pi}_{des} L^{\pi}_i> = \Sigma_j Q_{ij} c_j$ for all i. (Recall that $Q_{ij} \equiv <L^{\pi}_i L^{\pi}_j>$.) The least mean square, or LMS, algorithm of Widrow and Hoff (1960;1985) uses an estimate of the gradient of MSE′ and in effect performs gradient descent to compute the optimal c values. An ensemble-averaged form of the algorithm can be written as:

$$\dot{c}_i \propto <L^{\pi}_i (M^{\pi}_{des} - \Sigma_j L^{\pi}_j c_j)>. \tag{15}$$

Equations 14 and 15 give an objective function to be minimized and an algorithm for a *supervised* learning process. Both the inputs and the desired outputs are presented to the cell, and the error term $(M^{\pi}_{des} - M^{\pi})$ – the amount by which the actual output differs from the desired output – is fed back to change the c values until the mean square error is minimized.

Our optimal inference criterion, namely, the minimization of the objective function of equation 12, and a Hebb-type rule that implements it (equation 9) are formally similar to equations 14 and 15. But the optimal inference criterion provides a method for *unsupervised* learning. The criterion does not make any use of a desired output; it simply states that the M cell should have the property that knowing its output activity value allows one to infer the input activities with greatest possible accuracy.

Information Theory and the Principle of Maximum Information Preservation

For a single M cell receiving inputs from a given set of L cells, we have seen that, for a particular Hebb rule (equation 9), knowledge of the output activity value allows inference of the input values with greatest accuracy, in the sense of minimum mean squared error. For more general Hebb-type rules, we found that the variance of the output activity was maximized subject to various constraints. This result led us to suggest that, at least in an intuitive sense, a Hebb rule may act to generate an M cell whose output activity *preserves maximal information* about the input activities, subject to constraints.

We will now make this notion of maximum information preservation more precise, and will extend it to the case of an entire layer of M cells, by introducing some concepts from information theory. We will then derive the consequences of the resulting principle for some simple cases involving a single M cell and a coupled pair of M cells. We will also discuss qualitatively what the principle implies for the development of each layer (containing many cells) of a perceptual system. That is, given the statistical properties of the ensemble of input patterns at layer L, and certain constraints, what particular processing functions do the connections from layer L to layer M, and within layer M, develop to implement?

Shannon Information We will regard each presentation of real-valued inputs $L=(L_1,L_2,...,L_N)$ as a message, where L_i denotes the activity of the i'th L cell in the layer. We omit the π superscripts for clarity. Strictly speaking, even one real number carries an infinite amount of information. To avoid encountering expressions of the form $(\infty - \infty)$, and because infinite precision is physically and biologically meaningless, we will think of the N-dimensional space of the L vectors as being divided into small boxes. Two messages are regarded as identical if they lie in the same box. Each box is labeled by its location L. In the end, we will pass to the continuum limit, and the sums will become integrals.

Given an ensemble of messages, let P(L) be the probability that a randomly chosen message lies in box L. In a classic paper, Shannon (1948) showed that the information conveyed by sending a message that lies in box L is $I(L) = [-\ln P(L)]$. The average information conveyed per message is $\langle[-\ln P(L)]\rangle = -\Sigma_L P(L) \ln P(L)$, where $\langle ... \rangle$ is the usual ensemble average. If the base-2 (rather than natural) logarithm were used here, the information would be measured in bits.

Now suppose each input presentation L generates a set of output values, denoted by the vector M, via some known computation.

Suppose that we are told the value of M - or, more strictly, which discrete box M lies in. In general, M will not be uniquely determined by L because noise may be introduced in the computation of M. (More about this shortly.) How much additional information would we need to reconstruct the input message L that gave rise to M? (Shannon (1948) calls the ensemble average of this amount of additional information the "equivocation.") The answer is $I_M(L) = [-\ln P(L|M)]$, where $P(L|M)$ is the conditional probability of the input message lying in box L given that the output lies in box M. Therefore the amount of information that knowing M conveys about L is the difference, $I(L)-I_M(L) = \ln [P(L|M)/P(L)]$. The ensemble average of this quantity is the rate R, per message, of transmission of information from the cell's inputs to its output. This is the average amount of information that knowing M conveys about L. We have

$$R = \langle \ln [P(L|M)/P(L)] \rangle = \Sigma_{L,M} P(L,M) \ln [P(L,M)/P(L)P(M)]. \quad (16)$$

We have a standard identity $P(L|M)P(M) = P(L,M) = P(M|L)P(L)$, where $P(L,M)$ is the joint probability that the input lies in box L and the output lies in box M. Using this gives

$$R = \langle \ln [P(M|L)/P(M)] \rangle = \\ -\langle \ln P(M) \rangle + \langle \ln P(M|L) \rangle \equiv \langle I(M) \rangle - \langle I_L(M) \rangle. \quad (17)$$

The right-hand side is the ensemble average of the total information conveyed by M, minus the information that M conveys to one who already knows L. This second term is the "information" that M conveys about the processing noise, rather than about the signal L.

The information rate R is also referred to as the *mutual information* between the input message L and the output message M.

Maximum Information Preservation Let us now state the proposed *principle of maximum information preservation*, which we shall also refer to as the *infomax principle,* for each layer (or processing stage) of a perceptual network: Given a layer L of cells, and the (stationary) ensemble statistical properties of the signal activity values in this layer, and given that layer L is to provide input to another cell layer M, the transformation of activity values from L to M is to be chosen such that the rate R of information transmission from L to M is maximized, subject to constraints and/or additional cost terms. These constraints or costs may reflect, for example, biochemical and anatomical limitations on the formation of connections, or on the character of the allowed transformations.

The formulation of this principle was motivated by studying Hebb-type rules and recognizing certain optimization properties to which they lead for single M cells. Once formulated, however, the principle is independent of any particular local algorithm, whether Hebb-related or otherwise, that may be found to implement it. Let us explore: (1) the consequences of the principle for some simple cases; (2) how the principle might be implemented; and (3) how it may fit within a broader view of neural development.

Case of a Single M Cell Under certain conditions, maximizing the output activity variance of the M cell maximizes the Shannon information rate R. We illustrate this for a particularly simple but instructive case. The argument can be made somewhat more general than this, but it is not true that maximum information rate and maximum activity variance coincide when the probability distribution of signal values is arbitrary.

Suppose the M cell receives inputs from a set of L cells $L_1, L_2, \ldots, L_N$, and that the M cell's output in the presence of processing noise has the form

$$M^{\pi} = (\Sigma_i L_i^{\pi} c_i) + \nu^{\pi}. \tag{18}$$

Here π indexes the particular set of input and output values, so that if L is repeated but the output M is different, owing to noise, this counts as a different set of input-output values. The quantity ν^{π} is the noise, a random variable differing from one presentation to the next. Suppose that

1. M has a Gaussian distribution (with variance denoted by V);
2. ν has a Gaussian distribution with a mean of zero and variance denoted by B; and
3. ν is uncorrelated with any of the input components (that is, $\langle \nu L_i \rangle = 0$ for all i).

Then, omitting the details, we find that the information rate is

$$R = (\tfrac{1}{2}) \ln (V/B). \tag{19}$$

For a given noise variance B, this rate is maximized by maximizing the output variance V of the M cell. Note that V/B is essentially a signal-to-noise ratio.

Suppose that the noise model consists instead of independent Gaussian noise, ν_i, being introduced on each input line i, where each ν_i has variance B. Then $M^{\pi} = \Sigma_i (L_i^{\pi} + \nu_i^{\pi}) c_i$, and the information rate is found to be $R = (1/2)\ln [V/(B\Sigma c_i^2)]$. In that case, R is maximized (for fixed B) when $(V/\Sigma c_i^2)$ is maximized – that is, when the connection strengths are chosen so as to perform principal component analysis on the cell's inputs.

Why an Information-Theoretic Principle? As we have just noted, the infomax principle was motivated by studying Hebb-type rules and the optimization properties, such as PCA, to which they lead. The infomax principle gives the same result as PCA for certain simple cases, such as the case of a single linear cell with Gaussian signal and noise structure just illustrated. However, in general, infomax and PCA are *not* equivalent. For example, the two methods do not give the same results for arbitrary probability distributions. To illustrate this, consider a non-Gaussian distribution of signal activities that consists of two sharp peaks separated by a large interval. The variance of this distribution is large because of the large separation between the two peaks, but the information conveyed by a message selected from this distribution is small because each peak is narrow. A hypothetical M cell having this output activity distribution would tend to be selected by PCA, but not by the infomax principle.

My view is that the principle of maximum information preservation is more fundamental than PCA or the principle of optimal inference discussed above. There are several reasons for this view.

First, the infomax principle deals with the informational content of signals, rather than just with their activity values. Suppose one trivially rescales the value of a signal activity; that is, multiplies it and its associated noise by some factor. The information conveyed by that signal is unchanged, and the predictions of infomax are consequently unchanged. However, the results obtained using PCA will in general be changed since the signal variance has been increased. (This is related to the fact that PCA takes no explicit account of the noise process.)

Second, the infomax principle, unlike PCA, can apply irrespective of whether the allowed transformations are linear.

Third, when there is more than one output cell, the prediction of feature-analyzing properties depends critically upon the proper treatment of processing noise. In particular, if we were to use PCA to predict the properties of a set of N linear output cells, we might conclude that the output activity vector captures the N highest-variance attributes of the input vector. (Technically, this corresponds to finding the N leading components in a Karhunen-Loeve expansion.) The infomax principle predicts a quite different result, owing to the presence of processing noise and the fact that a single cell can compute a function of its input activities with only limited accuracy. The effect that noise has on the properties of the output cells will be made clear in the next section for a simple case.

Fourth, we have discussed an optimal inference criterion that is closely related to PCA. This criterion minimizes the mean squared error with which the input can be reconstructed, given the output. But there is no reason to require that the input to a processing stage be *reconstructed* from the output, nor (even if there were) that the mean squared error is the proper criterion for the goodness of reconstruction. In contrast, the infomax principle optimizes the extent to which the output of a processing stage allows one to *discriminate* different input patterns from one another in the presence of noise.

For a concrete example of the distinction between optimal reconstruction and discrimination, consider the well-known fact that our visual system is relatively insensitive to the absolute level of average illumination (the "discounting of the illuminant"). An array of contrast-enhancing cells, as discussed above, has this property. If one were to attempt to reconstruct the input intensities from the outputs of such an array, one would do very poorly as far as the mean squared error between reconstructed and actual intensities is concerned. On the other hand, one could do very well from the standpoint of preserving information about the inputs, since the average illumination varies relatively slowly over the visual field and therefore is describable by only a small number of bits of information. Furthermore, if one were instead to transmit the average illumination value as a part of the activity of each output cell, the amount of information that each output cell could convey about intensity contrasts would be greatly reduced, since each cell has only a limited dynamic range of allowed activity values in a biological system. The infomax principle is in qualitative accord with these facts; a rule that prescribes simple minimization of a reconstruction error is not.

In summary, we have described how the emergence of feature-analyzing functions in a simple network led us to explore connections between the development rules and certain optimization properties such as PCA and optimal inference, and how those optimization properties in turn suggested a more general principle, that of maximum information preservation. In some simple cases, the more general principle agrees with PCA as shown. (For an analysis of some consequences of PCA in cases of biological interest, see Linsker (1990a).) For more complex and biologically relevant cases, we must explore what the infomax principle implies, and how it may be implemented. We shall now address these issues, starting with a relatively simple model for which the infomax principle leads to new consequences.

Redundancy and Diversity – the Case of Two Coupled M Cells
Suppose there is an arbitrary number of L cells but just two linear M cells. Each M cell's output is some linear combination of the L cells' activities:

$$M_1^\pi = (\Sigma_i t_{1i} L_i^\pi) + v_1^\pi; \tag{20}$$

$$M_2^\pi = (\Sigma_i t_{2i} L_i^\pi) + v_2^\pi. \tag{21}$$

We consider the simple case in which each noise term is Gaussian and of variance B, the noise terms for the two M cells are uncorrelated with each other, and each noise term is uncorrelated with any of the L cell activities. We treat the case in which M_1 and M_2 have Gaussian distributions, with $\langle M_1^\pi \rangle = \langle M_2^\pi \rangle = 0$. Our task is to determine what values of the t_{ni}'s lead to the maximum information being preserved during the processing of L-cell activities to give M-cell output activities.

Note that the t_{ni}'s do not in general stand for the strengths of particular connections. There may be both feedforward and lateral (M-to-M) connections whose joint effect, possibly over several time steps, is to produce the M-cell outputs of equations 20 and 21. Our concern here is not with the particular connection strengths, nor with the development rule that may implement them, but rather with understanding what cell response properties – what t_{ni} values – are induced by the principle of maximum information preservation.

The information rate for this case, omitting details of the proof, is

$$R = (1/2)\ln(\text{Det}\, Q^M) - \ln B, \tag{22}$$

where the elements of the 2×2 covariance matrix Q^M are $Q^M_{nm} \equiv \langle M^\pi_n M^\pi_m \rangle$ and "Det" denotes the determinant. We find

$$\text{Det}\, Q^M = B^2 + B(W_1 + W_2) + W_1 W_2 (1 - \rho_{12}^2) \tag{23}$$

where W_n is the output variance of cell M_n in the absence of noise, and ρ_{12} is the correlation coefficient of the activities of M cells 1 and 2, also in the absence of noise.

To maximize R, given B, we must maximize $\text{Det}\, Q^M$. When B is large, the third term on the right-hand side of equation 23, which is independent of B, is small compared with the second term, which is of order B. In that case, maximizing $\text{Det}\, Q^M$ means maximizing $(W_1 + W_2)$. If no constraint prevents us, we can achieve this maximization by maximizing W_1 and W_2 separately. But this means constructing each M cell so that its output variance, which is W_n in the absence of noise, or $W_n + B$ in the presence of noise, is maximized. This is exactly what we found to be the optimum solution when there was only one M cell. (See equation 19.)

If the noise B is smaller, then the third term becomes relatively more important. The rate R is then maximized by making an optimal tradeoff between keeping W_1 and W_2 large, and making the responses of the two M cells uncorrelated.

We have thus found that, depending upon the noise level, there is competition between the value of having redundant M cell responses, which mitigate the information-destroying effects of noise, and the informational value of having different cells extract different linear combinations of the input. A high noise level favors redundancy. In this case, both M cells compute the same linear combination of inputs, if there is only one such combination that yields maximum output activity variance. A lower noise level favors diversity of response. In that case, the M cells compute different linear combinations of the L cell activities, even though each M cell's output variance may be reduced as a result of this choice.

To make this more concrete, consider a simple example. There are two L cells, and the Q matrix for L cell activity has $Q_{11} = Q_{22} = 1$

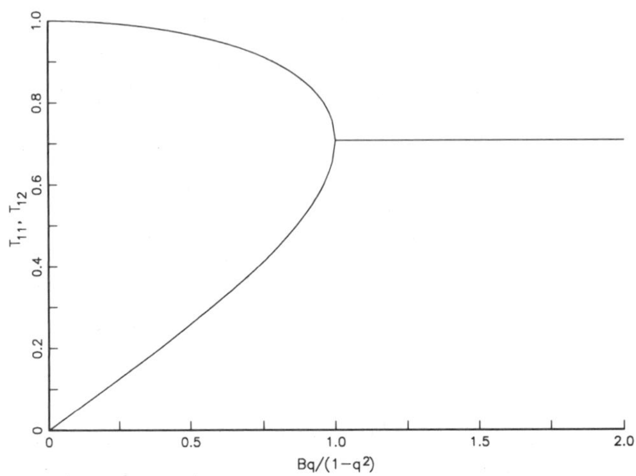

Figure 6: Values of the coefficients t_{ni} that maximize the preservation of information from layer L to M, for a simple case with two L and two M cells. Each M cell output (equations 20 and 21) includes random noise having variance B, and q is the correlation (or covariance) of the activities of the two L cells. For $x \equiv Bq/(1-q^2) \geq 1$, both M cells redundantly compute the same linear combination of the L cell activities (all $t_{ni} = 1/\sqrt{2}$). For $x<1$, the optimal t values satisfy $t_{11} = t_{22}$ and $t_{12} = t_{21}$, where the upper curve gives t_{11} and the lower curve gives t_{12}, or the reverse. The curves for $x<1$ are given by $y = (1/2)[(1+x)^{1/2} \pm (1-x)^{1/2}]$; this is derived by maximizing Det Q^M (equation 23) for this example.

and $Q_{12} = Q_{21} = q$ with $0<q<1$. We arbitrarily impose the constraint that $t_{n1}^2 + t_{n2}^2 = 1$ for each M cell ($n = 1,2$).

The solution that maximizes the preservation of information has $t_{11} = t_{22}$ and $t_{12} = t_{21}$, and the values of t_{11} and t_{12} are given in figure 6 as a function of B and q. For large B, both M cells receive the same linear combination of inputs: $(L_1+L_2)/\sqrt{2}$. For smaller B, the cells measure different linear combinations of L_1 and L_2. In the limit that B approaches zero, one M cell receives input only from cell L_1 and the other only from L_2.

An Entire Layer of M cells, with Nonlinearity and Lateral Connections What does the infomax principle imply qualitatively in this more general case? Maximizing R means that we attempt to (1) maximize the total information conveyed by the output message M, and (2) minimize the information that M conveys to one who already knows the input message L.

These criteria are related, but not equivalent, to the property of encoding signals so as to reduce redundancies present among the inputs to the perceptual system. The general idea that information theory may be useful for understanding perception is an old one (see Linsker (1990b) for a review). Significant contributions were made by Attneave (1954), Barlow (1961;1969), and Marr (1970). Much of this work has focused on the role of redundancy reduction. This property is one, but only one, aspect of the infomax principle. For example, we have seen that infomax also leads to the *introduction* of redundancy when this is useful in countering the effects of noise.

I have analyzed the qualitative consequences of the infomax principle in some simple models (Linsker, 1988b). The results show that the principle can, under certain conditions, lead to L-to-M transformations having the following properties:

- Topographic mapping from layer L to layer M, when the spatial extent of lateral connections within layer M is assumed to be limited. That is, near-neighbors in L tend to map to near-neighbors in M.
- Map distortions, in which a greater number of M cells tend to represent the types of layer-L excitation patterns that occur more often. This suggests a possible role for infomax in explaining certain map distortions, or "cortical magnification" effects, that are found in biological systems.
- The infomax principle selects which features of the input signals are represented in layer M. Features having relatively high signal-to-noise ratios are favored. This is the extension of our previous redundancy-diversity result to the full-layer case.
- Orientation-selective cells, and the arrangement of such cells in orientation columns, can emerge for some simple types of model input.
- When time-delayed information is made available to the layer, the infomax principle can cause M cells to extract and encode temporal correlations, in a manner similar to the extraction of spatial correlations.

I emphasize that much work is required to determine the

consequences of the infomax principle for cases involving more biologically realistic patterns of activity.

Implications

From a simplified set of assumptions – linear summation response, a simple Hebb-type rule having a covariance form, and feedforward connections only – we derived an optimization principle for the development of a single cell. This principle states that the mature M cell is such that its output activity variance is maximized subject to constraints. More generally, one can have cost terms instead of, or in addition to, constraints. Then the function that is maximized involves both the variance and the additional cost function.

This led us to infer a proposed principle of maximum information preservation, subject to constraints. It is equivalent to variance maximization in some simple cases, but it has a much broader scope. For example, it can be applied to cases in which a layer of L cells provides input to an entire layer of M cells, with lateral as well as feedforward connections. It can likewise apply to cases in which the response function is not necessarily linear. We discussed some important ways in which the infomax principle differs from the simpler variance-maximization ideas.

The consequences of this proposed principle are only beginning to be explored. One set of issues that needs clarification is the choice of biologically appropriate constraints and cost terms. A second, related, issue involves the choice of algorithms, whether of Hebb type or otherwise, that control the development of feedforward and lateral connections so as to implement the optimization principle. While much work needs to be done, I suggest that the infomax principle, or something like it, may play an important role in determining the character of perceptual processing at least in its early stages, where there is a chance that feedback may not affect development of feature-analyzing function in an essential way.

I expect that the infomax principle may play a role even in the presence of significant feedback, but it is not clear at this time how best to analyze this case. Feedback allows the output of a processing stage to affect the statistical properties of the input to that stage. Therefore applying the infomax principle to each stage in turn is not in itself sufficient to determine the processing function of each stage in the presence of significant feedback.

Also, just as feedback can modulate attentional mechanisms in biological systems, it is quite plausible that feedback signals may cause different portions of a network to weight information of different types, or from different sources, differently. A simple way of thinking about this is that feedback could modify the biological system parameters that correspond to the noise process in our model. (The significance of the noise process in our model is that it influences how different two input patterns have to be, in order to be discriminable from one another at the output of the processing stage.) Thus infomax could be one among a number of organizing principles that interact in determining the character of a perceptual system. The strategy we have chosen is to explore the consequences of "pure" infomax in order to determine both its predictive power and its limitations.

What might we expect to be the character of a layer of cells developing according to the principle of maximum information preservation? Although the necessary calculations for sufficiently realistic cases have not yet been carried out, we can speculate on the outcome.

Suppose there is a constraint on the distance within layer M over which the activity of one M cell can affect another. There might be, for example, a constraint on the length of lateral connections. Suppose also that each region of layer M "sees," or receives input from, only a limited region of layer L, and that nearby regions of M "see" nearby regions of L. Then, if the noise variance B is large, and there are not many M cells that "see" the same set of L cells, we might find that each M cell develops so as to maximize its activity variance, and performs processing that is redundant with that of many of its neighbors.

On the other hand, if B is smaller, or there are a large number of M cells that "see" the same L region, we may expect that the M cells in a region do not all perform the same processing function on the inputs from the L layer. Instead, they might span a range of feature-analyzing properties, each of which has a moderately high variance.

In the visual pathway of cat and monkey, there are multiple layers of center-surround cells, followed by layers of orientation-selective cells. The onset of orientation selectivity occurs at a different layer in cat than in monkey. It is possible that, in response to the ensemble of inputs seen by a particular layer, the layer can develop either center-surround or orientation-selective cells, as occurred in our previous model simulations (Linsker, 1986). Perhaps a parameter such as the noise level B "tunes" for redundancy or diversity of response. Redundancy could favor center-surround cell formation, with many cells performing substantially the same processing function. Diversity could favor the formation of orientation-selective cells spanning the entire range of orientation preferences within each region of the layer. (The point is that, of a group of cells comprising all orientation preferences, only a certain fraction will fire when presented with an oriented edge of illumination.) Hubel and Wiesel discovered (1977; 1979) that orientation-selective cells are arranged, within a cortical layer, so that each small ($\approx 1 \times 1$mm) region of cortex contains the "machinery" for analyzing substantially all edge orientations seen by either eye within a small region of visual space. Perhaps the principle of maximum information preservation, combined with limits on lateral interaction distance, can account for this efficient organization.

Local Algorithms The infomax principle is stated in terms of maximizing a complicated expression (equation 16). Is there an algorithm or process that deals with much simpler quantities and computations – local to each cell or pair of connected cells in a network – and yet implements the infomax principle, at least approximately?

I have found (Linsker, 1988b) that, for some simple cases, a Hebb-related algorithm developed by Kohonen for different purposes implements some of the qualitative features required by the infomax principle. Kohonen's (1984; 1988) algorithm was developed to show how lateral connections can induce topographic order in a simple model, and makes no reference to noise or information content. It is an extension of an earlier "k-means" technique in statistics.

The results obtained using Kohonen's (1984; 1988) algorithm are not in quantitative agreement with the requirements of the infomax principle. For example, different "cortical magnification" or map distortion effects are obtained. I have recently devised a local algorithm that implements the infomax principle in a quantitatively correct way, when certain simplifying conditions are satisfied

(Linsker, 1989). Work is in progress on local algorithms that embody the requirements of the infomax principle under more general conditions.

There is a complementary relationship between the infomax principle and an algorithm that implements it. The principle can suggest what the algorithm, and the lateral connections that it describes, are "for" – that is, what role such connections and processes may serve in the construction of a perceptual system. The algorithm can show how a complex optimization principle may be implemented by a network of cells individually having relatively little computational power.

Although I have focused on algorithms that perform activity-dependent modification of connections, other types of mechanisms may be used to implement a given optimization principle. Biochemical cell-cell adhesion markers, chemical or other gradients that may help to establish topographic maps, particular cell types that implement complex types of connectivity (as in the retina), and other mechanisms may all play a role. An organizing principle by itself does not specify the many design details that a particular system – biological or synthetic – may use to implement it.

Infomax and the Use of Perceptual Data Why might it be important for a perceptual system to maximize the amount of information that is preserved from one layer to the next?

I suggest that one goal of a perceptual system is to provide the brain with the means of discriminating different environmental situations that may demand different responses by the animal.

For a very simple network with only a few layers of processing from environmental input to motor output, we could imagine using some sort of supervised learning mechanism. The mechanism would pair inputs with the desired output responses and adjust the connection strengths accordingly. However, such a process involving more than a few layers appears biologically implausible, and moreover its performance may scale poorly as the number of layers is increased.

In a complex network, or an animal's brain, it is totally unclear how a component layer is to "decide" what transformation its connections should perform – *if* we assume that the layer needs to "know" what environmental features are important for the animal to respond to. This is the classic "credit assignment" problem of artificial intelligence: if the final output from a complex system is correct, which connections should be "rewarded" or strengthened?

The approach we propose avoids this problem. Instead of demanding that a connection or layer "know about" the ultimate goals of the animal, we use only local criteria for modifying connections. The information that reaches a layer is processed so that the maximum amount of information is preserved. We have seen that this does not in general lead to a trivial one-to-one identity mapping, in which each M cell receives input from only one L cell. In general, the identity mapping is not a solution that maximally preserves information, owing to the role of noise in our model. Instead, the M cells tend to respond to features that are among the statistically and information-theoretically most significant ones. Applying the principle of maximum information preservation to each layer of processing in turn, results in the emergence of a sequence of feature-analyzing functions.

The following analogy may help you to see intuitively how the process works. Imagine a layered organization of workers. No worker is ever told explicitly what his job is. Each worker is instead told: "Each week a large amount of data will pass across your desk. Your job is to prepare a summary of this data each week, subject to various constraints (for example, on the length of the summary). Your goal is to produce a summary that conveys the largest possible amount of information about the data you have seen that week." Furthermore, each worker may be allowed to communicate with a number of his colleagues within his layer. In that case the criterion is broadened so that the goal is to produce a set of summaries (as output from the worker's layer) that jointly conveys the largest possible amount of information about the input data to the entire layer.

The type of data each worker receives depends upon the environment external to the organization, the structure of the organization (what "layer" he is part of), and various constraints.

What will such a worker do? First suppose that the data received by the worker is a completely random-looking string of numbers; the worker cannot identify any structure or regularity in the data. Then the best the worker can do is to copy the data onto his summary sheet, and stop when he has run out of space. He has conveyed very little information in his summary, but that is the best he can do with unstructured input.

Now suppose that the worker receives numerical data that he finds has a regular pattern; for example, the data might consist of pairs of numbers with the second number of each pair being twice the first. Now the worker can write on his summary, "y = 2x, x = ... ," followed by a list of x values (until he runs out of space). By recognizing this regularity, the worker has been able to convey two pages of input data using one page of summary.

For more complex and subtle types of regularities in the input data, the basic idea remains the same: if the worker is able to identify the regularity, he should exploit it in order to produce a more informative summary. He is essentially performing a type of data compression on the input. Thus, over time, the worker will find that particular representations of information – for example, graphical plots involving certain combinations of variables - serve him best in preparing his summary.

The worker will be guided by a second consideration as well. If he receives data of differing reliabilities (for example, some of the data comes over a noisy telephone line), then – other things being equal – he should prefer to summarize information about the less noisy data. The reason is that otherwise he will be wasting a part of his summary on data that is more likely to be incorrect.

If each worker is working independently of his colleagues, we see from this analogy that two rules will guide the worker:

- Identify and exploit regularities (structure) in the input data, to produce a more compressed summary.
- Give greater weight to conveying information about data that has a higher signal-to-noise ratio.

If workers are allowed to collaborate, then two more rules emerge:

- If one worker is reliably summarizing certain features of the data, his colleagues should not duplicate the same features in their summaries.
- If the workers are unreliable (a "high processing noise level"), then several workers should redundantly perform the same, or similar, processing of the data – to increase the chance of certain data being conveyed accurately.

These rules illustrate the "redundancy/diversity" tradeoff that we discussed earlier in a more mathematical way.

There are other considerations that emerge as well from applying the infomax principle to this example, but the main point is that the workers will end up processing the data in specific ways, even though they have been told neither what the goals of the organization are, nor what their supervisors are interested in knowing, nor anything about what their specific tasks should be.

In both the organizational analogy and the real network, there is no need for any higher layer to attempt to *reconstruct* the raw data from the summary. The point is rather to enable the higher layers to *discriminate* different environmental situations from one another, in order to discriminate the relative value of different actions. If the information needed to make these discriminations has been lost at intermediate stages, it cannot be used. If a local optimization principle is to be used – one that does not attempt to take account of remote high-level goals – then the processing elements (cells or workers) at each stage will not "know" which particular information is going to be needed at high levels, and which information they can afford to discard. The virtue of the infomax principle is that it causes the processing elements to preserve as much information as possible within the imposed constraints. Although the principle of maximum information preservation was initially a far from obvious choice for an organizing principle, this discussion suggests that it is indeed a natural and attractive one to use in the construction of a layered perceptual system.

Evolution and Infomax Let us consider three aspects of the relationship between an organizing principle, such as infomax, and biological evolution:

1. What is the role of an organizing principle in the context of systems that evolve through mutation and selection?
2. What role might the infomax principle play in the overall structuring of a perceptual network?
3. Why might a *generic* organizing principle for processing information have adaptive value?

How can an organizing principle become established in a

biological system? I do *not* take the view that the principle is somehow specified to the system, and an algorithm then found for implementing it. Instead, through mutation and natural selection, nature may "try out" various ways of constructing networks – different synaptic modification rules, types of lateral interaction, and so on – some of which are found to be more successful than others. In our attempt to understand what it is about certain such ways that make them more successful, we are led to formulate possible organizing principles.

It is as if we knew a great deal about the structure of an eye but very little about the principles of optics. We would not know why the eye evolved as it did. Some might then wonder whether there were any underlying principles of ocular organization at all. They might suppose that, through mutation and selection, a large number of individual adaptations emerged that jointly resulted in excellent vision, and that our search for any integrative explanation of why this set of adaptations works is fruitless. This view of an extreme multiplicity of components and processes that are not reducible to a smaller number of explanatory principles has been expressed by some workers with respect to brain function. In contrast, I regard the search for such underlying principles – particularly in the face of the observed biological complexity – as a fruitful path toward greater neuroscientific understanding.

The infomax principle may determine what transformation each layer of a given network will implement. However, it does not specify the "gross architecture" of the entire network; that is, which layers provide input to which other layers. Nor does it specify the various parameters that may affect layer development, such as the noise level, the number of cells per layer, and so on. These aspects of the design, as determined by the course of biological evolution, may involve organizing principles not yet identified. It is possible that more powerful methods for implementing the infomax principle – that is, maximizing mutual information between input and output – across several processing stages may be developed and used to determine what architectures are infomax-optimal. Even if this is true, other criteria, not derivable from infomax alone, must play a role in the overall design of the network.

For an analogy, think of an electronic circuit designer who is not free to modify the properties of the components he uses, but who can connect them to form a variety of circuits. In the case of our proposed principle, each "component" is an entire cell layer, and the infomax principle determines that layer's behavior given a particular gross architecture or "circuit design." Thus evolution can "close the loop" on the design process, by experimenting with different "circuit designs" and favoring the survival of organisms whose perceptual systems are well-adapted to their environment.

There is a different, and important, evolutionary function that a *generic* principle for the development of a perceptual network layer – whether it be infomax or some other principle – can serve. Suppose that an evolutionary mutation produces a modified eye, or merges auditory signals into the visual pathway at some new point. If there were no generic principle for layer development, we might imagine that mutations would have to occur simultaneously in the processing function of several layers, for those layers to be able to use the novel input properly. But if there is such a generic principle – that applies to each layer regardless of what type of input reaches it – then the novel input will automatically be processed in accordance with that principle. This suggests that the existence of a generic principle may greatly increase the likelihood of a mutation being adaptive.

A Broader Context Other complex systems, besides neural networks, pose challenges similar to those we have discussed. How might complex structures and behaviors that may appear goal-oriented emerge from relatively simple local rules? We have seen that a local dynamical rule of Hebb type, acting at synapses, leads to an optimization principle – variance maximization – at the level of the whole cell. This led us to propose an optimization principle – maximum information preservation – that may apply at the level of an entire layer. From the information-theoretic standpoint advocated here, we may find that the immune response system and biological evolution, as well as other complex systems, have certain abstract similarities to the process of neural development and plasticity, although the dynamical rules and the substrates upon which they act are quite different.

A great deal of work remains to be done, if we are to take this or some other proposed organizing principle, extract testable predictions from it, and determine its scope and limitations. We need to identify and test such principles, in order to complement and help to focus the enormous amount of detail being revealed by progress in experimental neuroscience. The study of such principles may also provide the understanding needed to develop synthetic perceptual systems that require no explicit programming.

Acknowledgements

A version of this paper appeared in a special issue of IEEE Computer Magazine on neural networks, March 1988.

References

Attneave, F. (1954). Some informational aspects of visual perception. Psychol. Rev. 61:183-93.

Barlow, H. B. (1961). Possible principles underlying the transformations of sensory messages. In Sensory Communication, ed. W. A. Rosenblith, pp. 217-34. Cambridge, MA: M. I. T. Press and New York / London: John Wiley & Sons.

Barlow, H. B. (1969). Trigger features, adaptation and economy of impulses. In Information Processing in the Nervous System, ed. K. N. Leibovic, pp.209-26. New York: Springer-Verlag.

Blasdel, G. & Salama, G. (1986). Voltage-sensitive dyes reveal a modular organization in monkey striate cortex, Nature 321:579-585.

Hebb, D.O. (1949). The Organization of Behavior. New York: John Wiley & Sons.

Hopfield, J.J. (1982). Neural networks and physical systems with emergent collective computational abilities, Proc. Natl. Acad. Sci. USA 79: 2554-2558.

Hubel, D.H., and Wiesel, T.N. (1977). Functional architecture of macaque monkey visual cortex (Ferrier lecture), Proc. Royal Soc. London B198: 1-59.

Hubel, D.H., and Wiesel, T.N. (1979). Brain Mechanisms of Vision, Scientific American 241: September, 150-162.

Huber, P. (1985). Projection pursuit, Ann. Stat. 13: 435-475.

Kohonen, T. (1984; 1988). Self-Organization and Associative

Memory. New York: Springer-Verlag.

Liebelt, P.B. (1967). An Introduction to Optimal Estimation, Reading, MA: Addison-Wesley.

Linsker, R. (1986). From basic network principles to neural architecture (series), Proc. Natl. Acad. Sci. USA 83: 7508-7512, 8390-8394, 8779-8783.

Linsker, R. (1988a). Development of feature-analyzing cells and their columnar organization in a layered self-adaptive network. In R. Cotterill, Computer Simulation in Brain Science. New York: Cambridge University Press, 416-431.

Linsker, R. (1988b). Towards an organizing principle for a layered perceptual network. In D. Z. Anderson, Neural Information Processing Systems. New York: Amer. Insti. Phy., 485-494.

Linsker, R. (1989). How to generate ordered maps by maximizing the mutual information between input and output signals, Neural Computation 1: 396-405.

Linsker, R. (1990a). Designing a sensory processing system: What can be learned from principal components analysis?, Proc. Internat. Joint Conf. on Neural Networks, Winter 1990, in press.

Linsker, R. (1990b). Perceptual neural organization: some approaches based on network models and information theory, Annu. Rev. Neurosci. 13: 257-281, (in press).

Marr, D. (1970). A theory for cerebral neocortex. Proc. Roy. Soc. Lond. B 176:161-234.

Oja, E. (1982). A simplified neuron model as a principal component analyzer, J. Math. Biol. 15: 267-273.

Shannon, C.E. (1948). A mathematical theory of communication, Bell Syst. Tech. J. 27: 623-656.

Widrow, B., and M. E. Hoff, Jr. (1960). Adaptive switching circuits. In 1960 IRE WESTCON Conv. Record, Part 4, pp. 96-104.

Widrow, B., and Stearns, S. D. (1985). Adaptive Signal Processing. Englewood Cliffs, NJ: Prentice-Hall.

and there will be a luminance-brightness mismatch in two types of cases. They correspond to the two ways in which a one-to-one relation such as the power function can be violated (Todorović, 1987). On the one hand there are cases in which different luminances can induce the same brightness percept. On the other hand, there are converse cases in which surfaces of the same luminance can appear differently bright. These two types of cases are best exemplified by brightness contrast and brightness constancy, two classical phenomena of brightness perception.

Brightness constancy refers to the fact that surfaces of the same reflectance (same ratio of reflected to incoming light) tend to be perceived as approximately equally bright, even under unequal illumination (Arend and Reeves, 1986; Katz, 1935). For example, two identical pieces of gray paper are perceived as approximately equally bright, whether in sunlight or in shadow, whether directly under a lamp or in a dark corner of the room. Being differently illuminated, but having the same reflectance, the two pieces of paper have different luminance (which is the product of illumination and reflectance). Their approximately equal brightness is often thought to be due to their equal reflectance. It is assumed that in the percept of brightness the visual system in some way discounts the illumination and recovers the reflectance. Perceptual brightness constancy, then, is a reflection of physical reflectance constancy: the ratio of reflected to incoming light does not depend on the intensity of incoming light. In this approach, the brightness-from-luminance problem reduces to the reflectance-from-luminance problem (Horn, 1974; Hurlbert, 1986; Land, 1977, 1986).

A difficulty for such reflectance theories of brightness is that if they are completely successful (in recovering reflectance), then they must be wrong. This is because surfaces of equal reflectance do not always appear equally bright, and any theory which predicts that they do misses an important aspect of brightness perception. Brightness contrast is the prime example of such phenomena: put two equal pieces of gray paper under equal illumination but place them on differently luminant backgrounds. Then the gray piece surrounded by the black background will look brighter than the piece on the white background. This effect of contextual dependence of brightness, where surfaces of equal luminance and reflectance look differently bright, was already known to Leonardo da Vinci, and was seriously

CHAPTER 11

Solving the Brightness-From-Luminance Problem: A Neural Architecture for Invariant Brightness Perception

S. Grossberg and D. Todorović

Introduction

The spatial distribution of light that constitutes the input to our eyes is the foundation of all visual functions, such as perception of brightness, color, texture, form, and 3-D organization. The perception of brightness may perhaps appear to be the simplest of all functions: the most natural initial explanation of why surface A appears brighter than surface B is that more light arrives into our eyes from surface A than from B. However, as we will show in the following, the relation of luminance (which is a physical variable involving the amount of light energy arriving at the retina) and brightness (which is a psychological variable denoting perceived intensity of light) is much more complicated.

The brightness-from-luminance problem is the following: find the mapping that transforms any given spatial distribution of luminance into the corresponding spatial distribution of brightness. The problem is generally solved for the simple visual situation involving a bright patch on a dark background. Increasing the luminance of the patch causes it to look increasingly brighter, but in a nonlinear manner. The function is negatively accelerated, and the debate is only whether it is mathematically better described as a logarithm (Fechner, 1889) or as a power function (Stevens, 1957).

In more complicated visual situations containing several surfaces, their brightnesses may be predicted by taking logarithms, or power functions, of their luminances. This prediction will be wrong,

Figure 1: The Craik-O'Brien-Cornsweet Effect (COCE). The left rectangle looks uniformly brighter than the right one. They have identical luminance, except for the cusp-shaped profile of their shared vertical border (from Todorović, 1987).

studied in the nineteenth century. More recently, a host of related phenomena were discovered, such as the Hermann grid (Spillmann and Levine, 1971), the Koffka-Benussi ring (Koffka, 1935), brightness assimilation (Helson, 1963), the Wertheimer-Benary figure (Benary, 1938), and others.

An attractive phenomenon of this type is the Craik-O'Brien-Cornsweet Effect (COCE) (Cornsweet, 1970; see Todorović, 1987, for a review). One version of the COCE is presented in figure 1. Readers unfamiliar with this effect might suppose that, since the left rectangle is brighter than the right rectangle, it is also the more luminant one. However, the luminance of the two rectangles is actually identical, except for a luminance cusp overshoot at the left flank and a luminance cusp undershoot at the right flank of the midline (figure 6a). The illusory nature of the phenomenon is most easily demonstrated by the occlusion of the contour region. Placing a pencil or a piece of wire vertically across the midline in figure 1 causes the two rectangles to appear equally bright. To summarize, there are at least two factors that make the relation of brightness and luminance a problem: illumination discounting and contextual dependence. We will now present a neural network architecture that deals with both issues. The model generalizes to two dimensions the types of processes that Cohen and Grossberg (1984) used to simulate 1-D brightness phenomena. This generalization conjoins processing concepts and mechanisms from Cohen and Grossberg (1984) and those from Grossberg and Mingolla (1985, 1987). For a detailed presentation of the theory and simulations please consult Grossberg and Todorović (1988). The theory suggests that two parallel contour-sensitive processes interact to generate a brightness percept. The Boundary Contour (BC) System, comprised by several interacting networks, synthesizes an emergent boundary segmentation from combinations of oriented and unoriented scenic elements. The Feature Contour (FC) System triggers a diffusive filling-in of featural quality within perceptual domains whose boundaries are determined by output signals from the BC System. Neurophysiological and anatomical data from the lateral geniculate nucleus and visual cortex which have been analysed and predicted by the theory are summarized in Grossberg (1987a, b).

The Model

Figure 2 provides an overview of the neural network model that we have analyzed. The model has six levels depicted as thick-bordered rectangles numbered from one to six. Levels 1 and 2 are preprocessing levels prior to the BC and FC Systems. Output signals from level 2 generate inputs to both of these systems. Levels 3--5 are processing stages within the BC System. Level 6, which models the FC System, receives inputs from both level 2 and level 5. Each level contains a different type of neural network. The type of network is indicated by the symbol inside the rectangle. The symbols provide graphical mnemonics for the processing characteristics at a given level, and are used in the figures that present the computer simulations of the 2-D implementation of the model. The arrows connecting the rectangles depict the flow of processing between the levels. The type of signal processing between different levels is indicated inside thin-bordered insets attached by broken lines to appropriate arrows, and coded by letters A through E. The sketch inside the inset coded F depicts the

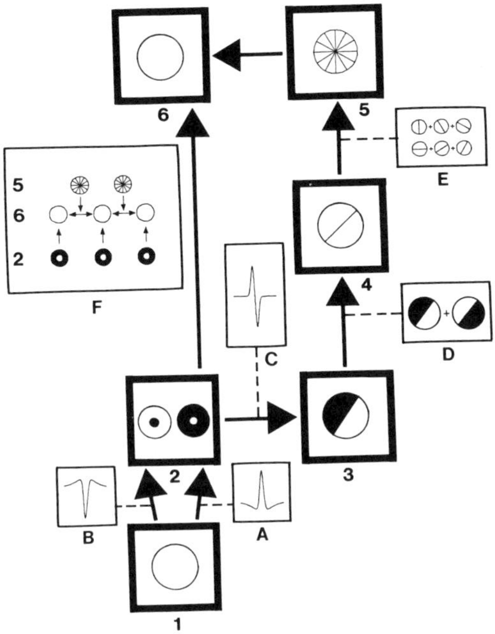

Figure 2: Overview of the model. The thick-bordered rectangles numbered from 1 to 6 correspond to the levels of the system. The symbols inside the rectangles are graphical mnemonics for the types of computational units residing at the corresponding model levels. The arrows depict the interconnections between the levels. The thin-bordered rectangles coded by letters A through E represent the type of processing between pairs of levels. Inset F illustrates how the activity at level 6 is modulated by outputs from level 2 and level 5. See the text for additional details.

complex interactions between levels 2, 5, and 6. The mathematical structure of the model is presented in the appendix.

The first level of the model consists of a set of units that sample the luminance distribution. In the 1-D version of the model the units are arranged on a line; in the 2-D version they form a square grid. Level 2 contains two networks with units that model on-cells and off-cells. These are neurons with concentric antagonistic receptive fields found at early levels of the visual system. In figure 2 the on-cells are symbolized with a white center and a black annulus, and the off-cells with a black center and a white annulus. The 1-D cross-sections of these fields are sketched in insets A and B of figure 2. In two dimensions, these profiles have the shapes of sombreros for on-units, and inverted sombreros for off-units. The activity level of such cells correlates with the size of the center-surround luminance contrast. Due to the postulated shunting interaction (see Appendix), the cells are sensitive to relative contrast in a manner approximating a Weber law (Grossberg, 1983). In addition, the cells are tuned to display non-negligible activity levels even for homogeneous stimulation, as do retinal ganglion cells (Enroth-Cugell and Robson, 1984). This property enables such a cell to generate output signals that are sensitive to both excitatory and inhibitory inputs.

Level 3 consists of units that share properties with cortical simple cells. The symbol for these units in figure 2 expresses their sensitivity to luminance contrast of a given orientation and a given direction of contrast. Inset C depicts the 1-D cross-section of the receptive field of such units, taken with respect to the network of on-cells. In our 2-D simulations, the function we used to generate this receptive field profile was the difference of two identical bivariate Gaussians whose centers were shifted with respect to each other. A similar formalization was used by Heggelund (1981a, 1981b, 1985). In our current implementation, Level 3 units are activated by Level 2 on-units.

Level 3 units are sensitive to oriented contrasts in a specific direction-of-contrast, as are cortical simple cells. However, complex cells sensitive to contrasts of specific orientation, regardless of polarity, are also well known to occur in striate cortical area 17 of monkeys (Hubel and Wiesel, 1968) and cats (Hubel and Wiesel, 1962). See Grossberg (1987a) for a review of relevant data and related models. Units fulfilling the above criteria populate level 4 of our network. Inset D in figure 2 depicts the construction of level 4 cells out of level 3 cells. The mathematical specification is similar to the one used by Grossberg and Mingolla (1985a, 1985b). The symbol for level 4 units expresses their sensitivity to oriented contrasts of either direction. Each level 4 unit at a particular location is excited by two level 3 units at the corresponding location having the same axis of orientation but opposite direction preference. Thus the 12 level 3 types of units give rise to six level 4 unit types. Interestingly, several physiological studies have found that the simple cells outnumber the complex cells in a ratio of approximately 2 to 1, and that complex cells have a higher spontaneous activity level than simple cells (Kato, Bishop, and Orban, 1978). Both of these properties are consistent with the proposed circuitry.

In the simulations presented in this paper, we have used a simplified version of the BC System. The final output of this system is found at level 5 of the model. A unit at a given level 5 location can be excited by any level 4 unit located at the position corresponding to the position of the level 5 unit. A level 4 unit excites a level 5 unit only if its own activity exceeds a threshold value. The pooling of signals sensitive to different orientations is sketched in inset E and expressed in the symbol for level 5 in figure 2. This pooling may, in principle, occur entirely in convergent output pathways from the BC System to the FC System, rather than at a separate level of cells within the BC System.

Network activity at level 6 of our model corresponds to the brightness percept. Level 6 is part of the FC System, which is composed of a syncytium of cells. This is a regular array of intimately connected cells such that contiguous cells can easily pass signals between each other's compartment membranes, possibly via gap junctions (Piccolino, Neyton, and Gerschenfeld, 1984). Due to the syncytial coupling of each cell with its neighbors, the activity can rapidly spread to neighboring cells, then to neighbors of the neighbors, and so forth. Because the spreading, or filling-in, of activation occurs via a process of diffusion, it tends to average the activation that is triggered by an FC input from level 2 across the level 6 cells that receive this spreading activity. The inset labeled F in figure 2 summarizes the three factors that influence the magnitude of activity of units at level 6. First, each unit receives bottom-up input from level 2, the field of concentric on-cells. Second, there are lateral connections between neighboring units at level 6 that define the syncytium that supports within-network spread of activation, or filling-in. Third, this lateral spread is modulated by inhibition from level 5 in the form of BC signals capable of decreasing the magnitude of mutual influence between neighboring level 6 units. The net effect of these interactions is that the FC signals generated by the concentric on-cells are diffused and averaged within boundaries generated by BC signals.

The idea of a filling-in process has been invoked in various forms by several authors in discussions of different brightness phenomena (Davidson and Whiteside, 1971; Fry, 1948; Gerrits and Vendrick, 1970; Hamada, 1984; Walls, 1954). In the present model, this notion is fully formalized, related to a possible neurophysiological

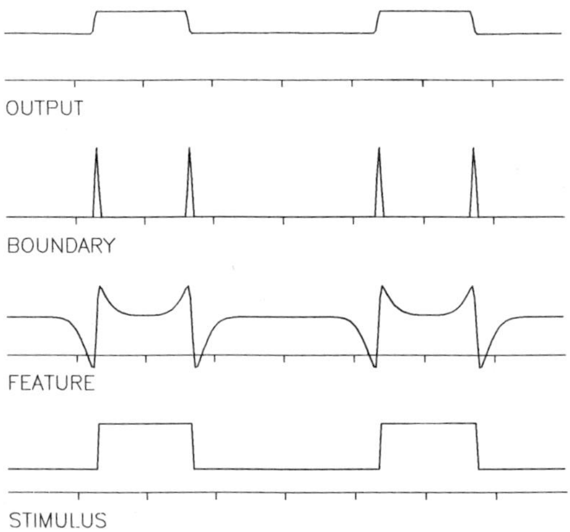

Figure 3: One-dimensional simulation of an evenly illuminated scene. In this and the following 1-D simulations the four graphs, from bottom to top respectively, refer to the level 1 stimulus distribution (labeled Stimulus), the level 2 on-cell distribution (labeled Feature), the level 5 BC output (labeled Boundary), and the level 6 filled-in syncytium (labeled Output). The parameters used in the simulations are listed in the appendix.

foundation, tied in with other mechanisms as a part of a more general vision theory, and applied in a systematic way to a variety of brightness phenomena.

1-D Simulations The model described in the preceding section was implemented in a 1-D and a 2-D version. Simulations from both versions will be presented. All graphical depictions of the 1-D simulations contain four distributions: the stimulus luminance distribution (level 1), the on-unit distribution (level 2), the output of the BC System (level 5), and the syncytium distribution (level 6) which corresponds to the predicted brightness distribution. Cohen and Grossberg (1984) presented their simulation of various brightness phenomena in a similar format. The graphs of the four distributions were scaled separately; that is, each was normalized with respect to its own maximum.

We begin with the simulation of a simple visual situation whose purpose is to set the context for the following simulations. The level 1 luminance distribution, labeled Stimulus, is presented in the bottom graph of figure 3. It portrays the horizontal cross-section of an evenly illuminated scene containing two equally luminant homogeneous patches on a less luminant homogeneous background. The level 2 reaction of on-units to such a stimulation, labeled Feature, illustrates the cusp-shaped profiles that correspond to luminance discontinuities. The four boundary contours formed at level 5 of the system are labeled Boundary. Finally, the top graph, labeled Output, presents the level 6 filled-in activity profile that embodies the prediction of a brightness distribution qualitatively isomorphic with the luminance distribution. This percept contains two homogeneous, equally bright patches on a darker, homogeneous background.

What happens when the two-patch scene is unevenly illuminated? Figure 4 presents a luminance distribution that mimics the effects of a light source off to the right side of the scene. The luminance profile is now tilted, and the right patch has more average luminance than the left patch. Inspection of the Output reveals that the model exhibits brightness constancy. It predicts a percept whose structure is very similar to the one in the preceding, evenly illuminated scene. One factor that contributes to this outcome is the ratio-processing characteristic of the level 2 units. Although the absolute luminance values in the stimulus distribution in figures 3 and 4 are different, the ratio of the lower to higher luminance across all edges in both distributions is 1:3. Therefore, the activity profiles of level 2 on-units are very similar in both cases, as is the activity in all subsequent processing stages. The consequence is that the illuminant is effectively discounted.

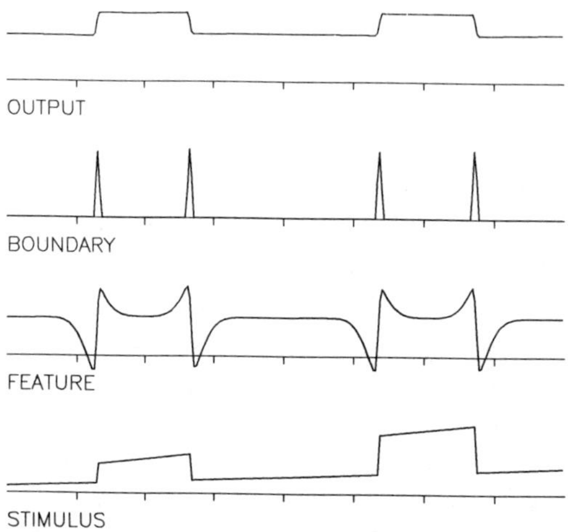

Figure 4: Brightness constancy: The same scene as in figure 3, but now unevenly illuminated. Although the two Stimulus distributions in figures 3 and 4 are different, the final output distributions are very similar.

The importance of luminance ratios for brightness perception was stressed by Wallach (1948, 1976). He found that if one region was completely surrounded by another, the brightness of the inner region was predominantly influenced by the size of the ratio of its luminance to the luminance of the surround. Our model provides a mechanical explanation of why the ratio principle is effective in such situations. In addition, the model is applicable to more general visual situations in which multiple regions have multiple neighbors, and it provides perceptually correct predictions in situations in which the ratio principle fails.

Figure 5 shows how the model handles brightness contrast. This luminance profile characterizing the favorite textbook example of this phenomenon is depicted as the Stimulus. The luminance distribution is similar to the one in figure 3 in that it contains two patches of medium luminance level. However, the left patch is positioned on

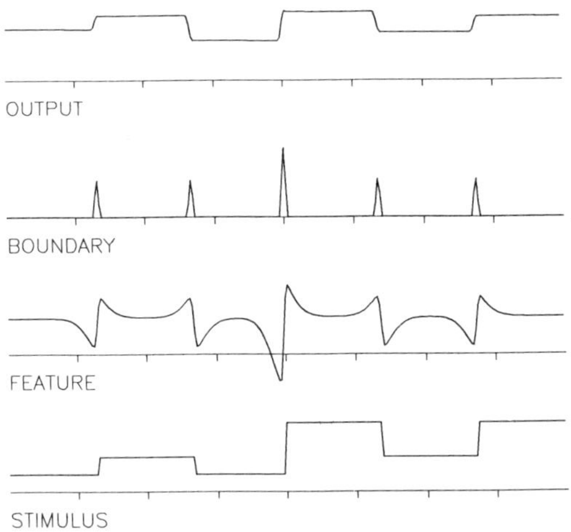

Figure 5: Brightness contrast: The stimulus contains two medium luminance patches, the left one on a low luminance background, and the right one on a high luminance background. The Output predicts the left patch to look brighter than the right patch. In contrast, in the Feature profile, the central activity levels corresponding to the two patches are equal. Thus brightness contrast cannot be explained solely by contour generated activity, but a filling-in process is also necessary.

a lower luminance background and the right patch on a higher luminant one. Inspection of the Output in figure 5 reveals that the prediction of the model is in accord with the perceptual fact that the patch on the dark background looks brighter than the patch on the bright background. Note that the central portions of the on-units profiles (level 2) that correspond to the stimulus patches in figure 5 have the same activity magnitude. Hence, these activity profiles cannot account for the difference in the appearance of the patches. However, the filled-in activity patterns within each region of the level 6 output in figure 5 are different and homogeneous.

In addition to the classical brightness phenomena, the model also explains a variety of more recently studied effects. Grossberg and Todorović (1988) presented 1-D simulations of experimental findings by Arend, Buehler, and Lockhead (1971), Arend and Goldstein (1987), Shapley (1986), and Shapley and Reed (1986). In all cases, brightness relationships found in the psychophysical experiments match those predicted by the model.

2-D Simulations Although a 1-D model suffices for some brightness effects, others can only be profitably studied and simulated by means of a 2-D architecture. The graphical depictions of our simulations consist of 30x30 or 40x40 arrays of circular symbols of different types. As noted, the type of a symbol serves as a mnemonic of the type of the unit it represents. The size of the radius of a symbol represents the magnitude of its activity. The particular sizes of the circular symbols on the printed page were chosen according to the following scaling procedure: the unit or units with the maximum activity are represented with circles whose radius is equal to half the distance between the centers of two neighboring units on the grid; the remaining circles are scaled proportionally.

Figure 6 shows how the model handles the COCE. Figure 6a is a 2-D stimulus representation (level 1) depicting the standard case of the COCE presented in figure 1. Figure 6b describes the activity pattern across the field of circular concentric on-units (level 2). Note that the middle portions of the left and right region corresponding to the two rectangles have approximately the same level of activity. However, there is an overshoot at all four edges of the left region, but only at three edges of the right region. Thus, on average there is more activity within the left than within the right region. Figure 6c describes the activity pattern across the field of boundary contour units which delineate two rectangular compartments (level 5). The activity pattern at level 2 generates a filling-in reaction at level 6 within these boundary compartments. Figure 6d, which should be compared with the percept in figure 1, presents this final filled-in activity pattern. The activity is uniformly higher in the left rectangle than in the right one, because on the left side there is a larger total amount of level 2 activity than on the right side, but they diffuse within the same-sized compartments. See Grossberg and Todorović (1988) for simulations of several variations of COCE presented by Todorović (1987).

a
b
c
d
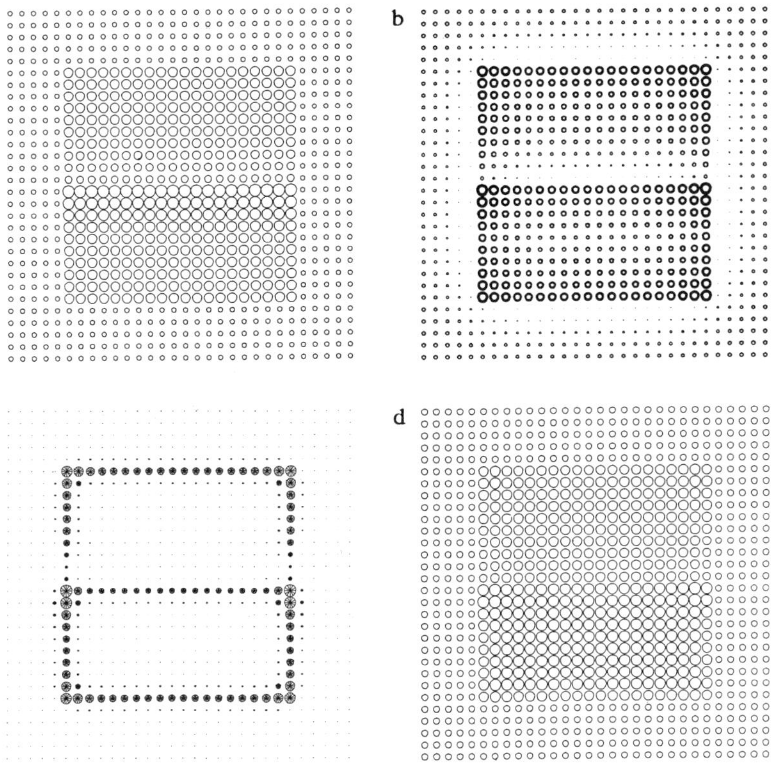

Figure 6: The COCE. (a) The stimulus distribution. (b) The on-cell activity profile. (c) The output of the BC System. (d) The filled-in syncytium, which predicts the brightness of the stimulus and should be compared with figure 1. The parameters for this and the following 2-D simulations are listed in the

The interactions of the BC System and the filling-in process are well illustrated in figure 7, which presents the simulation of the Koffka-Benussi ring (Berman and Leibowitz, 1965; Koffka, 1935). The version that we simulate uses a rectangular annulus. The annulus has an intermediate luminance level and is superimposed upon a bipartite background of the same type as in the classical brightness contrast condition, with one half having a high luminance level and the other half a low luminance level (fig. 7a). The percept of such a stimulus is that the annulus is approximately uniform in brightness, although the right and the left halves of the annulus exhibit some brightness contrast. This percept corresponds to the level 6 activity profile in figure 7b.

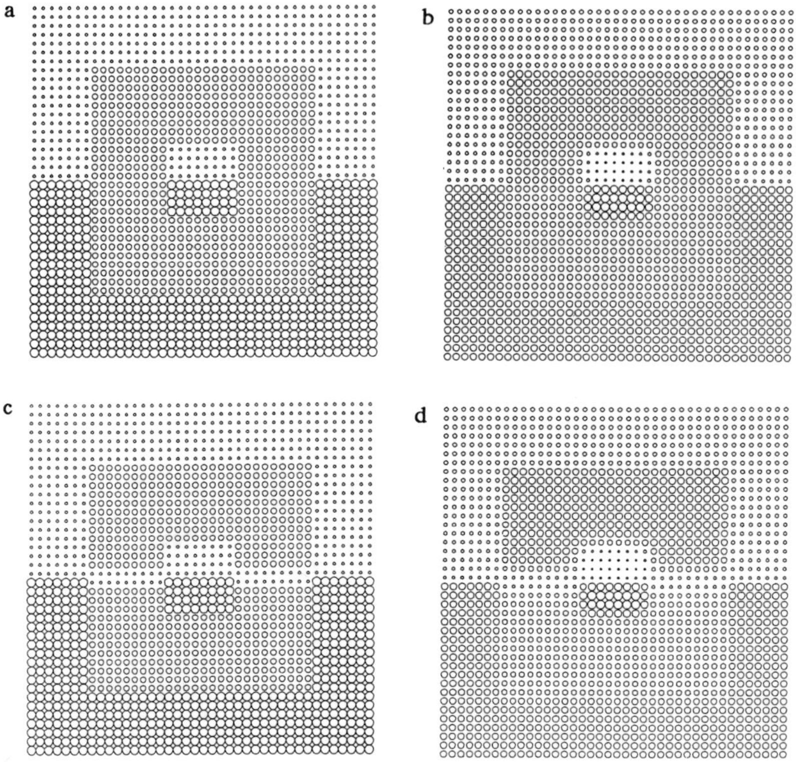

Figure 7: The Koffka-Benussi ring. (a) The stimulus distribution corresponding to the homogeneous undivided square annulus of medium luminance on a bipartite background. (b) The filled-in output corresponding to the stimulus in (a). (c) The same stimulus distribution as in (a), except that the annulus is here divided by vertical short lines into two equiluminous halves. (d) The filled-in output corresponding to the stimulus in (b). The two halves of the annulus are homogeneous and have different brightness levels.

The brightness distribution in the percept can be changed by the introduction of a narrow black line dividing the stimulus vertically into two halves. Figure 7c presents the new stimulus distribution. In the percept, as in the level 6 activity profile (fig. 7d), the annulus is now divided into two regions with homogeneous but different brightnesses that are in accord with brightness contrast. These effects depend critically upon interactions between contrast, boundaries, and filling-in in the model. In the unoccluded Koffka-Benussi ring, the annular region at Level 6 is a single connected compartment within which diffusion of activity proceeds freely. The opposite contrast due to the two halves of the background are effectively averaged throughout the annular region, although a residual effect of opposite contrast remains. The introduction of the occluding boundary (fig. 7c) divides the annulus into two smaller compartments (fig. 7d). The different contrasts are now constrained to diffuse within these compartments, generating two homogeneous regions of different brightness. See Grossberg and Todorović (1988) for additional 2-D simulations of brightness phenomena of the Hermann grid and the Kanizsa-Minguzzi (1986) anomalous brightness differentiation.

The last set of simulations that we present here was done on a set of images popularized by E. Land, who named them after Piet Mondrian, the Dutch painter. They consist of randomly arranged collages of homogeneous surfaces of different reflectances. We use here an example similar to the one presented by Shapley (1986). Consider the two squares in figure 8a, the first near the top left corner, and the second near the bottom right corner, that have the same size and luminance. Despite these equalities, the filled-in activity profile of the upper square is more intense than that of the lower square, corresponding to the percept that the upper square is brighter. This brightness difference is due to the following combination of factors in our model. The luminance of the regions surrounding the two squares were chosen such that, on average, the upper square is more luminant than its surround, and the lower square is less luminant than its surround. In consequence, as can be seen in figure 8b, more level 2 on-unit activity is present within the region corresponding to the upper square. The on-unit activity diffuses within the compartments delineated by the BCs (fig. 8c). Thus, in the filled-in upper square of figure 8d, a larger amount of activity is spread across the same area as in the lower square, thereby explaining the final brightness difference.

Figure 9a presents a Mondrian that is illuminated by a gradient of light that decreases linearly across space from the lower right corner of the figure. The upper square now exhibits, on average, less luminance than the lower square. Despite this fact, the filled-in activity profile of the upper square at level 6 is more intense than that of the lower square (fig. 9d). Figures 8b and 9b, 8c and 9c, and 8d and

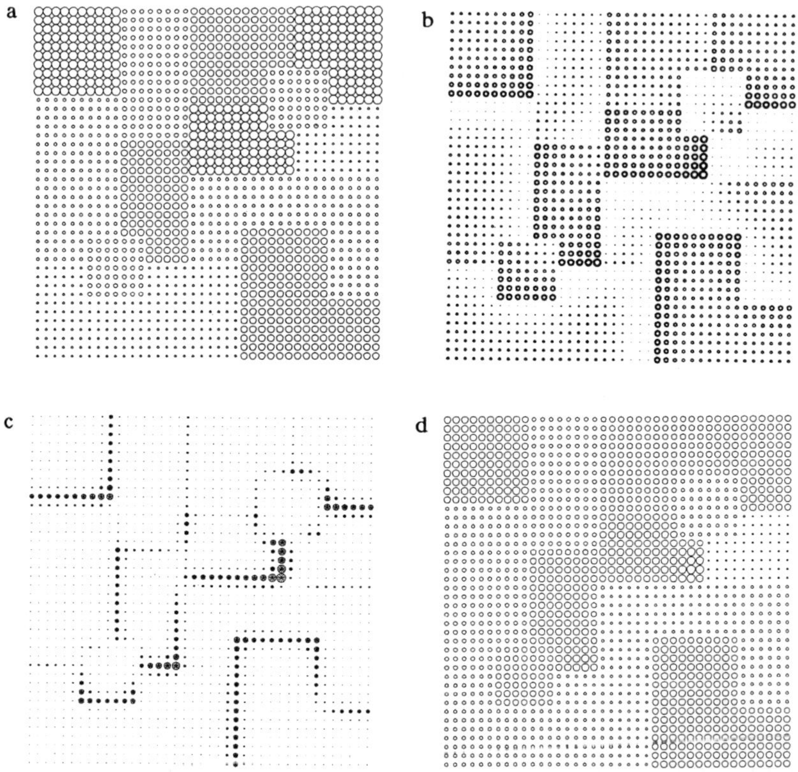

Figure 8: The evenly illuminated Mondrian. (a) The stimulus distribution consists of 13 homogeneous polygons with 4 luminance levels. Note that the square in the upper left portion of the stimulus has the same luminance and size as the square in the lower right portion. However, the average luminance of the regions surrounding the lower square is higher than the corresponding average luminance for the upper square. (b) The on-cell distribution. The amount of on-cell activity within the upper square is higher than that within the lower square. (c) The BC output. (d) The filled-in syncytium. The upper square is correctly predicted to look brighter than the lower square.

9d are, in fact, virtually indistinguishable. Thus this simulation exhibits brightness constancy by effectively discounting the illuminant, while at the same time retaining the effect of generalized

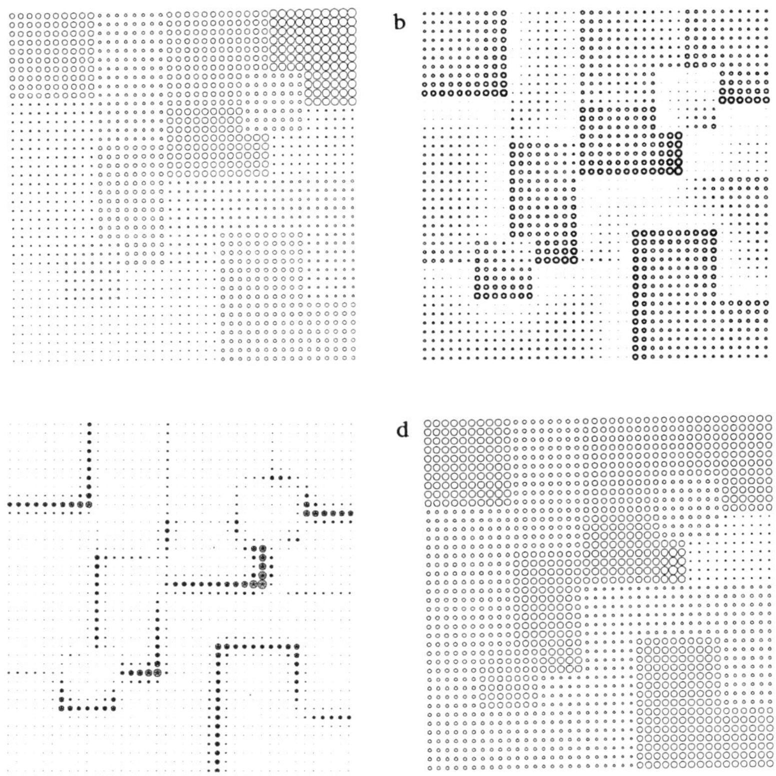

Figure 9: The unevenly illuminated Mondrian. (a) The stimulus distribution simulates the transformation of figure 8a caused by the presence of a light source whose intensity decreases linearly from the lower right corner toward the upper left corner of the stimulus. The lower square is now more luminant than the upper square. (b) The on-cell distribution. (c) The BC output. (d) The filled-in syncytium. Figures 9b, 9c, and 9d are very similar to the corresponding figures for the evenly illuminated Mondrian (fig. 8). This illustrates the model's discounting of the illuminant. In addition, the upper square is still predicted to appear brighter than the lower square, a case of contrast constancy.

brightness contrast. We call this *contrast constancy*, an effect which, to our knowledge, has not yet been psychophysically tested. This result does not, however, imply that complete discounting will occur in response to all combinations of achromatic and chromatic images, illuminants, and bounding regions (Arend and Reeves, 1986). The systematic analysis of all these factors remains for future research.

Discussion

The computer implementation of the model described in this paper has a limited domain of application since it deals only with monocular achromatic brightness effects. Extensions into the chromatic and the binocular domains have been described in Grossberg (1987a, b). Brightness can also be influenced by emergent segmentations that are not directly induced by image contrasts, as in Kanizsa's (1986) illusory triangle, the Ehrenstein illusion, and neon color spreading effects. These and related grouping and segmentation effects have been discussed by Grossberg and Mingolla (1985a, 1985b). Their implementation includes a version of the BC System in which emergent segmentations can be generated through lateral interactions between oriented channels. Such interactions may play a role in the orientation-sensitive brightness effects reported by McCourt (1982), Sagi and Hochstein (1985), and White (1979). The implementation in this paper also omits possible effects of multiple scale processing, as the receptive fields of all units within a network were assumed to have a single receptive field size. Units of multiple sizes may be involved in the explanation of classical brightness assimilation (Helson, 1963). Grossberg and Mingolla (1987) have studied the role of multiple scales in the perception of 3-D smoothly curved and shaded objects. A number of depth-related effects, such as the phenomena of transparency (Metelli, 1974) and proximity-luminance covariance (Dosher, Sperling, and Wurst, 1986) have been discussed by Grossberg (1987a, 1987b). The model in its current form also does not treat the temporal variation of brightness due to image motion (Cavanagh and Anstis, 1986; Todorović, 1983, 1987) or stabilization (Krauskopf, 1963; Pritchard, 1961; Yarbus, 1967). Finally, the application of the model to natural noisy images has yet to be accomplished.

In sum, the system described in this paper does not attempt to explain the complete gamut of brightness phenomena. These limitations are not, however, insurmountable obstacles; rather, they point to natural extensions of the model, many of which have been discussed and implemented in related work. However, even the processing of brightness in monocular, achromatic, static, noise-free images is full of surprising complexities. Only a model capable of handling these basic phenomena can be a foundation upon which still more complex effects can be explained. Not much computationally oriented work has been devoted to these fundamental aspects of visual perception. Several contemporary algorithms were influenced by Land's seminal work (Blake, 1985; Frisby, 1979; Horn, 1974; Hurlbert, 1986). Other computational models have provided alternative approaches to the analysis of filling-in (Arend and Goldstein, 1987; Hamada, 1984). Our model has been used to simulate a much larger set of brightness data, and includes mechanistic explanations of classical longstanding phenomena described in every review of brightness processing, recently discovered but unexplained data, and predictions of yet untested phenomena, including predictions of testable patterns of physiological activation.

Appendix

The equations underlying the model are based on and are an extension of work by Grossberg (1983), Cohen and Grossberg (1984) and Grossberg and Mingolla (1985b, 1986a). The exposition follows the description of levels in figure 2. Only the two-dimensional versions of the equations are presented. The one-dimensional forms can be derived by straightforward simplifications. The two-dimensional simulations were performed on a 30x30 lattice or a 40x40 lattice. The one-dimensional simulations involve 256 units.

Gray-Scale Image Description

We denote by I_{ij} the value of the stimulus input at position (i,j) in the lattice. In all simulations these values varied between 1 and 9. In order to compute the spatial convolutions of level 2 cells without causing spurious edge effects at the extremities of the luminance profile, the luminance values at the extremities were continued outward as far as necessary.

Level 2: Shunting On-Center Off-Surround Network for Discounting Illuminants and Extracting FC Signals The activity x_{ij} of a level 2 on-cell at position (i,j) of the lattice obeys a membrane equation

$$\frac{d}{dt}x_{ij} = -Ax_{ij} + (B - x_{ij})C_{ij} - (x_{ij} + D)E_{ij} \qquad \text{(A1)}$$

where C_{ij} (E_{ij}) is the total excitatory (inhibitory) input to x_{ij}. Each input C_{ij} and E_{ij} is a discrete convolution with Gaussian kernel of the inputs I_{pq}:

$$C_{ij} = \sum_{(p,q)} I_{pq}\, C_{pqij} \qquad \text{(A2)}$$

and

$$E_{ij} = \sum_{(p,q)} I_{pq} E_{pqij} \qquad \text{(A3)}$$

where

$$C_{pqij} = C \exp\left\{ -\alpha^{-2} \log 2[(p-i)^2 + (q-j)^2] \right\} \qquad \text{(A4)}$$

and

$$E_{pqij} = E \exp\left\{ -\beta^2 \log 2[(p-i)^2 + (q-j)^2] \right\}. \qquad \text{(A5)}$$

Thus, the influence exerted on the level 2 potential x_{ij} by input I_{pq} diminishes with increasing distance between the two corresponding locations. The decrease is isotropic, inducing the circular shape of the receptive fields. To achieve an on-center off-surround anatomy, coefficient C of the excitatory kernel in (A4) is chosen larger than coefficient E of the inhibitory kernel in (A5), but α, the radius of the excitatory spread at half strength in (A4), is chosen smaller than β, its inhibitory counterpart in (A5). In the simulations, this equation is solved at equilibrium. Then $\frac{d}{dt}x_{ij}=0$, so that

$$x_{ij} = \frac{\sum_{(p,q)} (BC_{pqij} - DE_{pqij})I_{pq}}{A + \sum_{(p,q)} (C_{pqij} + E_{pqij})I_{pq}}. \tag{A6}$$

The denominator term normalizes the activity x_{ij}.

The off-cell potential $\bar{x}_{ij}$ at position (i,j) also obeys a membrane equation with an equilibrium value of the same form

$$\bar{x}_{ij} = \frac{\sum_{(p,q)} (B\bar{C}_{pqij} - D\bar{E}_{pqij})I_{pq}}{A + \sum_{(p,q)} (\bar{C}_{pqij} + \bar{E}_{pqij})I_{pq}} \tag{A7}$$

The duality between on-cell and off-cell receptive fields was achieved by setting

$$\bar{C}_{pqij} = E_{pqij} \tag{A8}$$

and

$$\bar{E}_{pqij} = C_{pqij}. \tag{A9}$$

The output signal from level 2 is the nonnegative, or rectified, part of x_{ij}:

$$X_{ij} = \max(x_{ij}, 0). \tag{A10}$$

Levels 3--5 compute the Boundary Contour signals used to contain the featural filling-in process.

Level 3: Simple Cells The potential y_{ijk} of the cell centered at position (i,j) with orientation k obeys an additive equation

$$\frac{d}{dt} y_{ijk} = -y_{ijk} + \sum_{(p,q)} X_{pq} F^{k,}_{pqij} \tag{A11}$$

which is computed at equilibrium:

$$y_{ijk} = \sum_{(p,q)} X_{pq} F^{(k)}_{pqij} \tag{A12}$$

in all our simulations. In order to generate an oriented kernel $F^{(k)}_{pqij}$ as simply as possible, let $F^{(k)}_{pqij}$ be the difference of an isotropic kernel G_{pqij} centered at (i,j) and another isotropic kernel $H^{(k)}_{pqij}$ whose center $(i+m_{k,} j+n_{k)}$ is shifted from (i,j) as follows:

$$F^{(k)}_{pqij} = G_{pqij} - H^{(k)}_{pqij} \tag{A13}$$

where

$$G_{pqij} = \exp\left\{ -\gamma^{-2}[(p-i)^2 + (q-j)^2] \right\} \tag{A14}$$

and

$$H^{(k)}_{pqij} = \exp\left\{ -\gamma^{-2}[(p-i-m^2_{k)} + (q-j-n^2_{k)}] \right\} \tag{A15}$$

with

$$m_k = \sin 2\pi \frac{k}{K} \tag{A16}$$

and

$$n_k = \cos 2\pi \frac{k}{K}. \tag{A17}$$

In the 2-D simulations, the number K is 12, whereas for the 1-D simulations it is 2.

The output signal from level 3 to level 4 is the nonnegative, or rectified, part of y_{ijk}, namely

$$Y_{ijk} = \max(y_{ijk}, 0). \tag{A18}$$

Level 4: Complex Cells Each level 4 potential z_{ijk} with position (i,j) and orientation k is made sensitive to orientation but insensitive to direction-of-contrast by summing the output signals from the appropriate pair of level 3 units with opposite contrast sensitivities; viz.,

$$z_{ijk} = Y_{ijk} + Y_{ij(k+\frac{K}{2})}. \tag{A19}$$

An output signal Z_{ijk} is generated from level 4 to level 5 if the activity z_{ijk} exceeds the threshold L:

$$Z_{ijk} = \max(z_{ijk} - L, 0). \tag{A20}$$

Level 5: Boundary Contour Signals A level 5 signal z_{ij} at position (i,j) is the sum of output signals from all level 4 units at that position; viz.,

$$Z_{ij} = \sum_k Z_{ijk}. \tag{A21}$$

Level 6 computes the filling-in process, which is initiated by Feature Contour inputs from level 2 and contained by Boundary Contour inputs from level 5.

Level 6: Filling-In Process Each potential S_{ij} at position (i,j) of the syncytium obeys a nonlinear diffusion equation

$$\frac{d}{dt}S_{ij} = -MS_{ij} + \sum_{(p,q)\in N_{ij}} (S_{pq} - S_{ij})P_{pqij} + X_{ij} \tag{A22}$$

The diffusion coefficients that regulate the magnitude of cross influence of location (i,j) with location (p,q) depend on the Boundary Contour signals Z_{pq} and Z_{ij} as follows:

$$P_{pqij} = \frac{\delta}{1 + \varepsilon(Z_{pq} + Z_{ij})} \tag{A23}$$

The set N_{ij} of locations comprises only the lattice nearest neighbors of (i,j):

$$N_{ij} = \left\{ (i,j-1), (i-1,j), (i+1,j), (i,j+1) \right\}. \tag{A24}$$

At lattice edges and corners, this set is reduced to the set of existing neighbors. According to equation (A22), each potential S_{ij} is activated by the on-cell output signal X_{ij} and thereupon engages in passive decay (term $-MS_{ij}$) and diffusive filling-in with its four nearest neighbors to the degree permitted by the diffusion coefficients P_{pqij}. At equilibrium, each S_{ij} is computed as the solution of a set of simultaneous equations

$$S_{ij} = \frac{X_{ij} + \sum_{(p,q)\in N_{ij}} S_{pq}P_{pqij}}{M + \sum_{(p,q)\in N_{ij}} P_{pqij}} \tag{A25}$$

which is compared with properties of the brightness percept.

In all simulations the following parameter values were used: $A=1, B=90, D=60, \gamma=1$. All two-dimensional simulations shared the following parameters: $C=18, M=1, \alpha=.25, \varepsilon=1, E=.5, \beta=3, \delta=300, L=10$. All one-dimensional simulations used $C=4$, $M=10$, $\alpha=1$, $\varepsilon=100$, $E=.5$, $\beta=8$, $\delta=100{,}000$, $L=5$.

Acknowledgements

Supported in part by the Air Force Office of Scientific Research (AFOSR F49620-86-C-0037 and AFOSR F49620-87-C-0018), the Army Research Office (ARO DAAL03-88-K-0088), and the National Science Foundation (NSF IRI-87-16960). We wish to thank Cynthia Suchta and Carol Yanakakis for their valuable assistance in the preparation of the manuscript.

References

Arend, L.E., Buehler, J.N., and Lockhead, G.R. (1971). Difference information in brightness perception. Perception and Psychophysics 9: 367-370.

Arend, L.E., and Goldstein, R.E. (1987). Lightness models, gradient illusions, and curl. Perception and Psychophysics 42: 65-80.

Arend, L.E., and Reeves, A. (1986). Simultaneous color constancy. Journal of the Optical Society of America A3: 1743-1751.

Benary, W. (1924). Beobachtungen zu einem Experiment ueber Helligkeitskontrast. Psychologische Forschung 5: 131-142. Translated as "The influence of form on brightness contrast." In W.D. Ellis, A source book of Gestalt psychology. London: Routledge and Kegan Paul, 1938.

Berman, P.W., and Leibowitz, H.W. (1965). Some effects of contour on simultaneous brightness contrast. Journal of Experimental Psychology 69: 251-256.

Blake, A. (1985). Boundary conditions for lightness computation in Mondrian world. Computer Vision, Graphics, and Image Processing 14: 314-327.

Cavanagh, P., and Anstis, S.M. (1986). Brightness shift in drifting ramp gratings isolates a transient mechanism. Vision Research 26: 899-908.

Cohen, M.A., and Grossberg, S. (1984). Neural dynamics of brightness perception: Features, boundaries, diffusion, and resonance. Perception and Psychophysics 36: 428-456.

Cornsweet, T.N. (1970). Visual perception. New York: Academic Press.

Davidson, M., and Whiteside, J.A. (1971). Human brightness perception near sharp contours. Journal of the Optical Society of

America 61: 530-536.
Dosher, B.A., Sperling, G., and Wurst, S. (1986). Tradeoffs between stereopsis and proximitiy luminance covariance as determinants of perceived 3-D structure. Vision Research 26: 973-979.
Enroth-Cugell, C., and Robson, J.G. (1984). Functional characteristics and diversity of cat retinal ganglion cells. Investigative Ophthalmology and Visual Science 25: 250-267.
Fechner, G.T. (1889). Elemente der Psychophysik (2nd edition). Leipzig: Breitkopf und H"artel.
Frisby, J.L.I. (1979). Seeing: Illusion, Brain, and Mind. Oxford: Oxford University Press.
Fry, G.A. (1948). Mechanisms subserving simultaneous brightness contrast. American Journal of Optometry and Archives of the American Academy of Optometry 25: 162-178.
Gerrits, H.J.M., and Vendrick, A.J.H. (1970). Simultaneous contrast, filling-in process and information processing in man's visual system. Experimental Brain Research 11: 411-430.
Grossberg, S. (1983). The quantized geometry of visual space: The coherent computation of depth, form, and lightness. Behavioral and Brain Sciences 6: 625-692.
Grossberg, S. (1987a). Cortical dynamics of three-dimensional form, color, and brightness perception, I: Monocular theory. Perception and Psychophysics 41: 87-116.
Grossberg, S. (1987b). Cortical dynamics of three-dimensional form, color, and brightness perception, II: Binocular theory. Perception and Psychophysics 41: 117-158.
Grossberg, S., and Mingolla, E. (1985a). Neural dynamics of form perception: Boundary completion, illusory figures, and neon color spreading. Psychological Review 92: 173-211.
Grossberg, S., and Mingolla, E. (1985b). Neural dynamics of perceptual grouping: Textures, boundaries, and emergent segmentations. Perception and Psychophysics 38: 141-171.
Grossberg, S., and Mingolla, E. (1987). Neural dynamics of surface perception: Boundary webs, illuminants, and shape-from-shading. Computer Vision, Graphics, and Image Processing 37: 116-165.
Grossberg, S., and Todorović, D. (1988). Neural dynamics of 1-D and 2-D brightness perception: A unified model of classical and recent phenomena. Perception and Psychophysics 43: 241-277.

Hamada, J. (1984). A multistage model for border contrast. Biological Cybernetics 51: 65-70.
Heggelund, P. (1981a). Receptive field organisation of simple cells in cat striate cortex. Experimental Brain Research 42: 89-98.
Heggelund, P. (1981b). Receptive field organisation of complex cells in cat striate cortex. Experimental Brain Research 42: 99-107.
Heggelund, P. (1985). Quantitative studies of enhancement and suppression zones in the receptive field of simple cells in cat striate cortex. Journal of Physiology 373: 293-310.
Helson, H. (1963). Studies of anomalous contrast and assimilation. Journal of the Optical Society of America 53: 179-184.
Horn, B.K.Ll. (1974). Determining lightness from an image. Computer Graphics and Image Processing 3: 277-299.
Hubel, D.H., and Wiesel, T.N. (1962). Receptive fields, binocular interaction and functional architecture in the cat's visual cortex. Journal of Physiology 160: 106-154.
Hubel, D.H., and Wiesel, T.N. (1968). Receptive fields and functional architecture of monkey striate cortex. Journal of Physiology 195: 215-243.
Hurlbert, A. (1986). Formal connections between lightness algorithms. Journal of the Optical Society of America A3: 1684-1693.
Kanizsa, G., and Minguzzi, G.F. (1986). An anomalous brightness differentiation. Perception 15: 223-226.
Kato, H., Bishop, P.O., and Orban, G.A. (1978). Hypercomplex and simple/complex cell classifications in cat striate cortex. Journal of Neurophysiology 41: 1071-1095.
Katz, D. (1935). The World of Colour. London: Kegan Paul, Trench, Trubner and Co. Ltd.
Koffka, K. (1935). Principles of Gestalt Psychology. New York: Harcourt and Brace.
Krauskopf, J. (1963). Effect of retinal image stabilization on the appearance of heterochromatic targets. Journal of the Optical Society of America 53: 741-744.
Land, E. (1977). The Retinex theory of color vision. Scientific American 237: 108-128.
Land, E. (1986). Recent advances in Retinex theory. Vision Research 26: 7-21.
McCourt, M.E. (1982). A spatial frequency dependent grating-

induction effect. Vision Research 22: 119-134.

Metelli, F. (1974). The perception of transparency. Scientific American 230: 90-98.

Piccolino, M., Neyton, J., and Gerschenfeld, H.M. (1984). Decrease of gap junction permeability induced by dopamine and cyclic adenosine 3′:5′-monophosphate in horizontal cells of turtle retina. Journal of Neuroscience 4: 2477-2488.

Pritchard, R.M. (1961). Stabilized images on the retina. Scientific American 204: 72-78.

Sagi, D., and Hochstein, S. (1985). Lateral inhibition between spatially adjacent spatial-frequency channels? Perception and Psychophysics 37: 315-322.

Shapley, R. (1986). The importance of contrast for the activity of single neurons, the VEP and perception. Vision Research 26: 45-61.

Shapley, R. and Reid, R.C. (1986). Contrast and assimilation in the perception of brightness. Proceedings of the National Academy of Sciences USA 82: 5983-5986.

Spillmann, L., and Levine, J. (1971). Contrast enhancement in a Hermann grid with variable figure-ground ratio. Experimental Brain Research 13: 547-559.

Stevens, S.S. (1957). On the psychophysical law. Psychological Review 64: 153-180.

Todorović, D. (1983). Brightness perception and the Craik-O'Brien-Cornsweet effect. Unpublished M.A. Thesis. Storrs: University of Connecticut.

Todorović, D. (1987). The Craik-O'Brien-Cornsweet effect: New varieties and their theoretical implications. Perception and Psychophysics 42: 545-560.

Wallach, H. (1948). Brightness constancy and the nature of achromatic colors. Journal of Experimental Psychology 38: 310-324.

Wallach, H. (1976). On perception. New York: Quadrangle.

Walls, G. (1954). The filling-in process. American Journal of Optometry 31: 329-340.

White, M. (1979). A new effect of pattern on perceived lightness. Perception 8: 413-416.

Yarbus, A.L. (1967). Eye movements and vision. New York: Plenum Press.

INDEX